UNIPA Springer Series

Editors-in-chief

Carlo Amenta, Dept. of Economics, Management and Statistics Sciences,
University of Palermo, Italy

Series editors

Sebastiano Bavetta, Dept. of Economics, University of Palermo, Italy
Calogero Caruso, Dept. of Pathobiology, University of Palermo, Italy
Gioacchino Lavanco, Dept. of Psychology, University of Palermo, Italy
Bruno Maresca, Dept. of Pharmaceutical Sciences, University of Salerno, Italy
Andreas Öchsner, Dept. of Engineering and Information Technology,
Griffith University, Australia
Mariacristina Piva, Dept. of Economic and Social Sciences,
Catholic University of the Sacred Heart, Italy
Roberto Pozzi Mucelli, Dept. of Diagnostics and Public Health,
University of Verona, Italy
Antonio Restivo, Dept. of Mathematics and Computer Science,
University of Palermo, Italy
Norbert M. Seel, Dept. of Education, University of Freiburg, Germany
Gaspare Viviani, Dept. of Engineering, University of Palermo, Italy

More information about this series at http://www.springer.com/series/13175

Tommaso Angelone • Maria Carmela Cerra
Bruno Tota

Editors

Chromogranins: from Cell Biology to Physiology and Biomedicine

Editors
Tommaso Angelone
Department of Biology, Ecology
and Earth Sciences
University of Calabria
Arcavacata di Rende, Cosenza, Italy

Maria Carmela Cerra
Department of Biology, Ecology
and Earth Sciences
University of Calabria
Arcavacata di Rende, Cosenza, Italy

Bruno Tota
Department of Biology, Ecology
and Earth Science
University of Calabria
Arcavacata di Rende, Cosenza, Italy

ISSN 2366-7516 ISSN 2366-7524 (electronic)
UNIPA Springer Series
ISBN 978-3-319-58337-2 ISBN 978-3-319-58338-9 (eBook)
DOI 10.1007/978-3-319-58338-9

Library of Congress Control Number: 2017956128

© Springer International Publishing AG 2017
This work is subject to copyright. All rights are reserved by the Publisher, whether the whole or part of the material is concerned, specifically the rights of translation, reprinting, reuse of illustrations, recitation, broadcasting, reproduction on microfilms or in any other physical way, and transmission or information storage and retrieval, electronic adaptation, computer software, or by similar or dissimilar methodology now known or hereafter developed.
The use of general descriptive names, registered names, trademarks, service marks, etc. in this publication does not imply, even in the absence of a specific statement, that such names are exempt from the relevant protective laws and regulations and therefore free for general use.
The publisher, the authors and the editors are safe to assume that the advice and information in this book are believed to be true and accurate at the date of publication. Neither the publisher nor the authors or the editors give a warranty, express or implied, with respect to the material contained herein or for any errors or omissions that may have been made. The publisher remains neutral with regard to jurisdictional claims in published maps and institutional affiliations.

Printed on acid-free paper

This Springer imprint is published by Springer Nature
The registered company is Springer International Publishing AG
The registered company address is: Gewerbestrasse 11, 6330 Cham, Switzerland

Contents

History and Perspectives 1
Karen B. Helle

**Secretogranin II: Novel Insights into Expression and Function
of the Precursor of the Neuropeptide Secretoneurin** 21
Reiner Fischer-Colbrie, Markus Theurl, and Rudolf Kirchmair

**Chromogranins as Molecular Coordinators at the Crossroads
between Hormone Aggregation and Secretory Granule Biogenesis** 39
O. Carmon, F. Laguerre, L. Jeandel, Y. Anouar, and M. Montero-Hadjadje

**Involvement of Chromogranin A and Its Derived Peptides
to Fight Infections** 49
Marie-Hélène Metz-Boutigue and Francis Schneider

**Conserved Nature of the Inositol 1,4,5-Trisphosphate Receptor
and Chromogranin Coupling and Its Universal Importance
in Ca^{2+} Signaling of Secretory Cells** 69
Seung Hyun Yoo

Chromogranin A in Endothelial Homeostasis and Angiogenesis 83
Flavio Curnis, Fabrizio Marcucci, Elisabetta Ferrero, and Angelo Corti

**Full Lenght CgA: A Multifaceted Protein in Cardiovascular
Health and Disease** 99
Bruno Tota and Maria Carmela Cerra

**Cardiac Physio-Pharmacological Aspects of Three Chromogranin
A-Derived Peptides: Vasostatin, Catestatin, and Serpinin** 113
Tommaso Angelone, Bruno Tota, and Maria Carmela Cerra

**Comparative Aspects of CgA-Derived Peptides in Cardiac
Homeostasis** 133
Alfonsina Gattuso, Sandra Imbrogno, and Rosa Mazza

**Molecular and Cellular Mechanisms of Action of CgA-Derived
Peptides in Cardiomyocytes and Endothelial Cells** 149
Giuseppe Alloatti and Maria Pia Gallo

**Chromogranin A-Derived Peptides in Cardiac Pre- and
Post-conditioning**. 169
Claudia Penna and Pasquale Pagliaro

**Naturally Occurring Single Nucleotide Polymorphisms in Human
Chromogranin A (*CHGA*) Gene: Association with Hypertension
and Associated Diseases** 195
Nitish R. Mahapatra, Sajalendu Ghosh, Manjula Mahata,
Gautam K. Bandyopadhyay, and Sushil K. Mahata

Serpinin Peptides: Tissue Distribution and Functions 213
Y. Peng Loh, Niamh Cawley, Alicja Woronowicz, and Josef Troger

**Action and Mechanisms of Action of the Chromogranin A
Derived Peptide Pancreastatin**. 229
N. E. Evtikhova, A. Pérez-Pérez, C. Jiménez-Cortegana,
A. Carmona-Fernández, T. Vilariño-García, and V. Sánchez-Margalet

Chromogranins and the Quantum Release of Catecholamines 249
Leandro Castañeyra, Michelle Juan-Bandini, Natalia Domínguez,
José David Machado, and Ricardo Borges

Index. 261

Contributors

Giuseppe Alloatti Department of Life Sciences and Systems Biology, University of Turin, Turin, Italy
National Institute for Cardiovascular Research (INRC), Bologna, Italy

Tommaso Angelone Department of Biology, Ecology and Earth Sciences, University of Calabria, Arcavacata di Rende (CS), Italy

Y. Anouar Inserm U1239, University of Rouen-Normandy, UNIROUEN, Institute for Research and Innovation in Biomedicine, Mont-Saint-Aignan, France

Gautam K. Bandyopadhyay Department of Medicine, University of California, San Diego, CA, USA

Ricardo Borges Unidad de Farmacología, Facultad de Medicina, Universidad de La Laguna, Tenerife, Spain

O. Carmon Inserm U1239, University of Rouen-Normandy, UNIROUEN, Institute for Research and Innovation in Biomedicine, Mont-Saint-Aignan, France

A. Carmona-Fernández Department of Clinical Biochemistry, Virgen Macarena University Hospital, School of Medicine, University of Seville, Seville, Spain

Leandro Castañeyra Unidad de Farmacología, Facultad de Medicina, Universidad de La Laguna, Tenerife, Spain

Niamh Cawley Section on Cellular Neurobiology, Eunice Kennedy Shriver National Institute of Child Health and Human Development, National Institutes of Health, Bethesda, MD, USA

Maria Carmela Cerra Department of Biology, Ecology and Earth Sciences, University of Calabria, Arcavacata di Rende (CS), Italy

Angelo Corti Division of Experimental Oncology, San Raffaele Scientific Institute, via Olgettina, Milan, Italy

Flavio Curnis Division of Experimental Oncology, San Raffaele Scientific Institute, via Olgettina, Milan, Italy

Natalia Domínguez Unidad de Farmacología, Facultad de Medicina, Universidad de La Laguna, Tenerife, Spain

VU Medical Center, Clinical Genetics, HV Amsterdam, Netherlands

N.E. Evtikhova Department of Clinical Biochemistry, Virgen Macarena University Hospital, School of Medicine, University of Seville, Seville, Spain

Elisabetta Ferrero Division of Experimental Oncology, San Raffaele Scientific Institute, via Olgettina, Milan, Italy

Reiner Fischer-Colbrie Department of Pharmacology, Medical University Innsbruck, Innsbruck, Austria

Maria Pia Gallo Department of Life Sciences and Systems Biology, University of Turin, Turin, Italy

Alfonsina Gattuso Department of Biology, Ecology, and Earth Science, University of Calabria, Arcavacata di Rende, Italy

Sajalendu Ghosh Postgraduate Department of Zoology, Ranchi College, Ranchi, India

Karen B. Helle Department of Biomedicine, University of Bergen, Bergen, Norway

Sandra Imbrogno Department of Biology, Ecology, and Earth Science, University of Calabria, Arcavacata di Rende, Italy

L. Jeandel Inserm U1239, University of Rouen-Normandy, UNIROUEN, Institute for Research and Innovation in Biomedicine, Mont-Saint-Aignan, France

C. Jiménez-Cortegana Department of Clinical Biochemistry, Virgen Macarena University Hospital, School of Medicine, University of Seville, Seville, Spain

Michelle Juan-Bandini Unidad de Farmacología, Facultad de Medicina, Universidad de La Laguna, Tenerife, Spain

Rudolf Kirchmair Cardiology & Angiology, Internal Medicine III, Medical University Innsbruck, Innsbruck, Austria

F. Laguerre Inserm U1239, University of Rouen-Normandy, UNIROUEN, Institute for Research and Innovation in Biomedicine, Mont-Saint-Aignan, France

José David Machado Unidad de Farmacología, Facultad de Medicina, Universidad de La Laguna, Tenerife, Spain

Nitish R. Mahapatra Department of Biotechnology, Bhupat and Jyoti Mehta School of Biosciences, Indian Institute of Technology Madras, Chennai, India

Manjula Mahata Department of Medicine, University of California, San Diego, CA, USA

Contributors

Sushil K. Mahata VA San Diego Healthcare System, San Diego, CA, USA
Metabolic Physiology & Ultrastructural Biology Laboratory, Department of Medicine, University of California, San Diego, CA, USA

Fabrizio Marcucci Department of Pharmacological and Biomolecular Sciences, University of Milan, Milan, Italy

Rosa Mazza Department of Biology, Ecology, and Earth Science, University of Calabria, Arcavacata di Rende, Italy

Marie-Hélène Metz-Boutigue Inserm U1121, Université de Strasbourg, FMTS, Faculté de Chirurgie Dentaire Hôpital Civil, Porte de l'hôpital, Strasbourg, France

M. Montero-Hadjadje Inserm U1239, University of Rouen-Normandy, UNIROUEN, Institute for Research and Innovation in Biomedicine, Mont-Saint-Aignan, France

A. Pérez-Pérez Department of Clinical Biochemistry, Virgen Macarena University Hospital, School of Medicine, University of Seville, Seville, Spain

Pasquale Pagliaro Department of Clinical and Biological Sciences, University of Turin, Orbassano (TO), Italy

Y. Peng Loh Section on Cellular Neurobiology, Eunice Kennedy Shriver National Institute of Child Health and Human Development, National Institutes of Health, Bethesda, MD, USA

Claudia Penna Department of Clinical and Biological Sciences, University of Turin, Orbassano (TO), Italy

V. Sánchez-Margalet Department of Clinical Biochemistry, Virgen Macarena University Hospital, School of Medicine, University of Seville, Seville, Spain

Francis Schneider Inserm U1121, Université de Strasbourg, FMTS, Service de Réanimation Médicale, Hôpital Hautepierre, Avenue Molière, Strasbourg, France

Markus Theurl Cardiology & Angiology, Internal Medicine III, Medical University Innsbruck, Innsbruck, Austria

Bruno Tota Department of Biology, Ecology and Earth Sciences, University of Calabria, Arcavacata di Rende (CS), Italy

Josef Troger Department of Ophthalmology, Medical University of Innsbruck, Innsbruck, Austria

T. Vilariño-García Department of Clinical Biochemistry, Virgen Macarena University Hospital, School of Medicine, University of Seville, Seville, Spain

Alicja Woronowicz Section on Cellular Neurobiology, Eunice Kennedy Shriver National Institute of Child Health and Human Development, National Institutes of Health, Bethesda, MD, USA

Seung Hyun Yoo Gran Med Inc., Incheon, South Korea

History and Perspectives

Karen B. Helle

Abstract Research on chromaffin cells dates back to 1856 when the venous outflow of chemical substances from the adrenal medulla into the circulation was first described. The discovery of the chromaffin granules for storage of catecholamines in 1953 was the next major break-through. Soon thereafter the co-storage of catecholamines, ATP and uniquely acidic proteins was established, together making up the isotonic storage complex within elements of the diffuse sympathoadrenal system. The core proteins constitute a family of eight genetically distinct, uniquely acidic proteins, characterized by numerous pairs of basic residues and collectively named granins. A prohormone concept was formulated when the insulin-release inhibiting peptide, pancreastatin, was identified as the mid sequence of porcine chromogranin A. Subsequently, processing resulted in a range of peptides with antifungal and antibacterial potencies, predominantly from chromogranin A, a few from chromogranin B and one from secretogranin II. A wide range of biological activites has since been documented, notably for the chromogranin A –derived peptides, affecting endothelial stability, myocardial contractility, angiogenesis, cell adhesion and tumor progression. A physiological role for full-length chromogranin A and vasostatin-I as circulating stabilizers of endothelial integrity is now evident, while the high circulating levels of chromogranin A in neuroendocrine tumors and inflammatory diseases remain an unsolved and challenging puzzle for future research.

Abbreviations

bCgA	bovine CgA_{1-431}
CA	Catecholamines
CgA	Chromogranin A
CgB	Chromogranin B
GE-25	$bCgA_{367-391}$
PN-1	Protease nexin-1
PTH	Parathyroid hormone

K.B. Helle (✉)
Department of Biomedicine, University of Bergen, Bergen, Norway
e-mail: Karen.Helle@biomed.uib.no

PTX   Pertussis toxin
SgII   Secretogranin II
VIF   Vasoinhibitory factor – CgA_{79-113}
VS-I   Vasostatin I (CgA_{1-76})
VS-II Vasostatin II (CgA_{1-113})
WE-14 $bCgA_{316-330}$

1 History

1.1 The First Hundred Years of the Chromaffin Cells

Research on chromaffin cells and granins can be traced back to the mid nineteenth century when Vulpian (1856) described the venous outflow of chemical substances from the adrenal medulla into the circulation. Half a century later the strong cardiovascular effects of the adrenomedullary substances (Oliver and Schäfer 1895) led to the chemical identification and synthesis of the first hormones, adrenaline and noradrenaline (Stoltz 1904). We owe the first identification of catecholamines (CA) to the function of the adrenergic neuron to Loewi, who in Loewi 1921 described the so-called Accellerans-Stoff or Sympathin and its stimulating activity on the denervated frog heart. Twenty five years later, Sympathin E was identified as noradrenaline (Von Euler 1946).

1.2 The First Decade of the Chromaffin Granules

The discovery of the subcellular organelles responsible for the storage of CA in the adrenomedullary chromaffin cells, i.e. the chromaffin granules, was a major break-through (Blaschko and Welsch 1953, Hillarp et al. 1953). Soon thereafter the chromaffin granules were shown to be electron-dense, membrane-limited granules of 150–300 mµ diameter (Lever 1955, Welzstein 1957, Hagen and Barnett 1960, Coupland 1968). In parallel, the vesicles related to the storage of noradrenaline in the adrenergic fibres (Von Euler and Hillarp 1956, Von Euler 1958, Dahlstrøm 1966) were demonstrated to be smaller and of varying size and electron density both in the axons and in the terminals (De Robertis and Pellegrino de Iraldi 1961). Biochemical studies, on the other hand, revealed that both types of organelles bore a number of similarities, such as storing the respective CA together with the energy-rich nucleotide ATP in a molar ratio of CA: ATP of close to 4:1 in the adrenomedullary (Blaschko et al. 1956, Falck et al. 1956) and of 5:1 in the adrenergic nerve granules (Schümann 1958; Banks et al. 1969). Moreover, in the adrenomedullary chromaffin cells these low molecular weight constituents were stored intragranularly at concentrations of about 0.55 and 0.13 M for CA and ATP respectively, i.e. strongly hypertonic if osmotically active. This phenomenon led Hillarp in 1959 to the postulation of a third component involved in the storage complex, possibly a protein, which could be responsible for holding CA and ATP in an isotonic, non-diffusible form until discharge from the stimulated cell.

1.3 The First Thirty Five Years of the Granins

The search for a specific macromolecule involved in the isotonic retention of CA and ATP within the storage organelles was immediately directed to the core proteins in the bovine adrenomedullary chromaffin granules (Helle 1966a, Smith and Winkler 1967, Smith and Kirshner 1967). By means of an immunological identification method (Helle 1966b) it was established that the enzymatically inactive protein, subsequently named chromogranin (Blaschko et al. 1967), was exocytotically discharged from the stimulated adrenal gland in parallel with the co-stored CA and ATP both in vitro (Banks and Helle 1965) and in vivo (Blaschko et al. 1967). Due to the easy access from local slaughterhouses the bovine adrenals soon became a convenient source of chromaffin cells and chromaffin granules (Smith and Winkler 1967), notably for research on the structural, chemical and functional properties of the family of chromogranins, i.e. the granins (Huttner et al. 1991; Winkler and Fischer-Colbrie 1992).

1.3.1 Glucose Homeostasis, Pancreastatin and the Prohormone Concept

The first chromogranin A (CgA) peptide to be recognized for its regulatory potency was named pancreastatin due to its ability to inhibit the rapid phase of insulin release from the glucose-stimulated porcine pancreas (Tatemoto et al. 1986; Efendic et al. 1987). When identified as the mid-section of porcine and human CgA (Huttner and Benedum 1987; Konecki et al. 1987), a novel concept was coined, namely of the granins as putative prohormones for biologically active peptides with regulating potentials (Eiden 1987). Subsequently, pancreastatin was shown to be involved as a regulator of insulin action not only of glucose but also of lipid and protein metabolism (Sanchez-Margalet and Gonzalez-Yanes 1998). In rat hepatoma cells also the cell growth was inhibited, depending on the availability of nitric oxide (NO) production (Sanchez-Margalet et al. 2001). The accumulated literature supports the original observation of pancreastatin as an anti-insulin agent, impairing glucose homeostasis by diminishing insulin sensitivity (see review by Valicherla et al. 2013).

1.3.2 Calcium Homeostasis and the N-Terminus of CgA

In the parathyroid gland CgA was originally described as parathyroid secretory protein-I (Cohn et al. 1981), co-secreted from the gland with the parathyroid hormone (PTH), i. e. the primary regulator of serum calcium concentrations. Peptides containing the N-terminal sequence of CgA (CgA_{1-76}) inhibit PTH-secretion as effective as high physiological concentrations of calcium (Fasciotto et al. 1990). Pancreastatin ($bCgA_{248-293}$) and parastatin ($bCgA_{347-419}$) have also been shown to inhibit PTH secretion, but not yet detected in the effluents from the parathyroid cells in vivo. On the other hand, CgA_{1-76} was detected both in the medium of cultured parathyroid cells (Angeletti et al. 2000) and in the adrenomedullary effluents (Metz-Boutigue et al. 1993). A binding to a 78 kDa protein was identified on the parathyroid cell surface, and the blockade by pertussis toxin indicates a G-protein-coupled receptor. Moreover, the loop sequence CgA_{16-40} was required for inhibition of PTH secretion (Angeletti et al. 1996). Thus, inhibition of PTH secretion by CgA predominantly involves CgA_{1-76}, occuring either by an autocrine mechanism or via the circulating concentrations of the processed peptide.

1.4 The Granins and their Derived Peptides

Detailed investigations of the eight members of the granin family, i.e. CgA, chromogranin B (CgB), secretogranin II (SgII) and secretogranins III-VII, have since documented that these proteins are widely distributed in distinct patterns within the diffuse neuroendocrine system of vertebrates (Helle 2004). Stimuli for release of the granins derive from a wide range of environmental and intrinsic paths, raising the concentrations of the intact prohormones and processed peptides in the extracellular space and ultimately in the circulation. The degree of processing is extensive in the adrenomedullary storage granules (Metz-Boutigue et al. 1993; Strub et al. 1995) and gives rise to a wide range of peptides with a broad spectrum of biological potencies (Helle and Angeletti 1994). The peptides derived from CgA are the vasostatins I and II, chromofungin, chromacin, pancreastatin, catestatin, WE 14, chromostatin, GE25 and parastatin and, in addition, the two most resent arrivals on the scene, serpinin ($CgA_{403-428}$, Koshimizu et al. 2010) and the vasoconstriction-inhibiting factor (VIF, CgA_{79-113}, Salem et al. 2015). Vasostatin I (VS-I, CgA_{1-76}) and bovine catestatin ($bCgA_{344-364}$) were discovered and named according to their respective inhibitory potencies, on vasodilation (Aardal and Helle 1992) and on CA secretion (Mahata et al. 1997). Since then, notably VS-I and catestatin have been shown to be involved in regulation of a wide range of mechanisms, such as endothelial permeability, angiogenesis, myocardial contractility and innate immunity, however, in many tissue exhibiting oppositely directed activities (Helle et al. 2007; Helle 2010a, b; Mahata et al. 2010).

Peptides derived from CgB, being more extensively processed than CgA in most systems and species (Strub et al. 1995), may have specific regulatory functions yet to be unravelled. SgII, on the other hand, serves a prohormone for only one conspicuously active principle, secretoneurin (Kirchmair et al. 1993, Trudeau et al. 2012), nevertheless engaged in a wide range of modulating activities related to tissue repair (Helle 2010a). Stimulated polymorphonuclear neutrophils, when accumulated in response to invading microorganism, tissue inflammation and at sites of mechanical injury, represent a non-neuroendocrine source of CgA peptides that may affect a wide range of cells involved in inflammatory responses (Lugardon et al. 2000; Zhang et al. 2009). Among them we find the vascular endothelium, the endocardium and the epithelial cells, other leucocytes, fibroblasts, cardiomyocytes, vascular and intestinal smooth muscle cells (Helle et al. 2007; Helle 2010a, b). Taken together, the release of CgA-derived peptides from gland cells, nerve terminals and immunocytes would contribute to autocrine or paracrine modulations locally while endocrine effects would result from their subsequent overflow to the circulation.

1.4.1 The Antimicrobial Peptides and Innate Immunity

Antimicrobial activities of peptides derived from the matrix of secretory granules in the bovine adrenal medulla were first reported by Metz-Boutigue and colleagues in 1998. The first three peptides found to inhibit bacteria and fungal growth were derived from the N-terminal domain of CgA (VS-I), the C-terminal end of CgB (secretolytin) and the biphosphorylated C-terminal peptide of proenkephalin-A (enkelytin). These peptides are active in a diverse range of organisms, including prokaryotes, bivalves, frogs and mammals, suggesting an important role in innate immunity, a mechanism shared by all vertebrates and present at birth as an evolutionary ancient defence mechanism (Hoffmann et al. 1999; Metz-Boutigue et al. 2000). Another CgA peptide, catestatin, derived from CgA in keratinocytes, also possess antimicrobial activity against gram-positive and gram-negative bacteria, yeast and fungi, is active notably against skin pathogens and increases in skin in response to injury and infection (Radek et al. 2008). So far, no antimicrobial activity has been assigned to SN.

The innate immunity, independent of the adaptive immune responses, is used by vertebrates as a means for short term protection against pathogenic microorganisms. The need for new antimicrobial agents is now rapidly rising due to the fast growing number of antibiotica-resistant bacteria. Accordingly, the interest in antibacterial granin-derived peptides has grown exponentially. Their therapeutic potentials are now under intensive elucidation in immunodeficient patients, in chemotherapy, in organ grafting, and against antibiotica-resistant bacterial infections (Shooshtarizodeh et al. 2010).

1.5 Functional and Clinical Aspects

At the very end of the second millennium a large body of data had accumulated on the functional and clinical aspects of the granins and their derived peptides. As assessed in a range of comprehensive reviews appearing in the first book on chromogranins (Helle and Aunis 2000a), it was evident that granins were intimately involved not only in the intracellular sorting to the secretory granules (Gerdes and Glombik 2000) and release of the isotonic amine storage complex (Borges et al. 2000), but also in their transcription, expression and secretion (Taupenot et al. 2000; Anouar et al. 2000; Kähler and Fischer-Colbrie 2000). Notably, tissue-specific processing both within the core and in the extracellular space, rendered the granins, notably CgA, CgB and Sg II as the most conspicuous prohormones with widely different effects and targets for their derived peptides (Aunis and Metz-Boutigue 2000; Metz-Boutigue et al. 2000; Parmer et al. 2000; Portela-Gomes 2000; Curry et al. 2000; Ciesielski-Treska and Aunis 2000). Accordingly, the majority of properties assigned to the granins and their peptides up to the end of the twentieth century appeared to fit into patterns of modulating strategies which might be called upon when the organism was exposed to stressful situations requiring immediate protection via the vasculature, the heart, the pancreas, parathyroid and the innate immunity system (Helle and Aunis 2000b). Moreover, since the discovery of CgA as a circulating component in patients with phochromocytoma (O'Connor and Bernstein 1984), a large body of literature implicates granins, notably CgA, as markers for a variety of diseases, such as neuroendocrine tumors, chronic heart failure and brain disorders like Parkinson's and Alzheimer's (O'Connor et al. 2000).

Since the turn of the century the research interest in the granins, notably in CgA and its derived peptides, has surged, as indicated by the registered 460 reviews since year 2000 of a total of 630 on CgA since 1970. Similarly, the number of papers dealing with VS-I and catestatin has grown steadily since their respective discoveries in 1992 and 1997, reaching a total of 76 and 154 for VS-I and catestatin in 2016. The major achievements will be outlined in the following sections.

1.5.1 Vasostatins, Vasodilations, the Vascular Endothelium and Angiogenesis

The human internal thoracic artery and saphenous vein were the first targets to be examined for vascular responses to the N-terminal CgA_{1-76} and CgA_{1-113} (Aardal and Helle 1992; Aardal et al. 1993). The potent contractions to endothelin-1 (ET-1) were suppressed, affecting the maximal sustained tension response but not the potency for ET-1, independent of the endothelium and extracellular calcium. Accordingly, the term vasostatins was assigned to these two N-terminal CgA peptides, numbered according to length, i.e. as VS-I and VS-II. Moreover, the arterial dilatations were independent of other constrictors over a functional range of transmural pressures, and the intrinsic and concentration-dependent dilator effects persisted at moderately elevated extracellular [K^+] in both arteries (Brekke et al. 2000; Brekke et al. 2002). Thus, in pressure-activated bovine resistance arteries the naturally occurring VS-I appeared to have a direct dilator potential involving hyperpolarization, acting via the N-terminal, loop-containing domain. Moreover, as the dilator effect of CgA_{1-40} in the coronary artery was diminished by pertussis toxin (PTX) and abolished by antagonists to several subtypes of K^+ channels, the mechanism of action seemingly involves a $G\alpha i/o$ subunit and K^+ channel activation in the signal pathway. Significant species differences in vasoactivity were on the other hand apparent, as neither the rat betagranin peptide $rCgA_{7-57}$ nor the bovine chromofungin, $bCgA_{47-66}$, had vasodilator effects in the rat cerebral artery (Mandalà et al. 2005).

The vascular endothelium appears by itself to be a significant target for granin-derived peptides, e.g. VS-I (Ferrero et al. 2004; Blois et al. 2006a), catestatin (Theurl et al. 2010) and secretoneurin (Kähler et al. 2002). Bovine aorta endothelial cells internalizes bovine CgA (Mandalà et al. 2000) and both human CgA (Hsiao et al. 1990) and human STACgA1–78 (Roatta et al. 2011) are distributed across the vascular endothelium in two pools, a minor fraction in the blood and a major pool in the interstitium. Moreover, CgA and VS-I protect the endothelial barrier against the gap-forming, permeabilizing activity of TNFα (Ferrero et al. 2004) via a mechanism involving cytoskeletal reorganization and downregulation of the transmembrane protein intercellular VE-cadherin, responsible for the cell-cell adhesion (Ferrero et al. 2004). In contrast, catestatin (Theurl et al. 2010) as well as secretoneurin (Yan et al. 2006) impair the integrity of the endothelial barrier, however by different mechanisms. Other studies have shown that VS-1 also inhibits endothelial cell migration, motility, sprouting, invasion and capillary-like structure formation induced by vascular endothelial growth factor (VEGF) and basic fibroblast growth factor (bFGF) (Belloni et al. 2007).

The most recent newcomer among vasoactive CgA peptides corresponds to the C-terminal sequence of VS-II, CgA_{79-113}, and has vasodilatory properties (Salem et al. 2015). This peptide, named the vasoconstriction-inhibiting factor (VIF), acts as a cofactor for the angiotensin II type 2 receptor. As the plasma concentration of VIF was significantly increased in renal patients and patients with heart failure, it seems evident that yet another CgA-derived player and yet other targets may be involved in blood pressure regulation and vascular pathophysiology.

1.5.2 Vasostatins, Catestatin and Serpinin; Myocardial Contractility and Protection against Ischemia-Induced Injury

A large body of evidence suggests that CgA, either present in circulation or produced by the heart itself, is a novel regulator of the heart. Indeed, under normal and pathophysiological conditions alike, the heart is under constant exposure not only to CA but also to the circulating CgA originating from the sympathoadrenal system.

CgA may also derive locally from myocardial production, notably in ventricles of heart failure patients (Pieroni et al. 2007). The full-length CgA dilates coronaries and induce negative inotropism and lusitropism in the ex vivo perfused rat heart at 0.1–4 nM, but not at higher concentrations (Pasqua et al. 2013). Of note, analysis of the perfusates showed that exogenous CgA was not cleaved by the heart, suggesting that the myocardial effects were induced by the circulating, full-length protein. However, the same study demonstrated that physically and chemically stimulated rodent hearts could proteolytically process the intracardiac, endogenous CgA into fragments (Glattard et al. 2006). Moreover, the increased plasma levels in chronic heart failure (Ceconi et al. 2002), its over-expression in human dilated and hypertrophic cardiomyopathy (Pieroni et al. 2007) and the observation that the circulating CgA provide prognostic information on long-term mortality, independent of conventional risk markers in acute coronary syndromes (Jansson et al. 2009), all point to a significant role of CgA in human cardiovascular homeostasis. Hence, the systemic and intracardiac fates of full-length CgA and its fragments imply intriguing new aspects of the myocardial handling of CgA under normal and pathophysiological conditions.

To what extent the elevated circulating levels of CgA, VS-I and catestatin together are beneficial or detrimental to the failing heart, remains unanswered. Taking into account that an inflammatory response is caused by myocardial injury arising from ischemic reperfusion (Anaya-Prado and Toledo-Pereyra 2002), a link between plasma CgA and/or its fragments in cardioprotection seems plausible. For instance, it is well established that the human recombinant VS-1 (hrSTACgA$_{1-78}$) preconditions the rat heart against myocardial necrosis arising in response to reperfusion of the ischemia-injured tissue, presumably involving the endothelial/endocardial adenosine/nitric oxide signaling pathway (Cappello et al. 2007). In contrast, catestatin, being without pre-conditioning effects, may modulate reperfusion injury during the post-ischemic reperfusion period (Penna et al. 2010; Penna et al. 2014). Hence, it seems likely that N- and C-terminal CgA fragments arising from processing of the circulating and intracardiac pools of CgA in species-specific patterns, may exert beneficial effects, not only under experimental conditions in animal models (Pasqua et al. 2013), but also in the failing human heart in situ.

Although the two structurally different CgA peptides, VS-I and catestatin, both exert negative myocardial inotropy, non-competitively inhibiting the β-adrenenoceptor on cardiomyocytes (Tota et al. 2008; Angelone et al. 2008), these apparently converging effects on the heart may be less puzzling when realizing that these two peptides may not reach peak concentrations in the same frame of time (Crippa et al. 2013). The thrombin.induced C-terminal processing of the anti-angiogenic, full length CgA into a catestatin-containing angiogenic fragment point to a functional rationale, namely maintaining protection of the heart against excessive adrenergic stimulation by CA, whether by VS-I or catestatin, regardless of the quiescent or stimulus-activated state of the vasculature.

History and Perspectives

The C-terminal peptide serpinin ($CgA_{403-428}$, Koshimizu et al. 2010), is a novel CgA-derived factor in cardiovascular modulations (Tota et al. 2012). This fragment was first described for its ability to signal the increase in transcription of the serine protease inhibitor, protease nexin-1 (PN-1), a potent inhibitor of plasmin released during inflammatory processing causing cell death. Two other forms have since been identified, (pGlu)serpinin and serpinin-Arg-Arg-Gly (Koshimizu et al. 2011a). In addition to the serpinin-like effect on increasing the levels of PN-1, (pGlu)serpinin also excerts anti-apoptotic effects of relevance to protection of neurons in the central nervous system (Koshimizu et al. 2011b). Intriguingly, in the perfused rat heart both serpinin and (p-Glu)serpinin excert positive inotropic and lusitropic effects via a β1-adrenergic receptor/adenylate cyclase/cAMP/PKA pathway (Tota et al. 2012), thus contrasting the inhibitory effects of VS-I and catestatin on the cardiac β2-adrenoceptor mediated activations. It remains to be seen to what extent and at what stage in the C-terminal processing of the full-length CgA the concentrations of serpinin and pGly-serpinin may reach their functional maxima (Loh et al. 2012).

1.5.3 Angiogenesis, Cell Adhesion and Tumor Progression

CgA appears to regulate angiogenesis and tumor growth in several models of solid tumors (Corti 2010), affecting fibroblasts (Dondossola et al. 2010) and endothelial cells (Corti and Ferrero 2012) in the tumor microenvironment. Recent studies have reveled that the full-length CgA contains one anti-angiogenic site in the C-terminal region ($CgA_{410-439}$) (Crippa et al. 2013), and another site in a latent form in the N-terminal domain CgA_{1-76}. Proteolytic liberation is necessary for full activation of the anti-angiogenic property of VS-I. Intriguingly, further processing of VS-1 leads to the antimicrobial peptide CgA_{47-66}, originally named chromofungin (Lugardon et al. 2001). Even this degradation product is able to cause negative inotropic effects and, like the unprocessed VS-I, to elicit post-conditional protection against ischemia/reperfusion damage (Filice et al. 2015).

Given the potential ability of CgA and/or its fragments to regulate tumor vessel biology, these molecules might also contribute to inhibit tumor growth, as shown in mouse lymphomas (Bianco et al. 2016) and mammary adenocarcinomas genetically engineered to release CgA locally (Colombo et al. 2002). In animal models both CgA and VS-I reduced the trafficking of tumor cells from tumor-to-blood, from blood-to-tumor and from blood-to-normal tissues (i.e. the tumor "self-seeding" and metastasis processes), by enhancing the endothelial barrier function and reducing the trans-endothelial migration of cancer cells (Dondossola et al. 2012). In certain tumor patients the CgA plasma levels may reach up to 10–100-fold. Whether these high levels of circulating CgA may also affect the growth and progression of non-neuroendocrine tumors, remains a challenging question awaiting detailed analyses of plasma concentrations of full-length CgA and VS-1 in these patients.

1.5.4 A Physiological Role for the Circulating CgA

Fifty years ago when the exocytotic release of CgA into the effluents from the stimulated adrenal medulla was first reported (Banks and Helle 1965; Blaschko et al. 1967), no functional significance was assigned to the released protein. Nearly twenty years lapsed before the enzymatically inactive CgA was detected in the circulation of pheochromocytoma patients (O'Connor and Bernstein 1984). After another thirty years the N- and C-terminal domains in CgA have finally been quantified in normal plasma, thanks to highly refined immunochemical analyses, revealing subnanomolar levels of both full length CgA (0.1 nM) and VS-1 (0.4 nM) (Crippa et al. 2013). This report was also the first to show that full-length CgA and VS-1 exerted potent anti-angiogenic activity when performed with biologically relevant concentrations in the various in vitro and in vivo assays. Rather unexpectedly, the anti-angiogenic property of the intact CgA was converted to a potent pro-angiogenic fragment corresponding to the catestatin-containing fragment CgA_{1-373} upon blood coagulation in a thrombin-dependent manner (Crippa et al. 2013). Thus, the full length CgA, VS-1 and the catestatin-containing peptide seemingly form a balance of anti- and pro-angiogenic factors tightly regulated by proteolysis as a functional response to tissue injury when repair of the damaged tissue is called for (Crippa et al. 2013; Helle and Corti 2015). Hence, a physiological role is finally apparent for the anti-angiogenic, full-length CgA and its N-terminal peptide VS-I when circulating at normal concentrations, namely in maintaining the vascular endothelium in a quiescent state by protecting its structural integrity and, in addition, protecting the myocardium against excessive β-adrenergic stimulation and detrimental effects of ischemia-induced injury.

1.5.5 Circulating CgA as a Marker for Inflammatory Diseases

Inflammatory processes, in particular those involving the cardiovascular system, pose clinical challenges in diagnosing and therapy. For instance, the elevated plasma CgA in chronic heart disease is a strong indicator of a relationship between high plasma CgA and pro-inflammatory markers (Corti et al. 2000) as well as an independent marker of mortality (Pieroni et al. 2007). Vascular inflammation may induce pathological arterial changes and variable blood pressure. Moreover, endothelial dysfunction is now recognized as a crucial factor in hypertension, with endothelial NO production as essential for maintenance of vascular tone, being compromised as a result of systemic and localized inflammatory responses (Watson et al. 2008).

1.6 Putative Receptors for CgA and CgA-Derived Peptides

Classical, high-affinity cell surface receptors have not yet been identified for most of the CgA-derived peptides. The exception is the nicotinic acetylcholine receptor for catestatin in the sympatoadrenal system mediating the autocrine inhibitory effect of catestatin on CA secretion (Mahata et al. 1997). On the other hand, binding studies have shown that VS-I and chromofungin engage in electrostatic and hydrophobic interactions with membrane-relevant phospholipids at physiological conditions, particularly with phosphatidylserine (Blois et al. 2006b). Moreover, binding to membrane proteins with molecular weights 74 and 78 kDa were early findings for VS-I both in cultured calf smooth muscle and parathyroid cells, respectively (Angeletti et al. 1994; Russell et al. 1994). Similarly, a 70 kDa glycoprotein coupled to two different G-proteins was detected as the receptor for pancreastatin in adipocytes and hepatocytes (Sanchez-Margalet et al. 1996; Sanchez-Margalet et al. 2000). Also catestatin, eliciting histamine-release from rat mast cells, does so via its cationic and amphipatic properties (Krüger et al. 2003). Thus, analogous to the cell penetrating properties of cationic and amphipatic peptides in microorganisms (Metz-Boutigue et al. 2004), both VS-I and catestatin have been postulated to interact with and penetrate into mammalian cells via their cationic and amphipathic properties (Helle et al. 2007; Helle 2010b). Consistent with this hypothesis VS-I was reported to activate PI3K-dependent e-NOS phosphorylation via binding to a heparin sulphate proteoglycan, leading to caveolae endocytosis in bovine aortic endothelial cells (Ramella et al. 2010). A role for heparin sulphate proteoglycan as a cell surface endocytosis receptor entry of macromolecules in mammalian cells has recently gained strong support (Christianson and Belting 2014). A selective binding of CgA and VS-I to the epithelial integrin αvß6 was also recently demonstrated in a study of would healing in injured mice (Curnis et al. 2012). Integrin αvß6 belongs to a large family of heterodimeric transmembrane glycoproteins that attach cells to extracellular matrix proteins of the basement membrane. Notably, the interaction of the RGD/α-helix motif of CgA with avß6-integrin could regulate keratinocyte physiology in wound healing (Curnis et al. 2012). Although αvß6 is upregulated in tissue repair and in cancer (Bandyopadhyay and Raghavan 2009) it remains to be seen whether circulating CgA and VS-I bind to this integrin also in cancerous tissues. An indirect involvement of integrins was observed for VS-I via the phospholipid-binding amphiphilic α-helix within the chromofungin sequence CgA47-66 and the hydrophilic C-terminus CgA67-78 in murine and human dermal fibroblasts (Dondossola et al. 2010). This adhesion mechanism required cytoskeleton rearrangement but not protein synthesis, enhancing fibroblast adhesion to solid-phases.

G-protein-regulated signalling pathways coupled to Gαi/o subunits are commonly identified by their activation by PTX, the *Bordetella pertussis* toxin. So far, most reports on binding of CgA-derived peptides to membrane proteins refer on PTX-sensitive effects, suggesting coupling to G-proteins containing Gαi-subunits. Intriguingly, not only the glycoprotein receptor for parastatin in adipocytes and hepatocytes was sensitive to PTX (Sanchez-Margalet et al. 1996), but also the dilator effect of CgA_{1-40} in the coronary artery (Brekke et al. 2002) and the inhibitory effect of VS-I on gap-formation via a blockade of the activation of p38MAPK by PTX in pulmonary and coronary arterial endothelial cells (Blois et al. 2006a). Likewise, the catestatin induced release from rat mast cells was sensitive to PTX (Krüger et al. 2003). On the other hand, catestatin as well as VS-1 signal via AKT/PKB to eNOS mediating their inhibitory effects in the rat heart (Angelone et al. 2008; Tota et al. 2008). Thus, in the rat heart different G-proteins may be involved in the NO-production by VS-I and catestatin, both serving as non-competitive inhibitors of the β-adrenoceptor. Hence, not only pancreastatin, but also VS-I and catestatin appear to interact with membrane constituents via their membrane-penetrating properties, coupling to distinctly different G-protein-coupled pathways, in some tissues involving PTX-sensitive Gαi/o subunits (Helle 2010b).

2 Conclusions

While the research history of the adrenal chromaffin cells dates back to the mid 1850ies, our knowledge of the granins reflects research from the most recent six decades. The accumulated literature has unraveled that these unique proteins serve as prohormones for a range of regulatory peptides with widely different effects and target tissues. Moreover, their properties seemingly fit into patterns of functionally protective activities, e.g. in calcium, glucose and vascular homeostasis, in angiogenesis, tissue repair and heart physiology, with implications also for the diagnosis and treatment of a wide range of neuroendocrine tumors, inflammatory pathologies and cardiovascular diseases. Thus, the co-release of granins with CA and other biogenic amines opens for a novel concept for the diffuse sympathoendocrine system, namely that of buffering and counterbalancing the immediate responses to the stress-activated system.

A dual role for the circulating full-length CgA is now apparent, protecting the vascular endothelium by inhibiting angiogenesis under normal conditions, yet accelerating local angiogenesis in response to tissue damage, e.g. after C-terminal cleavage of the prohormone by thrombin. In additiom, the circulating pool of VS-I, which seemingly contributes to preservation of endothelial cell quiescence, may also serve to counter-balance the pro-angiogenic activity of catestatin-containing fragments when released in the systemic circulation from various sites of injury.

Although classical members of the high-affinity, transmembrane-spanning classes of receptors have yet to be linked to the effects of most of the CgA-derived peptides, other receptor classes have been implicated; in addition to G-proteins for VS-I, pancreastatin and catestatin, for VS-I also the cell surface endocytosis receptor heparin sulphate proteoglycan and the epithelial, transmembrane glycoprotein, the integrin $\alpha v \beta 6$.

3 Perspectives

The few reports on CgA processing in patients so far published, indicate complex and disease-related patterns of fragments. For instance, decreased levels of plasma catestatin are characteristic of patients with essential hypertension and also in normotensive subjects with a family history of hypertension and increased epinephrine secretion (O'Connor et al. 2002). On the other hand, elevated plasma levels of VS-I occur in critically ill patients (Schneider et al. 2012) and of catestatin in patients with coronary heart disease and after acute myocardial infarction (Liu et al. 2013; Meng et al. 2013). On the other hand, in patients with chronic kidney disease and heart failure a new fragment derived from VS-II (VIF), is elevated (Salem et al. 2015) while VS-I and fragments, lacking the anti-angiogenic C-terminal region of CgA were increased in patients suffering from a rare form of systemic, inflammatory large vessel vasculitis although the levels of the CgA fragments did not reflect disease activity or extent (Tombetti et al. 2016). Hence, research into the pathophysiological patterns of CgA and its processing in cardiovascular and inflammatory diseases and in tumors emerges as a major challenge in order to assess whether a given pattern of circulating CgA fragments is beneficiary or detrimental to the survival of the afflicted patient.

References

Aardal S, Helle KB (1992) The vasoinhibitory activity of bovine chromogranin A fragment (vasostatin) and its independence of extracellular calcium in isolated segments of human blood vessels. Regul Pept 41:9–18

Aardal S, Helle KB, Elsayed S, Reed RK, Serch-Hanssen G (1993) Vasostatins, comprising the N-terminal domain of chromogranin A, suppress tension in isolated human blood vessel segments. J Neuroendocrinol 5:105–112

Anaya-Prado R, Toledo-Pereyra LH (2002) The molecular events underlying ischemia/reperfusion injury. Transplant Proc 34:2518–2519

Angeletti RH, D'Amico T, Russell J (2000) Regulation of parathyroid secretion. Adv Exp Med Biol 482:217–223

Angeletti RH, Aardal S, Serch-Hanssen G, Gee P, Helle KB (1994) Vasoinhibitory activity of synthetic peptides from the amino terminus of chromogranin A. Acta Physiol Scand 132:11–19

Angeletti RH, Mints L, Aber C, Russell J (1996) Determination of residues in chromogranin A (16–40) required for inhibition of parathyroid hormone secretion. Endocrinology 137:2918–2922

Angelone T, Quintieri AM, Brar BK, Limchaiyawat PT, Tota B, Mahata SK, Cerra MC (2008) The antihypertensive chromogranin A peptide catestatin acts as a novel endocrine/paracrine modulator of cardiac inotropism and lusitropism. Endocrinology 149:4780–4793

Anouar Y, Desmoucelles C, Vaudry H (2000) Neuroendocrine cell-specific expression and regulation of the human secretogranin II gene. Adv Exp Med Biol 482:113–124

Aunis D, Metz-Boutigue M-H (2000) Chromogranins: current concepts. Structural and functional aspects. Adv Exp Med Biol 482:21–38

Bandyopadhyay A, Raghavan S (2009) Defining the role of integrin alphavbeta6 in cancer. Curr Drug Targets 10:645–652

Banks P, Helle KB (1965) The release of protein from the stimulated adrenal medulla. Biochem J 97:40C–41C

Banks P, Helle KB, Major D (1969) Evidence for the presence of a chromogranin-like protein in bovine splenic nerve granules. Mol Pharmacol 5:210–212

Belloni D, Scabini S, Foglieni C, Vescini L, Giazzon A, Colombo B et al (2007) The vasostatin-I fragment of chromogranin A inhibits VEGF-induced endothelial cell proliferation and migration. FASEB J 21:3052–3062

Bianco H, Gasparri A, Generoso L, Assi E, Colombo B, Scarfò L, Bertilaccio MT, Scieizo C, Ranghetti P, Dondossola E, Ponzoni M, Caligaris-Cappio F, Ghia P, Corti A (2016) Inhibition of chronic lymphocytic leucemia progession by full length chromogranin A and its N-terminal fragment in mouse models. Oncotarget 7:41725–41736. 9407.

Blaschko H, Welsch AD (1953) Localization of adrenaline in cytoplasmic particles of the bovine adrenal medulla. Arch Exp Path Pharmak 219:17–22

Blaschko H, Born GVR, D'Iorio A, Eade NR (1956) Observations on the distribution of catecholamines and adenosine phosphate in the bovine adrenal medulla. J Physiol Lond 183:548–557

Blaschko H, Comline RS, Schneider FH, Silver M, Smith AD (1967) Secretion of a chromaffin granule protein, Chromogranin, from the adrenal gland after splanchnic stimulation. Nature (Lond) 215:58–59

Blois A, Srebro B, Mandalà M, Corti A, Helle KB, Serck-Hanssen G (2006b) The chromogranin A peptide vasostatin-I inhibits gap formation and signal transduction mediated by inflammatory agents in cultured bovine pulmonary and coronary arterial endothelial cells. Regul Pept 135:78–84

Blois A, Holmsen H, Martino G, CortiA M-BM-H, Helle KB (2006a) Interactions of chromogranin A-derived vasostatins and monolayers of phosphatidyl serine, phosphatidylcholine and phosphatidylethanolamine. Regul Pept 134:30–37

Borges R, Machado JD, Alonso C, Brioso MA, Goméz JF (2000) Functional role of chromogranins: the intragranular matrix in the last phase of exocytosis. Adv Exp Med Biol 482:69–82

Brekke JF, Kirkeleit J, Lugardon K, Helle KB (2000) Vasostatins. Dilators of bovine resistance arteries. Adv Exp Med Biol 482:239–246

Brekke JF, Osol GJ, Helle KB (2002) N-terminal chromogranin derived peptides as dilators of bovine coronary resistance arteries in vitro. Regul Pept 105:93–100

Cappello S, Angelone T, Tota B, Pagliaro P, Penna C, Rastaldo R, Corti A, Losano G, Cerra MC (2007) Human recombinant chromogranin A-derived vasostatin-1 mimics preconditioning via an adenosine/nitric oxide signaling mechanism. Am J Physiol Heart Circul Physiol 293:H719–H727

Ceconi C, Ferrari R, Bachetti T, Opasich C, Volterrani M, Colombo B, Parrinello G, Corti A (2002) Chromogranin A in heart failure; a novel neurohumoral factor and a predictor for mortality. Eur Heart J 23:967–974

Christianson HC, Belting M (2014) Heparan sulphate proteoglycan as a cell surface endocytosis receptor. Matrix Biol 35:51–55

Ciesielski-Treska J, Aunis D (2000) Chromogranin A introduces a neurotoxic phenotype in brain microglial cells. Adv Exp Med Biol 482:291–298

Cohn DV, Morrisey JJ, Hamilton JW, Shofstall RE, Smardo FL, Chu LLH (1981) Biochemistry 20:4135–4140

Colombo B, Curnis F, Foglieni C, Monno A, Arrigoni G, Corti A (2002) Chromogranin a expression in neoplastic cells affects tumor growth and morphogenesis in mouse models. Cancer Res 62:941–946

Corti A (2010) Chromogranin A and the tumor microenvironment. Cell Mol Neurobiol 30:1163–1170

Corti A, Ferrero E (2012) Chromogranin A and the endothelial barrier function. Curr Med Chem 19:4051–4058

Corti A, Ferrari R, Ceconi C (2000) Chromogranin A and tumor necrosis factor-alpha (TNF) in chronic heart failure. Adv Exp Med Biol 482:351–359

Coupland RE (1968) Determining sizes and distribution of sizes of spherical bodies such as chromaffin granules in tissue sections. Nature (Lond) 217:384–388

Crippa L, Bianco M, Colombo B, Gasparri AM, Ferrero E, LohYP CF, Corti A (2013) A new chromogranin A-dependent angiogenic switch activated by thrombin. Blood 121:392–402

Curnis F, Gasparri A, Longhi R, Colombo B, D'Alessio S, Pastorino F, Ponzoni M, Corti A (2012) Chromogranin A binds to avb6-integrin and promotes wound healing in mice. Cell Mol Life Sci 69:2791–2803

Curry WJ, Norlen P, Barkatullah SC, Johnston CF, Håkanson R, Hutton JC (2000) Chromogranin A and its derived peptides in the rat and porcine gastro-entero-pancreatic system: expression, localization and characterization. Adv Exp Med Biol 482:205–213

Dahlstrøm A (1966) The intraneuronal distribution of noradrenaline and the transport and lifespan of aqmine storage granules in the sympathetic adrenergic neuron, a histological and biochemical study. Ivar Högströms Trykkeri AB, Stockholm

De Robertis E, Pellegrino de Iraldi A (1961) Plurovesicular secretory processes and nerve endings in the pineal gland of the rat. J Biophys Biochem Cytol 10:361–372

Dondossola E, Gasparri A, Bachi A, Longhi R, Metz-Boutigue MH, Tota B, Helle KB, Curnis F, Cort A (2010) Role of vasostatin-1 C-terminal region in fibroblast cell adhesion. Cellular and molecular life sciences : Cell Mol Life Sci 67:2107–2118

Dondossola E, Crippa L, Colombo B, Ferrero E, Corti A (2012) Chromogranin a regulates tumor self-seeding and dissemination. Cancer Res 72:449–459

Efendic S, Tatemoto K, Mutt V, Quan C, Chang D, Ostenson CG (1987) Pancreastatin and islet hormone release. Proc Natl Acad Sci U S A 84:7257–7260

Eiden L (1987) Is chromogranin A a prohormone? Nature 325:301

Von Euler US (1946) A specific sympathomimetic ergone in adrenergic nerve fibres (Sympathin) and its relations to adrenaline and noradrenaline. Acta Physiol Scand 12:73–97

Von Euler US (1958) The presence of the adrenergic neurotransmitter in intraaxonal structures. Acta Physiol Scand 43:155–166

Von Euler US, Hillarp N-Å (1956) Evidence of the presence of noradrenaline in submicroscopic structures of adrenergic axons. Nature (Lond) 177:44–45

Falck B, Hillarp N-Å, Högberg B (1956) Content and intracellular distribution of adenosine triphosphate in cow adrenal medulla. Acta Physiol Scand 36:360–376

Fasciotto BH, Gorr S-U, Bourdeau AM, Cohn DV (1990) Autocrine regulation of patathyroid secretion: inhibition of secretion by chromogranin A (secretory protein-I) and potentiation of secretion by chromogranin A and pancreastatin antibodies. Endocrinology 133:461–466

Ferrero E, Scabini S, Magni E, Foglieni C, Belloni D, Colombo B, Curnis F, Villa A, Ferrero ME, Corti A (2004) Chromogranin A protects vessels against tumor necrosis factor alpha-induced vascular leakage. FASEB J 18:554–555

Filice E, Pasqua T, Quintieri AM, Cantafio P, Scavello F, Amodio N, Cerra MC, Marban C, Schneider F, Metz-Boutigue M-H, Angelone T (2015) Chromofungin, CgA 47–66 –derived peptide, produces basal cardiac effects and postconditioning cardioprotective action during ischemia/reperfusion injury. Peptides 71:40–48

Gerdes H-H, Glombik MM (2000) Signal-mediated sorting of secretory granules. Adv Exp Med Biol 482:41–54

Glattard E, Angelone T, Strub JM, Corti A, Aunis D, Tota B, Metz-Boutigue MH, Goumon Y (2006) Characterization of natural vasostatin-containing peptides in rat heart. FEBS J 273:3311–3321

Hagen P, Barnett RJ (1960) The storage of amines in the chromaffin cell. In: Wolstenholme GEW, O'Connor MO (eds) Adrenergic Mechanisms. Ciba Found Symp, Churchill, London, pp 83–99

Helle KB (1966a) Some chemical and physical properties of the soluble protein fraction of bovine adrenal cgromaffin granules. Mol Pharmacol 2:298–310

Helle K (1966b) Antibody formation against soluble protein from bovine adrenal chromaffin granules. Biochim Biophys Acta 117:107–110

Helle KB (2004) The granin family of uniquely acidic proteins of the diffuse neuroendocrine asystem: comparative and functional aspects. Biol Rev Camb Philos Soc 79:769–794

Helle KB (2010a) Regulatory peptides from chromogranin A and secretogranin II. Putative modulators of cells and tissues involved in inflammatory conditions. Regul Pept 165:45–51

Helle KB (2010b) The chromogranin A-derived peptides vasostatin.I and catestatin as regulatory peptides for cardiovascular functions. Cardiovasc Res 85:9–16

Helle KB, Angeletti RH (1994) Chromogranin A: a multipurpose prohormone? Acta Physiol Scand 152:1–10

Helle KB, Aunis D (2000b) A physiological role for the granins as prohormones for homeostatically important regulary peptides? Adv Exp Med Biol 482:389–397

Helle KB, Aunis D (2000a) Chromogranins. Functional and Clinical Aspects. Adv. Exp. Med. Boil 482:1–405

Helle KB, Corti A (2015) Chromogranin A: a paradoxical player in angiogenesis and vascular biology. Cell Mol Life Sci 72:339–348

Helle KB, Corti A, Metz-Boutigue MH, Tota B (2007) The endocrine role for chromogranin A: a prohormone for peptides with regulatory properties. Cell Mol Life Sci 64:2863–2886

Hillarp N-Å, Lagerstedt S, Nilson B (1953) The isolation of a granular fraction from the subrarenal medulla, containing the sympathomimetic catecholamines. Acta Physiol Scand 29:251–263

Hillarp N-Å (1959) Further observations on the state of the chatecholamines stored in the adrenal medullary granules. Acta Physiol Scand 47:271–279

Hoffmann JA, Kafator FC, Janeway CA Jr, Ezkowitz RAB (1999) Phylogenetic perspectives in innate immunity. Science 284:1313–1318

Hsiao RJ, Seerger RC, Yyu A, O'Connor DT (1990) Chromogranin A in children with neuroblastoma: plasma concentration parallels disease stage and predicts survival. J Clin Invest 85:1555–1559

Huttner WB, Benedum UM (1987) Chromogranin A and pancreastatin. Nature 325:305

Huttner WB, Gerdes H-H, Rosa P (1991) Chromogranins/secretogranins-widespread consttituets of the secretory granule matrix in endocrine cells and neurons. In: Gratzl M, Langley K (eds) Markers for neural and endocrine cells. VCH, Weinheim, pp 93–133

Jansson AM, Rosjo H, Omland T, Karlsson T, Hartford M, Flyvbjerg A, Caidahl K (2009) Prognostic value of circulating chromogranin A levels in acute coronary syndromes. Eur Heart J 30:25–32

Kähler CM, Fischer-Colbrie R (2000) A novel link between the nervous and the immune system. Adv Exp Med Biol 482:279–290

Kähler CM, Kaufmann G, Kähler ST, Wiedermann CJ (2002) The neuropeptide secretoneurin stimulates adhesion of human monocytes to arterial and venous endothelial cells in vitro. Regul Pept 110:65–73

Kirchmair R, Hogue-Angeletti R, Guitirrez J, Fischer-Colbrie R, Winkler H (1993) Secretoneurin – a neuropeptide generated in brain, adrenal medulla, and other endocrine tissues by proteolytic processing of secretoneurin II (Chromogranin C). Neuroscience 53:359–365

Konecki DS, Benedum UM, Gerdes HH, Huttner WB (1987) The primary structure o human chromogranin A and pancreastatin. J Biol Chem 261:17026–17030

Koshimizu H, Kim T, Cawley NX, Loh YP (2010) Chromogranin A: a new proposal for traffiking, processing and induction of granule biogenesis. Regul Pept 160:153–159

Koshimizu H, Cawley NX, Kim T, Yergey AL, Loh YP (2011a) Serpinin: a novel chromogranin A-derived, secreted peptide up-regulates protease nex-1 expression and granule biogenesis in endocrine cells. Mol Endocrinol 25:732–744

Koshimizu H, Cawley NX, Yergey AL, Loh YP (2011b) Role of pGlu-serpinin; a novel chromogranin A-derived peptide in inhibition of cell death. J Mol Neurosci 45:294–303

Krüger P-G, Mahata SK, Helle KB (2003) Catestatin (CgA344–364) stimulates rat mast cell release of histamine in a manner comparable to mastoparan and other cationic charged neuropeptides. Regul Pept 114:29–35

Lever JD (1955) Electron microscopic observations on the normal and denervated adrenal medulla of the rat. Endocrinol 57:621–635

Liu L, Ding W, Zhao F, Shi L, Pang Y, Tang C (2013) Plasma levels and potential roles of catestatin in patients with coronary heart disease. Scand Cardiovasc J 47:217–224

Loewi O (1921) Über Humorale Übertragbarkeit der Herznervenwirkung. Pflüg Arch Ges Physiol 189:239–241

Loh YP, Koshimizu H, Cawley NX, Tota B (2012) Serpinins: role in granule biogenesis, inhibition of cell death and cardiac functions. Curr Med Chem 19:4086–4092

Lugardon K, Raffner R, Goumon Y, Corti A, Delmas A, Bulet P, Aunis D, Metz-Boutigue M-H (2000) Antibacterial and antifungal activities of vasostatin-1, the N-terminal fragment of chromogranin A. J Biol Chem 275:10745–10753

Lugardon K, Chasserot-Golaz S, Kiefler AE, Maget-Dana R, Nullans G, Kiefler B, Aunis D, Metz-Boutigue M-H (2001) Structural and biochemical characterization of chromofungin, the antifungal chromogranin A-(47–68)-derived peptide. J Biol Chem 276:35875–35882

Mandalà M, Stridsberg M, Helle KB, Serck-Hanssen G (2000) Endothelial handling of chromogranin A. Adv Exp Med Biol 482:167–178

Mandalà M, Brekke JF, Serck-Hanssen G, Metz-Boutigue MH, Helle KB (2005) Chromogranin A-derived peptides: interaction with rat posterior cerebral artery. Regul Pept 124:73–80

Mahata SK, O'Connor DT, Mahata M, Yoo SH, Taupenot L, Wu H, Gill BM, Parmer RJ (1997) Novel autocrine feedback control of catecholamine release. A discrete chromogranin a fragment is a noncompetitive nicotinic cholinergic antagonist. J Clin Invest 100:1623–1633

Mahata SK, Mahata M, Fung M, O'Connor DT (2010) Catestatin: a multifunctional peptide from chromogranin A. Regul Pept 162:33–43

Meng L, Wang J, Ding WH, Han P, Yang Y, Qi LT, Zhang BW (2013) Plasma catestatin level in patients with acute myocardial infarction and its correlation with ventricular remodelling. Postgrad Med J 89:193–196

Metz-Boutigue M-H, Garcia-Sablone P, Hogue-Angeletti R, Aunis D (1993) Intracellular and extracellular processing of chromogranin A: determination of cleavage sites. Europ J Biochem 217:247–257

Metz-Boutigue M-H, Goumon Y, Lugardon K, Stub JM, Aunis D (1998) Antibacterial peptides are present in chromaffin cell secretory granules. Cell Mol Neurobiol 18:249–266

Metz-Boutigue M-H, Lugardon K, Goumon Y, Raffner R, Strub J-M, Aunis D (2000) Antibacterial and antifungal peptides derived from chromogranins and proenkephalin.A: from structural to biological aspects. Adv Exp Med Biol 482:299–315

Metz-Boutigue M-H, Helle KB, Aunis A (2004) The innate immunity: roles for antifungal and antibacterial peptides secreted by chromffaffin granules from the adrenal medulla. Curr Med Chem-Immun, Endoc Metab Agents 4:169–177

O'Connor DT, Bernstein KN (1984) Radioimmunoassay of chromogranin A in plasma as a measure of exocytotic sympathoadrenal activity in normal subjects and patients with pheochromocytoma. N Engl J Med 311:764–770

O'Connor DT, Mahata SK, Taupenot L, Mahata M, Taylor CVL (2000) Chromogranin in human disease. Adv Exp Med Biol 482:377–388

O'Connor DT, Kailasam MT, Kennedy BP, Ziegler MG, Yanaihara N, Parmer RJ (2002) Early decline in the catecholamine release-inhibitory peptide catestatin in humans at genetic risk of hypertension. J Hypertens 20:1335–1345

Oliver G, Schäfer EA (1895) The physiological effects of extracts of the supraadrenal capsules. J Physiol Lond 18:230

Parmer RJ, Mahata M, Gong Y, Jiang Q, O'Connor DT, Xi XP, Miles LA (2000) Processing of chromogranin A by plasmin provides a novel mechanism for regulating catecholamine secretion. J Clin Invest 106:907–915

Pasqua T, Corti A, Gentile S, Pochini L, Bianco M, Metz-Boutigue MH, Cerra MC, Tota B, Angelone T (2013) Full-length human Chromogranin-A cardioactivity: myocardial, coronary and stimulus-induced processing evidence in normotensive and hypertensive male rat hearts. Endocrinology 154:3353–3365

Penna C, Alloatti G, Gallo MP, Cerra MC, Levi R, Tullio F, Bassino E, Dolgetta S, Mahata SK, Tota B et al (2010) Catestatin improves post-ischemic left ventricular function and decreases ischemia/reperfusion injury in heart. Cell Mol Neurobiol 30:1171–1179

Penna C, Pasqua T, Amelio D, Perrelli M-G, Angotti C, Tullio F, Mahata SK, Tota B, Pagliaro P, Cerra MC, Angelone T (2014) Catestatin increases the expression of anti-apoptotic and proangiogenetic factors in the post-ischemic hypertrophied heart of SHR. PLoS One 9:e102536. pp1–11

Pieroni M, Corti A, Tota B, Curnis F, Angelone T, Colombo B, Cerra MC, Bellocci F, Crea F, Maseri A (2007) Myocardial production of chromogranin A in human heart: a new regulatory peptide of cardiac function. Eur Heart J 28:1117–1127

Portela-Gomes GM (2000) Chromogranin A immunoreactivity in neuroendocrine cells in the human gastrointestinal tract and pancreas. Adv Exp Med Biol 482:193–204

Radek KA, Lopez-Garcia B, Hupe M, Niesman IR, Elias PM, Taupenot L, Mahata SK, O'Connor DT, Gallo RL (2008) The neuroendocrine peptide catestatin is a cutaneous antimicrobial and induced in the skin after injury. J Invest Dermatol 2008 128:1525–1534

Ramella R, Boero O, Alloatti G, Angelone T, Levi R, Gallo MP (2010) Vasostatin-1 activates eNOS in endothelial cells through a proteoglycan-dependent mechanism. J Cell Biochem 110:70–79

Roatta S, Passatore M, Novello M, Colombo B, Dondossola E, Mohammed M, Losano G, Corti A, Helle KB (2011) The chromogranin A-derived peptide vasostatin-I: in vivo effects on cardiovascular variables in the rabbit. Regul Pept 168:10–20

Russell J, Gee P, Liu SM, Angeletti RH (1994) Stimulation of parathyroid hormone secretion by low calcium is inhibited by amino terminal chromogranin peptides. Endocrinology 135:337–342

Salem S, Jankowski V, Asare Y, Liehn E, Welker P, Raya-Bermudez A, Pineda-Martos C, Rodrigues M, Muñoz-Castrañeda JR, Bruck H, Marx N, Machado FB, Straudt M, Heinze G, Zidek W, Jankowski J (2015) Identification of the vasoconstriction-inhibiting factor (VIF), a potent endogenous cofactor of angiotensin II acting on the angiotensin II type receptor. Circulation 131:1426–1434

Sanchez-Margalet V, Gonzalez-Yanes C (1998) Pancreastatin inhibits insulin action in rat adipocytes. Am J Phys 275:E1055–E1060

Sanchez-Margalet V, Gonzalez-Yanes C, Najib S (2001) Pancreastatin, a chromogranin A-derived peptide, inhibits DNA and protein synthesis by producing nitric oxide in HTC rat hepatpma cells. J Hepatol 35:80–85

Sanchez-Margalet V, Gonzalez-Yanes C, Santos-Alvarez J, Najib S (2000) Pancreastatin. Biological effects and mechanism of action. Adv Exp Med Biol 482:247–262

Sanchez-Margalet V, Lucas M, Goberna R (1996) Pancreastatin action in the liver: dual coupling to different G-proteins. Cell Signal 8:9–12

Schneider F, Bach C, Chung H, Crippa L, Lavaux T, Bollaert PE, Wolff M, Corti A, Launoy A, Delabranche X et al (2012) Vasostatin-I, a chromogranin A-derived peptide, in non-selected critically ill patients: distribution, kinetics, and prognostic significance. Intensive Care Med 38:1514–1522

Schümann HJ (1958) Über den Noradrenaline und ATP-gehalt sympatischer Nerven. Arch Exp Path Pharmak 233:296–300

Shooshtarizodeh P, Zhang D, Chich JF, Gasnier C, Schneider F, Aunis D, Metz-Boutique M-H (2010) The antimicrobial peptides derived from chromogranin/secretogranin family, new actors of innate immunity. Regul Pept 165:102–110

Smith AD, Winkler H (1967) Purification and properties of an acidic protein from chromaffin granules of bovine adrenal medulla. Biochem J 103:483–492

Smith WJ, Kirshner A (1967) A specific soluble protein from the catecholamine storage vesicles of bovine adrenal medulla. Mol Pharmacol 3:52–62

Stoltz F (1904) Über Adrenaline und Alkylamineacetobrenzcathechin. Ber Deutsch Chem Gesellsch 37:4149–4154

Strub JM, Garcia-Sablone P, Lønning K, Taupenot L, Hubert P, van Dorsselar A, Aunis D, Metz Boutigue MH (1995) Processing of chromogranin B in bovineadrenal medulla; identification of secretolytin, the endotenous C-terminal fragment of residues 614–626 with antibacterial activity. Eur J Biochem 229:356–368

Tatemoto K, Efendic S, Mutt V, Makk G, Feistner GJ, Barchas JD (1986) Pancreastatin, a novel pancreatic peptide that inhibits insulin secretion. Nature 324:476–478

Taupenot L, Mahata M, Mahata SK, Wu H, O'Connor DT (2000) Regulation of chromogranin A transcription and catecholamine secretion by the neuropeptide PACAP. Adv Exp Med Biol 482:97–112

Theurl M, Schgoer W, Albrecht K, Jeschke J, Egger M, Beer AG, Vasiljevic D, Rong S, Wolf AM, Bahlmann FH et al (2010) The neuropeptide catestatin acts as a novel angiogenic cytokine via a basic fibroblast growth factor-dependent mechanism. Circ Res 107:1326–1335

Tombetti E, Colombo B, Di Chio MC, Sartorelli S, Papa M, Salerno A, Bozzolo EP, Tombolini E, Benedetti G, Godi C, Lanzani C, Rovere-Querini P, Del Maschio A, Ambrosi A, De Cobelli F, Sabbadini MG, Baldisera E, Corti A, Manfredi AA (2016) Chromogranin-A production and fragmentation in patients with Takayasy arteritis. Arthritis Res Ther 18:187–200

Tota B, Angelone T, Mazza A, Cerra CM (2008) The chromogranin A derived vasostatins: new players in the endocrine heart. Curr. Med. Chem 15:1444–1451

Tota B, Gentile S, Pasqua T, Bassino E, Koshimizu H, Cawley NX, Cerra MC, Loh YP, Angelone T (2012) The novel chromogranin A-derived serpinin and pyroglutaminated serpinin peptides are positive cardiac beta-adrenergic-like inotropes. FASEB J 26:2888–2898

Trudeau VL, Martyniuk CJ, Zhao E, Hu H, Volkoff H, Decatur WA, Basak A (2012) Is secretoneurin a new hormone? Gen Comp Endocrinol 175:10–18

Valicherla GR, Hossain Z, Mahata SK, Gayen JR (2013) Pancreastatin is an endogenous peptide that regulates glucose homeostasis. Physiol Genomics 45:1060–1071

Vulpian M (1856) Note sur quelque reactions proper a la substances des capsules surrénales. C R Acad Sci (Paris) 43:663

Watson T, Goon PK, Lip GY (2008) Endothelial progenitor cells, endothelial dysfunction, inflammation, and oxidative stress in hypertension. Antioxid Redox Signal 10:1079–1088

Welzstein R (1957) Electronenmichroscopische Untersuchungen aus Nebennierenmark von Maus, Meehrschweinchen und Katze. Z Zellforsch 46:517–576

Winkler H, Fischer-Colbrie R (1992) The chromogranins A and B: the first 25 years and future perspectives. Neuroscience 49:497–528

Yan S, Wang X, Wang H, Yao Q, Chen C (2006) Secretoneurin increases monolayer permeability in human coronary artery endothelial cells. Surgery 140:243–251

Zhang D, Shooshtarizadeh P, Laventie B-J, Colin DA, Chich J-F, Vidic J, de Barry J, Chasserot-Golaz S, Delalande F, Van Dorsselaer A, Schneider F, Helle K, Aunis D, Prevost G, Metz-Boutigue M-H (2009) Two chromogranin A-derived peptides induce calcium entry in human neutrophils by calmodulin-regulated alcium independent phospholipase A2. PLoS One 4:e4501. pp1–14

Secretogranin II: Novel Insights into Expression and Function of the Precursor of the Neuropeptide Secretoneurin

Reiner Fischer-Colbrie, Markus Theurl, and Rudolf Kirchmair

Abstract SgII is an acidic secretory which belongs to the family of chromogranins. It is present in the large-dense cored vesicles of the regulated secretory pathway of many neurons and endocrine cells and it is well conserved during evolution. Like chromogranin A, SgII can induce granulogenesis in endocrine cells but also in cells typically lacking secretory vesicles like fibroblasts. In the secretory vesicles SgII is processed to smaller peptides, e.g. secretoneurin, EM66 and manserin. For secretoneurin several biological effects like induction of neurotransmitter release, chemotactic activity towards immune-, endothelial- and muscle cells, and potent angiogenic and vasculogenic properties have been established. Thus, SN displays potent hormonal and paracrine effects, which help to orchestrate development, maintenance, physiologic activity and repair of the surrounding tissue. In addition, SgII has been established as valuable biomarker for endocrine tumours and cardiovascular diseases.

1 Introduction

Secretogranin II (SgII) – also named chromogranin C - is an acidic secretory protein expressed in many neurons and endocrine tissues. It was initially isolated from rat PC12 and the soluble content of bovine chromaffin granules as third main secretory protein and belongs thus to the family of chromogranins. The chromogranins are typically widely expressed throughout neuronal and endocrine tissues. They are stored within the cell in large dense secretory vesicles and released via the regulated pathway in a calcium-dependent manner. The chromogranins represent

---

R. Fischer-Colbrie (✉)
Department of Pharmacology, Medical University Innsbruck,
Peter Mayr Str. 1a, A-6020 Innsbruck, Austria
e-mail: fischer-colbrie@i-med.ac.at

M. Theurl • R. Kirchmair
Cardiology & Angiology, Internal Medicine III, Medical University Innsbruck, Innsbruck, Austria

proteins of 50–120 kD, which share several features, such as acidic nature, heat-stability and proteolytic processing to smaller peptides. The so-called granin motif (D/E-S/N-L-S/A/N-X-X-D/E-X-D/E-L) present in the C-terminus of SgII represents a moderately homologous stretch of 10 amino acids also found in chromogranins A and B. It was postulated when the primary amino acid sequence of these proteins became available (Huttner et al. 1991) and is most likely without physiological relevance. This is underlined by the absence of this motif in secretogranins III and V as well as its presence in unrelated proteins like BRCA1, BRCA2, Golgin-245 or *trans*-Golgi p230, an acidic protein localized to the cytosolic side of Golgi membranes (Erlich et al. 1996). The chromogranins share some physico-chemical properties like acidic pI and binding of calcium with low affinity but high capacity, which is responsible for the electron-dense core of large dense secretory vesicles. They are cleaved at pairs of consecutive basic amino acids to multiple smaller peptides and thus give rise to the characteristic pattern of multiple immunoreactive bands of intermediate size seen in immunoblots with monospecific antisera. Chromogranins have recently been shown to be ultimately involved in vesicle formation or biogenesis, to contribute to packaging and sorting of hormones and enzymes into LDVs and they can function as precursors of small peptides generated by prohormone convertases from chromogranins A, B, and SgII. Furthermore, chromogranins have been established as valuable biomarkers for the characterisation of neuroendocrine tumours and cardiovascular diseases. Several excellent reviews have recently been published, which provide further details on the structure and physiologic aspects of this protein family (Bartolomucci et al. 2011; Conlon 2010; Fischer-Colbrie et al. 2005, 1995; Helle 2010; Huttner et al. 1991; Portela-Gomes et al. 2010; Stridsberg et al. 2008; Taupenot et al. 2003).

2 Structure & Posttranslational Modifications of Secretogranin II

Human and mouse SgII consist of 617 amino acids (aa) whereas the bovine homolog is 4 amino acids shorter. SgII contains 20% acidic amino acids in its primary amino acid sequence (Fischer-Colbrie et al. 1990), which causes its untypical physico-chemical properties like a random-coil structure and heat-stability. Even though a molecular weight of 67 kDa can be deduced from the primary sequence, a band with an apparent M_r of 86,000 is seen in SDS gels. This abnormal migration in SDS electrophoresis is probably due to the high amount of acidic amino acids found in the primary sequence. After co-translational cleavage of a 27 amino acid signal peptide at a typical cleavage site (Ala-X-Ala/Ala), (see (Fischer-Colbrie et al. 1990) for discussion), SgII is further posttranslationally modified in the Golgi apparatus (Fig. 1): it is sulphated at Tyr-151 and phosphorylated at Ser-532 (Lee et al. 2010). SgII is not or only marginally glycosylated. The primary amino acids sequence of bovine SgII contains no consensus sequence for N-glycosylation and SgII does not bind to concanavalin A lectin, in addition no significant O-glycosylation was detected

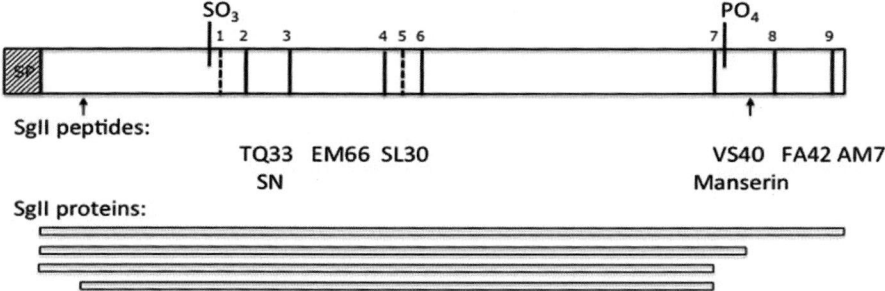

Fig. 1 Structure and processing of secretogranin II (*SgII*). Human SgII contains 617 amino acids including a 27 amino acid signal peptide (*SP*). One potential tyrosine sulfation and phosphorylation site is shown. Sites of two consecutive basic amino acids are labelled by numbers. These sites with the exception of site 1 and 5 are used by prohormone convertases (PCs) for endoproteolytic processing. Proteins and peptides generated by PCs from the SgII precursor are shown below. Two non-dibasic cleavage sites are indicated by arrows. The name of the peptides represent their first and last amino acid in the single letter code plus the total length. SN, secretoneurin

(Fischer-Colbrie et al. 1995). In the secretory granules SgII is processed to small peptides and proteins of intermediate size by prohormone convertases PC1 and PC2, two proteases belonging to the furin-like prohormone convertase group. SgII contains 9 pairs of basic amino acids (7 KR and 2 RK sites). Five peptides (secretoneurin (SN), EM66, SL30, manserin, FA42, AM7; see Fig. 1) flanked by KR sites have been isolated from different tissues and are thus generated in vivo (Anouar et al. 1996; Kirchmair et al. 1993; Lee et al. 2010; Tilemans et al. 1994; Vaudry and Conlon 1991; Yajima et al. 2004), whereas the RK sites are probably not used for processing. In addition to these small peptides four longer processing intermediates of 20–90 kDa size have been identified by mass spectroscopy (Fournier et al. 2011; Gupta et al. 2010). These proteins include full-length SgII, two proteins cleaved at the C-terminal end at position 547 and 524 and one cleaved N-terminally at position 59. Position 527 represents dibasic cleavage site #7 (see Fig. 1) whereas cleavage sites 59 and 547 are non-dibasic sites. Cleavage site 547 is located in the manserin peptide and represents an Ala-IsoLeu site, the second non-dibasic cleavage site is located between Ala-58/Leu-59 in the N-terminal fragment of SgII (see Fig. 1). The proteolytic enzymes cleaving at these sites have not been characterized yet.

Proteolytic processing of SgII is pronounced in most tissues, especially the central nervous system, resulting in more than 90% of SgII processed to small peptides including SN. In contrast, in the adrenal medulla, processing is more limited with the majority of SgII immunoreactivity existing as high molecular weight proteins of 20–86 kD (see Fig.1 in (Fischer-Colbrie et al. 1995). This limited processing is due to the high amount of catecholamines stored in adrenomedullary vesicles, which are potent inhibitors of prohormone convertases PC1 and PC2 (Wolkersdorfer et al. 1996). Also, in astrocytes SgII is mainly unprocessed (Fischer-Colbrie et al. 1993) reflecting the relative absence of prohormone convertase activity in these cells.

3 Subcellular Localization

In neurons and endocrine cells SgII is localized to the so-called large dense-cored synaptic vesicles (LDV). It is not found in the cytoplasm or other subcellular organelles like lysosomes. The subcellular localisation to the LDV was established by several means including subcellular fractionation techniques, as well as by immunohistochemistry and immune electron-microscopy. In fact, due to this well-established subcellular localisation SgII is considered and has been used in numerous studies as marker molecule for LDVs. In two studies SgII-immunoreactivity was also found in the nucleus by immune-electron microscopy and immunohistochemical techniques (Yajima et al. 2008; Yoo et al. 2007). It is currently not known whether epitopes shared by a nuclear protein cross-react or whether small amounts of SgII are indeed translocated to the nucleus. This for a secreted protein untypical nuclear localisation can only unequivocally be established when tissue from SgII knock-out mice becomes available. In any case the presence of a well-defined signal peptide rather argues against a nuclear form of SgII.

The N-terminal 27 amino acids of SgII comprise a signal peptide, which cause its co-translational translocation to the rough endoplasmatic reticulum. From there SgII is sorted to the LDV of the regulated pathway. Two regions of SgII, ie a putative -helix at the very N-terminus (SgII 25–41) and 16 amino acids in the middle region (SgII 334–348) have been identified as sorting motifs (Courel et al. 2008). Sorting of SgII depends on a saturable machinery which is greatly uncharacterized still. It has been suggested that sorting occurs via binding to secretogranin III (Hotta et al. 2009), another granin. However, in secretogranin III knock-down cells SgII was still secreted in a regulated manner (Sun et al. 2013) in line with a proposed sorting mechanism operating mainly by retention (Kuliawat and Arvan 1994; Tooze 1998).

4 Tissue Expression and Secretion from Cells

SgII is constitutively and abundantly expressed throughout the endocrine and nervous system. SgII has been identified in the adrenal medulla, all 3 lobes of the pituitary, the endocrine pancreas, in C-cells of the thyroid and endocrine cells of the whole gastrointestinal tract whereas the parathyroid gland is devoid of SgII (for details see (Fischer-Colbrie et al. 1995)). In the nervous system SgII is distributed in the phylogenetically older parts of the brain with high densities of immunoreactive terminals and fibers found in the hypothalamus, extended amygdala, hippocampus, lateral septum, medial thalamic nuclei, locus coeruleus, nucleus tractus solitarii and substantiae gelatinosae of the spinal cord (Fischer-Colbrie et al. 1995). In the peripheral nervous system SgII is expressed in sensory as well as sympathetic and parasympathetic neurons. In the eye SgII/SN immunoreactivity is expressed in amacrine cells of the retina (Overdick et al. 1996) and capsaicin-sensitive sensory neurons innervating the iris/ciliary complex (Troger et al. 2005).

Under pathological conditions SgII is induced in tissues, which *per se* do not express SgII. Ischaemia induces SgII expression in muscle cells of the hindlimb (Egger et al. 2007; Theurl et al. 2015) and the heart (Røsjø et al. 2012; Theurl et al. 2015). A similar phenomenon was seen in animal models of cerebral ischemia for neurons of the central nervous system (Kim et al. 2002; Marti et al. 2001). Furthermore, epithelial cells from adenocarcinomas of the prostate and the gastro-intestinal tract can acquire a so-called endocrine phenotype by the induction of SgII synthesis (Pruneri et al. 1998).

Like other neuroendocrine secretory proteins SgII is released *en bloc* with co-stored classical transmitters, other granins and neuropeptides and biosynthetic enzymes from endocrine cells and neurons into circulation or the synaptic cleft, respectively. This secretory cocktail, however, contains only small amounts of intact, full-length SgII whereas the majority of SgII immunoreactivity comprises small SgII-derived peptides, i.e. SN or EM66, generated in the vesicles by proteolytic processing prior to release. Following depolarisation of cells by action potentials or hormonal signals SgII-derived peptides are exocytotically released in a calcium-dependent manner (Troger et al. 1994).

SN released from endocrine cells and neurons was detected in several body fluids. Serum steady-state SN levels of 22 fmol/ml originate from enterochromaffin cells of the gastrointestinal tract since other sources like the adrenal medulla, pituitary and the endocrine pancreas contribute only little (Ischia et al. 2000a). This is corroborated by the fact that high-emetogenic chemotherapy, which potently stimulates secretion from gastro-intestinal enterochromaffin cells, leads to a 50% increase of SN serum levels (Ischia et al. 2000a). Similarily, serum chromogranin A, which is co-stored together with SgII in these vesicles, is increased 2.5-fold (Cubeddu et al. 1995). SN levels are elevated 5-fold in childhood and are positively correlated with serum creatinine suggesting an influence of renal clearance on SN serum levels. This can be readily explained by the low molecular weight of SN. The half-life of SN was established experimentally as 2.5 h by analysing its decline in serum following removal of SN secreting tumors (Stridsberg et al. 2008). In the urine 80 fmol/ml have been found.

SN is furthermore released in significant amounts into the cerebro-spinal fluid (1500 fmol/ml (Eder et al. 1998; Miller et al. 1996), the aqueous humor (500 fmol/ml (Stemberger et al. 2004)), synovia (16 fmol/ml (Eder et al. 1997)) and faeces (Wagner et al. 2013).

5 Phylogenetic Conservation of Secretogranin II

SgII is well conserved during evolution. It is expressed in the entire vertebrate lineage including mammalia, birds, reptilia, amphibia and fish. SN is even found in agnatha like the lamprey (Trudeau et al. 2012) but has not been detected in any invertebrates so far. In concordance with the teleost-specific whole-genome duplication, teleost bony fish contain two moderately different SgII gene products. In some teleost fish like salmon even 4 SgII gene products exist (unpublished observation).

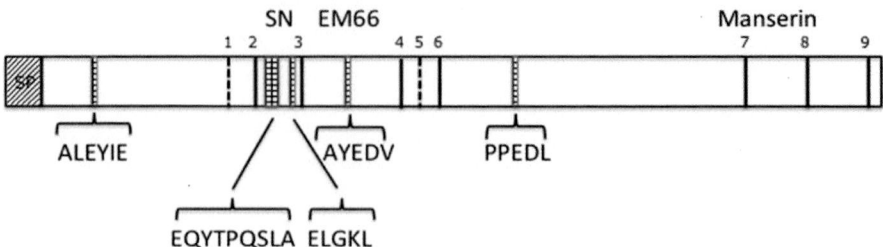

Fig. 2 Phylogenetic conservation of secretogranin II (*SgII*). The primary amino acid sequence of human and zebrafish SgII were compared. Homologous regions with a minimum of 5 identical amino acids are given in the single letter code. Only the region of SgII comprising the secretoneurin (*SN*) peptide is significantly conserved

Table 1 Conservation of SN

Human (1)	TNEIVEEQYTPQSLATLESVFQELGKLTGPNNQ
Primates (21)	TNEIV EQYTPQSLATLESVFQELGKL PN Q
Mammalia (84)	TNE EQYTPQ LATLESVFQELGK
Birds (62)	TNEIVEEQYTPQSLATLES FQELGK
Reptilia (19)	TNEI EEQYTPQSLATLESVFQELGK P
Amphibia (6)	T EIVE QYTPQ LATL SVF ELGK N
Cartilage fish (3)	TNEIVEEQYTPQSLATLES F ELG
Pre-Teleost bony fish (2)	TNEIVEEQYTPQSLATLESVF ELGKL P
Teleost fish – gene A(46)	E YTPQ LA L S F EL
Teleost fish – gene B(44)	E E YTPQ L S EL
Agnatha (1)	E E YTPQ LA L QELG

The primary amino acid sequence of human SN is given in the single letter code. For the various taxa shown only conserved amino acids are listed and variable ones are blanked out. The number of species analyzed per taxa is given in brackets

To date, SgII sequence information of up to 300 species has been deposited in the GenBank sequence database. The entire SgII precursor is highly conserved between mammalia, whereas between mammalia and fish only the region of SgII comprising the SN peptide is significantly homolog (see Kähler and Fischer-Colbrie 2000). Apart from SN only three additional regions with stretches of 5 or more amino acids conserved between human and zebrafish SgII can be identified (Fig. 2) suggesting that SN is indeed the physiological relevant peptide present within the SgII precursor.

Table 1 presents the high degree of phylogenetic conservation of SN peptide with 16 out of its 33 amino acids identical between human, mammalian, bird, reptile, amphibian, cartilage fish and pre-teleost bony fish. In teleost bony fish still 12 (gene A) and 10 (gene B) are identical with human homolog and SN is even found in agnatha like the lamprey. The middle part of SN comprises the best-conserved region, whereas the C-terminus varies greatly (see Table 1). It is interesting to note that the highly conserved part of SN is composed of 2 contiguous -helices as determined by two-dimensional ^{1}NMR analysis (Oulyadi et al. 1997).

6 Gene Organisation and Gene Regulation

The secretogranin II gene is organized into 2 exons only, which are separated by a 3 kb intron (Schimmel et al. 1992). It is located as a single copy gene on human chromosome 2q35-q36 and mouse chromosome 1, respectively (Mahata et al. 1996). The entire open reading frame plus 15 nt of the 5′ untranslated region are located on exon 2, exon 1 contains only 5′ untranslated region.

In the SgII promoter region several regulatory elements have been identified. A functional cyclic AMP response element (CRE) is located 74 bp upstream of the transcription site (Scammell et al. 2000). This site is conserved during evolution and found in the human, rat and mouse SgII promoter and confers induced gene-expression in response to several stimuli including nicotine and PACAP (Mahata et al. 1999), histamine (Bauer et al. 1993), gonadotropin-releasing hormone (GnRH) (Song et al. 2003) and NO (Li et al. 2008). In addition, the SgII promoter contains 2 TRE-like elements (Li et al. 2008) and a serum-response element (SRE) (Mahata et al. 1999), which is present in the mouse and rat promoter but not conserved in human. Nevertheless, the inactivation of this SRE decreased SgII expression in response to nicotine and PACAP (Mahata et al. 1999). Recent studies demonstrated that SgII is a genuine target gene for the RE-1 silencing transcription factor (REST) (Hohl and Thiel 2005; Watanabe et al. 2004). Since REST is significantly repressed by hypoxia (Liang et al. 2014; Lin et al. 2016) this might explain the potent up-regulation of SgII under ischemic conditions (Egger et al. 2007) despite the absence of a hypoxia-response element (HRE) in the SgII promoter.

7 Biomarker and Disease

7.1 Tumors

Twenty years after the detection of chromogranin A as main secretory protein of adrenergic chromaffin cells its pan-endocrine expression was discovered (Cohn et al. 1982; O'Connor et al. 1983). This led to its rapid application as marker for an endocrine phenotype of normal and malign tissues. In the following years this concept was extended to other members of the chromogranin family, namely chromogranin B and SgII.

SgII has been identified in pituitary adenomas, gastro-enterohepatic carcinomas, pheochromocytomas, Merckel cell carcinoma, midgut carcinoids and oat cell lung and prostate carcinomas (Conlon 2010; Guillemot et al. 2006; Ischia et al. 2000a, b; Portela-Gomes et al. 2010; Wiedenmann et al. 1988). In general, the expression of chromogranin B and SgII in tumors comprising an endocrine phenotype is more restricted than that of chromogranin A. Only in the appendix the majority of carcinoids (94%) stained positive for SgII-immunoreactivity, whereas chromogranin A was expressed less frequent there (83%, (Prommegger et al. 1998)).

In the serum, elevated levels of SgII can be expected if tumors secrete actively. In general however, serum levels vary greatly between patients and tumor types depending on their renal clearance and the degree of proteolytic processing and secretory capacity of the individual tumor cells. For two peptides generated from SgII, i.e. SN and EM66 data are available. In gut carcinoids and endocrine pancreatic tumors SN serum levels were elevated 20- and 15-fold, in pheochromocytomas, oat cell carcinomas of the lung and neuroblastomas up to 2.5–4.5-fold (Guillemot et al. 2006; Ischia et al. 2000a).

7.2 CNS Diseases

In the CSF SN levels are on average 70-fold higher compared to serum. However, there is a marked inter-individuality limiting its usefulness as potential biomarker for neurological and psychiatric diseases. On average, SgII levels are decreased by 15% in patients with multiple sclerosis (Mattsson et al. 2007) and not altered in schizophrenic patients (Miller et al. 1996) and patients with Parkinson's or Alzheimer's disease (Eder et al. 1998). Post-mortem studies demonstrated that SgII levels are decreased in brains of patients with tauopathies (Lechner et al. 2004) and amyotrophic laterals sclerosis due to a loss of presynaptic large dense core vesicles (Schrott-Fischer et al. 2009).

7.3 Heart Failure

Serum levels of SN were increased 2.4-fold in patients with cardiac arrest in the first 7 days and back to normal levels after another week (Hasslacher et al. 2014). These findings were corroborated in patients with acute heart failure although in this study the increase was less pronounced (1.2-fold, (Ottesen et al. 2015; Hasslacher et al. 2014)). In both studies SN serum levels were significantly correlated with a poor clinical outcome. The source of increased serum SN levels after heart failure has not unequivocally been established. It has been proposed that under severe ischemic conditions SN leeks from the cerebrospinal fluid into circulation due to an impaired tightness of the blood brain barrier (Hasslacher et al. 2014). In accordance, SN levels in umbilical cord blood was elevated in neonates with hypoxic-ischaemic encephalopathy (Wechselberger et al. 2016).

8 Physiologic Function of Secretogranin II

8.1 Biogenesis of Secretory Granules

Chromogranins induce biogenesis of secretory granule in neuronal but also non-neuronal cells. After the initial discovery that gene ablation of chromogranin A leads to a reduced granule biosynthesis (Kim et al. 2005) a similar function was established for SgII. Expression of full-length SgII stimulated biogenesis of secretory granules in fibroblast cells (Beuret et al. 2004) and a secretory-deficient PC12 cell-line (Courel et al. 2010). From these newly formed secretory granules cargo was released in a calcium dependent manner (Beuret et al. 2004). *Vice versa*, silencing of SgII expression led to a decrease in the number and size of large dense vesicles (Courel et al. 2010). Thus, induction of SgII in non-neuronal tissues typically lacking secretory granules sorting of SgII into the trans-Golgi network might first induce granule biogenesis whereas later on after storage and proteolytic processing in these vesicles SgII might function in addition as a neuropeptide precursor protein. It seems justified to speculate that this scenario is initiated by ischemic conditions in skeletal muscle cells (Egger et al. 2007) or under endocrine differentiation of various adenocarcinomas (Courel et al. 2014).

8.2 Sorting and Release of Secretory Proteins

SgII facilitates sorting of proopiomelanocortin (POMC) to secretory vesicles and its release from AtT-20 cells (Sun et al. 2013). It remains to be demonstrated whether SgII promotes sorting of LDV secretory proteins like POMC *per se* or if its established granulogenic activity alone is sufficient to mediate this effect. Also, SgII alters release of viral particles from infected cells. Depending on the type of virus, shRNA mediated down-regulation of SgII can either decrease or increase the amount of viral particles excreted into medium (Berard et al. 2015).

8.3 Precursor of the Neuropeptide/Cytokine Secretoneurin

In recent years, numerous important physiological functions of SN in the nervous, immune and endocrine system as well as on blood vessels were unravelled (for a review see (Fischer-Colbrie et al. 2005)). In the nervous system SN stimulates dopamine release from rat striatal slices and basal ganglia in vivo. It stimulates neurite outgrowth and survival of cerebellar granules cells (Fujita et al. 1999; Gasser et al. 2003). In concordance, SgII following up-regulation by REST promotes differentiation and maturation of adult hippocampal progenitor cells (Kim et al. 2015). In the immune system SN displays potent chemotactic activity toward monocytes, eosinophils and dendritic cells, which has also been shown for smooth muscle cells and fibroblasts (Dunzendorfer et al. 1998, 2001; Kähler et al. 1997a, b; Reinisch et al. 1993). In the endocrine system SN inhibits melatonin release from melanocytes and stimulates gonodotropin II release from goldfish pituitary as well as food intake (Blázquez et al. 1998; Mikwar et al. 2016).

8.3.1 Angiogenesis and Vasculogenesis

In 2004 a potent angiogenic and vasculogenic property of SN was discovered. Due to biologic actions of SN on vascular cells such as induction of chemotaxis in endothelial or vascular smooth muscle cells (Kähler et al. 1997a, b) and the fact that SN-containing nerve fibers are closely associated with blood vessels in the uterus (Collins et al. 2000) we hypothesized that SN might induce the growth of new blood vessels out of the pre-existing vasculature, a process called angiogenesis. We observed that SN indeed induced angiogenesis in the mouse corneal neovascularization assay and that newly generated vessels are covered by smooth muscle cells what might indicate durable, stable vessels. In vitro SN induced angiogenesis in a matrigel assay, stimulated proliferation and inhibited apoptosis of endothelial cells (ECs) and activated prominent intracellular signal transduction pathways like Akt or MAPK (Kirchmair et al. 2004a). Beside angiogenesis new blood vessels might also be generated by circulating progenitor cells, a process called vasculogenesis. We could show that SN also stimulates incorporation of these cells into new blood vessels and activates these cells in vitro (Kirchmair et al. 2004b).

Regulation by Hypoxia. Angiogenic factors typically are up-regulated by hypoxia to counteract lack of oxygen by generation of new blood vessels. In the case of SN it was already shown that induction of hypoxia in the central nervous system by ligation of the carotid artery leads to up-regulation of SN in neurons of the hippocampus or the cerebral cortex (Marti et al. 2001). We could show that also muscle cells (a cell type that normally doesn't produce SN) in the ischemic hindlimb after ligation of the femoral artery express SN. In vitro L6 myocytes also increased SN after prolonged hypoxia but this effect was indirect as the promoter region of the gene encoding the SN precursor secretogranin-II does not contain a hypoxia-responsive element and hypoxia did not increase SN in the absence of serum. We could show that SN increase by hypoxia was dependent on hypoxia-inducible factor 1 and basic fibroblast growth factor (FGF) in contrast to the direct regulation of vascular endothelial growth factor (VEGF) (Egger et al. 2007).

SN Gene Therapy. To explore if SN might possess therapeutic potential in the treatment of ischemic limb and heart disease we generated a plasmid gene therapy vector and could demonstrate biologic activity of recombinant SN by EC chemotaxis. After injection of plasmid into ischemic hindlimbs in mice SN improved clinical outcome (less limb necrosis) and increased tissue perfusion and density of capillaries and arteries compared to control plasmid. SN gene therapy additionally increased numbers endothelial progenitor cells in this ischemia model. These findings indicate induction of angiogenesis, arteriogenesis and vasculogenesis by SN gene therapy. We could also show that SN increased other potent angiogenic factors in ECs (basic fibroblast growth factor and platelet-derived growth factor B) and stimulated the nitric oxide pathway (Schgoer et al. 2009).

In the heart SN gene therapy improved systolic function and inhibited scar formation and remodelling of the left ventricle after an experimental myocardial infarction (permanent LAD ligation). Again, density of capillaries and arterioles/arteries in the infarct border zone was increased, consistent with induction of angiogenesis and arteriogenesis also in this ischemia model.

In arterial coronary ECs SN inhibited apoptosis, stimulated proliferation and in-vitro angiogenesis and activated Akt and MAPK. Interestingly, in vitro effects were blocked by a neutralizing antibody against VEGF, indicating that SN effects depend on VEGF. We indeed could demonstrate that SN stimulates activation of VEGF receptor-2 in a receptor tyrosine kinase (RTK) array. Further experiments elucidated that SN stimulates binding of VEGF to its co-receptors neuropillin-1 and heparan-sulfate proteoglycans. In RTK assays also activation of receptors for FGF and insulin-like growth factor-1 (IGF-1) was observed by SN. This activation of several potent angiogenic growth factor receptors by SN might be the reason for the robust effect of SN we observed on growth of new blood vessels as a complex biological effects like angiogenesis probably is mediated by different factors (Albrecht-Schgoer et al. 2012). A similar effect was shown recently in airway epithelial cells where it was demonstrated that SN stimulates mucus secretion by enhancing binding of epidermal growth factor (EGF) to neuropilin-1 (Xu et al. 2014).

These observations also might indicate that SN acts via binding to growth factors thereby stimulating binding of these factors to and activation of respective tyrosine kinase receptors instead of acting via an own specific cell surface receptor. Indeed, no specific SN receptor was detected so far. In this respect it is also of interest that SN was considered to act via a G-protein coupled receptor (GPCR) in cell migration experiments as effects were blocked by pertussis toxin. It will be interesting to elucidate if, in analogy to RTK, also GPCRs are activated by SN via classical chemotactic factors.

8.3.2 Wound Healing

To investigate potential effects of SN gene therapy we investigated wound healing in diabetic mice (db/db mice). Application of SN accelerated wound closure in this model and increased density of capillaries and arterioles in the wound. In microvascular dermal endothelial cells SN stimulated proliferation and in vitro angiogenesis in a basic-FGF dependent manner. We could show that FGF receptor-3 mediates SN-induced effects and that SN stimulates binding of basic FGF to heparan-sulfate proteoglycans on dermal endothelial cells (Albrecht-Schgoer et al. 2014). This finding corroborates our observation on coronary endothelial cells that SN stimulates receptors of potent angiogenic cytokines.

8.3.3 Effects of SN on Cerebral Ischemia

Systemic (intra-venous) application of SN in a rat model of cerebral ischemia reduced infarct area, enhanced motor performance and increased brain metabolic activity. In ischemic areas of the brain in this animal model as well as in human samples after stroke SN was increased in neuronal cells. In vitro SN inhibited apoptosis of neurons in cell culture after oxygen/glucose deprivation by stimulation of the Jak/Stat pathway. In this work it was also demonstrated that SN enhanced growth of blood vessels in the ischemic brain area and attracted neuronal stem cells (Shyu et al. 2008).

8.3.4 Effects of SN on Cardiomyocytes

Recently it was demonstrated that SN is taken up by cardiomyocytes and influences Calcium (Ca^{2+}) handling in these cells by reduction of Ca^{2+}/calmodulin (CaM)-dependent protein kinase II δ (CaMKIIδ) activity. SN binds to CaM and CaMKII and attenuates CaMKIIδ-dependent phosphorylation of the ryanodine receptor. SN also inhibits sarcoplasmic reticulum Ca^{2+} leak and augments sarcoplasmic reticulum Ca^{2+} content. SN also attenuates Ca^{2+} sparks and waves in cardiomyocytes (Ottesen et al. 2015). These findings indicate that SN might have a potential to inhibit arrythmias.

8.3.5 Effects of SN Gene Therapy on Animal Models of High Vascular Risk

Over the last two decades a variety of endothelial growth factors have been investigated as promising novel therapeutic agents for the treatment of peripheral arterial disease or myocardial ischemia. Unfortunately, the promising results obtained in animal models could not be verified in human trials so far. One reason might be that, in pre- clinical models, often young and healthy animals were studied, whereas in clinical trials patients with severe, long-lasting atherosclerosis and impaired vascular response were treated. Nevertheless, available data from clinical trials suggest that intramuscular injection of DNA might be safe and not associated with an increased rate of cancer. Due to promising results of SN gene therapy in hindlimb and myocardial ischemia we investigated the efficacy of SN gene transfer in two animal models (type I diabetes mellitus and hypercholesterolemia) with impaired angiogenic response (Schgoer et al. 2013; Theurl et al. 2015). In both animal models therapy of hindlimb ischemia with a SN-encoding plasmid resulted in significant improvement of limb reperfusion. Moreover, animals treated with the SN-plasmid showed a significant reduction in tissue defects and amputation rate. In the Apo E −/− mice, a model of severe hypercholesterolemia, we also evaluated the effect of SN-gene therapy in myocardial ischemia induced by permanent ligation of the left anterior descending artery. Similar to our data in rats SN treatment resulted in significant improvement of cardiac function as shown by echocardiographic parameters. Immunofluorescence staining revealed a significant increase of capillary and arteriole density as possible underlying mechanism for the favourable outcome of SN-treated animals.

Despite the negative influence of hypercholesterolemia and hyperglycemia on vascular cells, SN showed beneficial effects on EC function like proliferation, in vitro angiogenesis, or activation of the MAPK-ERK1/2 signaling cascade.

The local injection of the SN-plasmid did not result in elevated SN-serum levels and did not influence plaque progression as described for systemic administration of VEGF. Therefore, to our current knowledge, SN-gene therapy seems to be an effective and safe treatment strategy for hind limb and myocardial ischemia. Large animal models should be the next step to bring SN-gene therapy from bench to bedside.

References

Albrecht-Schgoer K, Schgoer W, Holfeld J, Theurl M, Wiedemann D, Steger C, Gupta R, Semsroth S, Fischer-Colbrie R, Beer AG, Stanzl U, Huber E, Misener S, Dejaco D, Kishore R, Pachinger O, Grimm M, Bonaros N, Kirchmair R (2012) The angiogenic factor secretoneurin induces coronary angiogenesis in a model of myocardial infarction by stimulation of vascular endothelial growth factor signaling in endothelial cells. Circulation 126:2491–2501

Albrecht-Schgoer K, Schgoer W, Theurl M, Stanzl U, Lener D, Dejaco D, Zelger B, Franz WM, Kirchmair R (2014) Topical secretoneurin gene therapy accelerates diabetic wound healing by interaction between heparan-sulfate proteoglycans and basic FGF. Angiogenesis 17:27–36

Anouar Y, Jégou S, Alexandre D, Lihrmann I, Conlon JM, Vaudry H (1996) Molecular cloning of frog secretogranin II reveals the occurrence of several highly conserved potential regulatory peptides. FEBS Lett 394:295–299

Bartolomucci A, Possenti R, Mahata SK, Fischer-Colbrie R, Loh YP, Salton SR (2011) The extended Granin family: structure, function, and biomedical implications. Endocr Rev 32:755–797

Bauer JW, Kirchmair R, Egger C, Fischer-Colbrie R (1993) Histamine induces a gene-specific synthesis regulation of secretogranin II but not of chromogranin a and B in chromaffin cells in a calcium-dependent manner. J Biol Chem 268:1586–1589

Berard AR, Severini A, Coombs KM (2015) Differential reovirus-specific and herpesvirus-specific activator protein 1 activation of secretogranin II leads to altered virus secretion. J Virol 89:11954–11964

Beuret N, Stettler H, Renold A, Rutishauser J, Spiess M (2004) Expression of regulated secretory proteins is sufficient to generate granule-like structures in constitutively secreting cells. J Biol Chem 279:20242–20249

Blázquez M, Bosma PT, Chang JP, Docherty K, Trudeau VL (1998) G-aminobutyric acid up-regulates the expression of a novel secretogranin-II messenger ribonucleic acid in the goldfish pituitary. Endocrinology 139:4870–4880

Cohn DV, Zangerle R, Fischer-Colbrie R, Chu LLH, Elting JJ, Hamilton JW, Winkler H (1982) Similarity of secretory protein I from parathyroid gland to chromogranin a from adrenal medulla. Proc Natl Acad Sci U S A 79:6056–6059

Collins JJ, Wilson K, Fischer-Colbrie R, Papka RE (2000) Distribution and origin of secretoneurin-immunoreactive nerves in the female rat uterus. Neuroscience 95:255–264

Conlon JM (2010) Granin-derived peptides as diagnostic and prognostic markers for endocrine tumors. Regul Pept 165:5–11

Courel M, El Yamani FZ, Alexandre D, El Fatemi H, Delestre C, Montero-Hadjadje M, Tazi F, Amarti A, Magoul R, Chartrel N, Anouar Y (2014) Secretogranin II is overexpressed in advanced prostate cancer and promotes the neuroendocrine differentiation of prostate cancer cells. Eur J Cancer 50(17):3039–3049

Courel M, Soler-Jover A, Rodriguez-Flores JL, Mahata SK, Elias S, Montero-Hadjadje M, Anouar Y, Giuly RJ, O'Connor DT, Taupenot L (2010) Pro-hormone secretogranin II regulates dense core secretory granule biogenesis in catecholaminergic cells. J Biol Chem 285:10030–10043

Courel M, Vasquez MS, Hook VY, Mahata SK, Taupenot L (2008) Sorting of the neuroendocrine secretory protein Secretogranin II into the regulated secretory pathway: role of N- and C-terminal alpha-helical domains. J Biol Chem 283:11807–11822

Cubeddu LX, O'Connor DT, Parmer RJ (1995) Plasma chromogranin A: a marker of serotonin release and of emesis associated with cisplatin chemotherapy. J Clin Oncol 13:681–687

Dunzendorfer S, Kaser A, Meierhofer C, Tilg H, Wiedermann CJ (2001) Peripheral neuropeptides attract immature and arrest mature blood-derived dendritic cells. J Immunol 166:2167–2172

Dunzendorfer S, Schratzberger P, Reinisch N, Kähler CM, Wiedermann CJ (1998) Secretoneurin, a novel neuropeptide, is a potent chemoattractant for human eosinophils. Blood 91:1527–1532

Eder U, Leitner B, Kirchmair R, Pohl P, Jobst KA, Smith AD, Mally J, Benzer A, Riederer P, Reichmann H, Saria A, Winkler H (1998) Levels and proteolytic processing of chromogranin a and B and secretogranin II in cerebrospinal fluid in neurological diseases. J Neural Transm 105:39–51

Eder U, Hukkanen M, Leitner B, Mur E, Went P, Kirchmair R, Fischer-Colbrie R, Polak JM, Winkler H (1997) The presence of secretoneurin in human synovium and synovial fluid. Neurosci Lett 224:139–131

Egger M, Schgoer W, Beer AG, Jeschke J, Leierer J, Theurl M, Frauscher S, Tepper OM, Niederwanger A, Ritsch A, Kearney M, Wanschitz J, Gurtner GC, Fischer-Colbrie R, Weiss G, Piza-Katzer H, Losordo DW, Patsch JR, Schratzberger P, Kirchmair R (2007) Hypoxia up-regulates the angiogenic cytokine secretoneurin via an HIF-1alpha- and basic FGF-dependent pathway in muscle cells. FASEB J 21:2906–2917

Erlich R, Gleeson PA, Campbell P, Dietzsch E, Toh BH (1996) Molecular characterization of trans-Golgi p230. A human peripheral membrane protein encoded by a gene on chromosome 6p12–22 contains extensive coiled-coil alpha-helical domains and a granin motif. J Biol Chem 271:8328–8337

Fischer-Colbrie R, Kirchmair R, Kähler CM, Wiedermann CJ, Saria A (2005) Secretoneurin: a new player in angiogenesis and chemotaxis linking nerves, blood vessels and the immune system. Curr Protein Pept Sci 6:373–385

Fischer-Colbrie R, Kirchmair R, Schobert A, Olenik C, Meyer DK, Winkler H (1993) Secretogranin II is synthesized and secreted in astrocyte cultures. J Neurochem 60:2312–2314

Fischer-Colbrie R, Gutierrez J, Hsu CM, Iacangelo A, Eiden LE (1990) Sequence analysis, tissue distribution and regulation by cell depolarization, and second messengers of bovine secretogranin II (chromogranin C) mRNA. J Biol Chem 265:9208–9213

Fischer-Colbrie R, Laslop A, Kirchmair R (1995) Secretogranin II: molecular properties, regulation of biosynthesis and processing to the neuropeptide secretoneurin. Prog Neurobiol 46:49–70

Fournier I, Gaucher D, Chich JF, Bach C, Shooshtarizadeh P, Picaud S, Bourcier T, Speeg-Schatz C, Strub JM, Van Dorsselaer A, Corti A, Aunis D, Metz-Boutigue MH (2011) Processing of chromogranins/secretogranin in patients with diabetic retinopathy. Regul Pept 167:118–124

Fujita Y, Katagi J, Tabuchi A, Tsuchiya T, Tsuda M (1999) Coactivation of secretogranin-II and BDNF genes mediated by calcium signals in mouse cerebellar granule cells. Mol Brain Res 63:316–324

Gasser MC, Berti I, Hauser KF, Fischer-Colbrie R, Saria A (2003) Secretoneurin promotes pertussis toxin-sensitive neurite outgrowth in cerebellar granule cells. J Neurochem 85:662–669

Guillemot J, Anouar Y, Montero-Hadjadje M, Grouzmann E, Grumolato L, Roshmaninho-Salgado J, Turquier V, Duparc C, Lefebvre H, Plouin PF, Klein M, Muresan M, Chow BK, Vaudry H, Yon L (2006) Circulating EM66 is a highly sensitive marker for the diagnosis and follow-up of pheochromocytoma. Int J Cancer 118:2003–2012

Gupta N, Bark SJ, Lu WD, Taupenot L, O'Connor DT, Pevzner P, Hook V (2010) Mass spectrometry-based neuropeptidomics of secretory vesicles from human adrenal medullary pheochromocytoma reveals novel peptide products of prohormone processing. J Proteome Res 9:5065–5075

Hasslacher J, Lehner GF, Harler U, Beer R, Ulmer H, Kirchmair R, Fischer-Colbrie R, Bellmann R, Dunzendorfer S, Joannidis M (2014) Secretoneurin as a marker for hypoxic brain injury after cardiopulmonary resuscitation. Intensive Care Med 40:1518–1527

Helle KB (2010) Chromogranins a and B and secretogranin II as prohormones for regulatory peptides from the diffuse neuroendocrine system. Results Probl Cell Differ 50:21–44

Hohl M, Thiel G (2005) Cell type-specific regulation of RE-1 silencing transcription factor (REST) target genes. Eur J Neurosci 22:2216–2230

Hotta K, Hosaka M, Tanabe A, Takeuchi T (2009) Secretogranin II binds to secretogranin III and forms secretory granules with orexin, neuropeptide Y, and POMC. J Endocrinol 202:111–121

Huttner WB, Gerdes HH, Rosa P (1991) The granin (chromogranin/secretogranin) family. TIBS 16:27–30

Ischia R, Gasser RW, Fischer-Colbrie R, Eder U, Pagani A, Cubeddu LX, Lovisetti-Scamihorn P, Finkenstedt G, Laslop A, Winkler H (2000a) Levels and molecular properties of secretoneurin-immunoreactivity in the serum and urine of control and neuroendocrine tumor patients. J Clin Endocrinol Metab 85:355–360

Ischia R, Hobisch A, Bauer R, Weiss U, Gasser RW, Horninger W, Bartsch G Jr, Fuchs D, Bartsch G, Winkler H, Klocker H, Fischer-Colbrie R, Culig Z (2000b) Elevated levels of serum secretoneurin in patients with therapy resistant carcinoma of the prostate. J Urol 163:1161–1164

Kähler CM, Kirchmair R, Kaufmann G, Kähler STEA, Reinisch N, Fischer-Colbrie R, Hogue-Angeletti R, Winkler H, Wiedermann CJ (1997a) Inhibition of proliferation and stimulation of migration of endothelial cells by secretoneurin in vitro. Arterioscler Thromb Vasc Biol 17:932–939

Kähler CM, Schratzberger P, Wiedermann CJ (1997b) Response of vascular smooth muscle cells to the neuropeptide secretoneurin. A functional role for migration and proliferation in vitro. Arterioscler Thromb Vasc Biol 17:2029–2035

Kähler CM, Fischer-Colbrie R (2000) Secretoneurin - a novel link between the nervous and the immune system. Conservation of the sequence and functional aspects. Adv Exp Med Biol 482:279–290

Kim HJ, Denli AM, Wright R, Baul TD, Clemenson GD, Morcos AS, Zhao C, Schafer ST, Gage FH, Kagalwala MN (2015) REST regulates non-cell-autonomous neuronal differentiation and maturation of neural progenitor cells via secretogranin II. J Neurosci 35:14872–14884

Kim T, Zhang CF, Sun Z, Wu H, Loh YP (2005) Chromogranin a deficiency in transgenic mice leads to aberrant chromaffin granule biogenesis. J Neurosci 25:6958–6961

Kim YD, Sohn NW, Kang C, Soh Y (2002) DNA array reveals altered gene expression in response to focal cerebral ischemia. Brain Res Bull 58:491–498

Kirchmair R, Egger M, Walter DH, Eisterer W, Niederwanger A, Woell E, Nagl M, Pedrini M, Murayama T, Frauscher S, Hanley A, Silver M, Brodmann M, Sturm W, Fischer-Colbrie R, Losordo DW, Patsch JR, Schratzberger P (2004a) Secretoneurin, an angiogenic neuropeptide, induces postnatal vasculogenesis. Circulation 110:1121–1127

Kirchmair R, Gander R, Egger M, Hanley A, Silver M, Ritsch A, Murayama T, Kaneider N, Sturm W, Kearny M, Fischer-Colbrie R, Kircher B, Gaenzer H, Wiedermann CJ, Ropper AH, Losordo DW, Patsch JR, Schratzberger P (2004b) The neuropeptide secretoneurin acts as a direct angiogenic cytokine in vitro and in vivo. Circulation 109:777–783

Kirchmair R, Hogue-Angeletti R, Gutierrez J, Fischer-Colbrie R, Winkler H (1993) Secretoneurin--a neuropeptide generated in brain, adrenal medulla and other endocrine tissues by proteolytic processing of secretogranin II (chromogranin C). Neuroscience 53:359–365

Kuliawat R, Arvan P (1994) Distinct molecular mechanisms for protein sorting within immature secretory granules of pancreatic beta-cells. J Cell Biol 126:77–86

Lechner T, Adlassnig C, Humpel C, Kaufmann WA, Maier H, Reinstadler-Kramer K, Hinterholzl J, Mahata SK, Jellinger KA, Marksteiner J (2004) Chromogranin peptides in Alzheimer's disease. Exp Gerontol 39:101–113

Lee JE, Atkins N, Hatcher NG, Zamdborg L, Gillette MU, Sweedler JV, Kelleher NL (2010) Endogenous peptide discovery of the rat circadian clock: a focused study of the suprachiasmatic nucleus by ultrahigh performance tandem mass spectrometry. Mol Cell Proteomics 9:285–297

Li L, Hung AC, Porter AG (2008) Secretogranin II: a key AP-1-regulated protein that mediates neuronal differentiation and protection from nitric oxide-induced apoptosis of neuroblastoma cells. Cell Death Differ 15:879–888

Liang H, Studach L, Hullinger RL, Xie J, Andrisani OM (2014) Down-regulation of RE-1 silencing transcription factor (REST) in advanced prostate cancer by hypoxia-induced miR-106b~25. Exp Cell Res 320:188–199

Lin TP, Chang YT, Lee SY, Campbell M, Wang TC, Shen SH, Chung HJ, Chang YH, Chiu AW, Pan CC, Lin CH, Chu CY, Kung HJ, Cheng CY, Chang PC (2016) REST reduction is essential for hypoxia-induced neuroendocrine differentiation of prostate cancer cells by activating autophagy signaling. Oncotarget 7:26137–26151

Mahata SK, Kozak CA, Szpirer J, Szpirer C, Modi WS, Gerdes HH, Huttner WB, O'Connor DT (1996) Dispersion of chromogranin/secretogranin secretory protein family loci in mammalian genomes. Genomics 33:135–139

Mahata SK, Mahata M, Livsey CV, Gerdes HH, Huttner WB, O'Connor DT (1999) Neuroendocrine cell type-specific and inducible expression of the secretogranin II gene: crucial role of cyclic adenosine monophosphate and serum response elements. Endocrinology 140:739–749

Marti E, Ferrer I, Blasi J (2001) Differential regulation of chromogranin a, chromogranin B and secretoneurin protein expression after transient forebrain ischemia in the gerbil. Acta Neuropathol 101:159–166

Mattsson N, Ruetschi U, Podust VN, Stridsberg M, Li S, Andersen O, Haghighi S, Blennow K, Zetterberg H (2007) Cerebrospinal fluid concentrations of peptides derived from chromogranin B and secretogranin II are decreased in multiple sclerosis. J Neurochem 103:1932–1939

Mikwar M, Navarro-Martin L, Xing L, Volkoff H, Hu W, Trudeau VL (2016) Stimulatory effect of the secretogranin-ll derived peptide secretoneurin on food intake and locomotion in female goldfish (Carassius Auratus). Peptides 78:42–50

Miller C, Kirchmair R, Troger J, Saria A, Fleischhacker WW, Fischer-Colbrie R, Benzer A, Winkler H (1996) CSF of neuroleptic-naive first-episode schizophrenic patients: levels of biogenic amines, substance P, and peptides derived from chromogranin a (GE-25) and secretogranin II (secretoneurin). Biol Psychol 39:911–918

O'Connor DT, Burton D, Deftos LJ (1983) Chromogranin a: immunohistology reveals its universal occurrence in normal polypeptide hormone producing endocrine glands. Life Sci 33:1657–1663

Ottesen AH, Louch WE, Carlson CR, Landsverk OJ, Kurola J, Johansen RF, Moe MK, Aronsen JM, Hoiseth AD, Jarstadmarken H, Nygard S, Bjoras M, Sjaastad I, Pettila V, Stridsberg M, Omland T, Christensen G, Rosjo H (2015) Secretoneurin is a novel prognostic cardiovascular biomarker associated with cardiomyocyte calcium handling. J Am Coll Cardiol 65:339–351

Oulyadi H, Davoust D, Vaudry H (1997) A determination of the solution conformation of secretoneurin, a neuropeptide originating from the processing of secretogranin II, by 1H-NMR and restrained molecular dynamics. Eur J Biochem 246:665–673

Overdick B, Kirchmair R, Marksteiner J, Fischer-Colbrie R, Troger J, Winkler H, Saria A (1996) Presence and distribution of a new neuropeptide, secretoneurin, in human retina. Peptides 17:1–4

Portela-Gomes GM, Grimelius L, Wilander E, Stridsberg M (2010) Granins and granin-related peptides in neuroendocrine tumours. Regul Pept 165:12–20

Prommegger R, Obrist P, Ensinger C, Schwelberger HG, Wolf C, Fischer-Colbrie R, Mikuz G, Bodner E (1998) Secretoneurin in carcinoids of the appendix – immunohistochemical comparison with chromogranins a, B and secretogranin II. Anticancer Res 18:3999–4002

Pruneri G, Galli S, Rossi RS, Roncalli M, Coggi G, Ferrari A, Simonato A, Siccardi AG, Carboni N, Buffa R (1998) Chromogranin a and B and secretogranin II in prostatic adenocarcinomas:

neuroendocrine expression in patients untreated and treated with androgen deprivation therapy. Prostate 34:113–120

Reinisch N, Kirchmair R, Kähler CM, Hogue-Angeletti R, Fischer-Colbrie R, Winkler H, Wiedermann CJ (1993) Attraction of human monocytes by the neuropeptide secretoneurin. FEBS Lett 334:41–44

Røsjø H, Stridsberg M, Florholmen G, Stenslokken KO, Ottesen AH, Sjaastad I, Husberg C, Dahl MB, Oie E, Louch WE, Omland T, Christensen G (2012) Secretogranin II; a protein increased in the myocardium and circulation in heart failure with cardioprotective properties. PLoS One 7:e37401

Scammell JG, Reddy S, Valentine DL, Coker TN, Nikolopoulos SN, Ross RA (2000) Isolation and characterization of the human secretogranin II gene promoter. Mol Brain Res 75:8–15

Schgoer W, Theurl M, Albrecht-Schgoer K, Jonach V, Koller B, Lener D, Franz WM, Kirchmair R (2013) Secretoneurin gene therapy improves blood flow in an ischemia model in type 1 diabetic mice by enhancing therapeutic neovascularization. PLoS One 8:e74029

Schgoer W, Theurl M, Jeschke J, Beer AG, Albrecht K, Gander R, Rong S, Vasiljevic D, Egger M, Wolf AM, Frauscher S, Koller B, Tancevski I, Patsch JR, Schratzberger P, Piza-Katzer H, Ritsch A, Bahlmann FH, Fischer-Colbrie R, Wolf D, Kirchmair R (2009) Gene therapy with the angiogenic cytokine secretoneurin induces therapeutic angiogenesis by a nitric oxide-dependent mechanism. Circ Res 105:994–1002

Schimmel A, Bräunling O, Rüther U, Huttner WB, Gerdes H-H (1992) The organisation of the mouse secretogranin II gene. FEBS Lett 3:375–380

Schrott-Fischer A, Bitsche M, Humpel C, Walcher C, Maier H, Jellinger K, Rabl W, Glueckert R, Marksteiner J (2009) Chromogranin peptides in amyotrophic lateral sclerosis. Regul Pept 152:13–21

Shyu WC, Lin SZ, Chiang MF, Chen DC, Su CY, Wang HJ, Liu RS, Tsai CH, Li H (2008) Secretoneurin promotes neuroprotection and neuronal plasticity via the Jak2/Stat3 pathway in murine models of stroke. J Clin Invest 118:133–148

Song SB, Rhee M, Roberson MS, Maurer RA, Kim KE (2003) Gonadotropin-releasing hormone-induced stimulation of the rat secretogranin II promoter involves activation of CREB. Mol Cell Endocrinol 199:29–36

Stemberger K, Pallhuber J, Doblinger A, Troger J, Kirchmair R, Kralinger M, Fischer-Colbrie R, Kieselbach G (2004) Secretoneurin in the human aqueous humor and the absence of an effect of frequently used eye drops on the levels. Peptides 25:2115–2118

Stridsberg M, Eriksson B, Janson ET (2008) Measurements of secretogranins II, III, V and proconvertases 1/3 and 2 in plasma from patients with neuroendocrine tumours. Regul Pept 148:95–98

Sun M, Watanabe T, Bochimoto H, Sakai Y, Torii S, Takeuchi T, Hosaka M (2013) Multiple sorting systems for secretory granules ensure the regulated secretion of peptide hormones. Traffic 14:205–218

Taupenot L, Harper KL, O'Connor DT (2003) The chromogranin-secretogranin family. N Engl J Med 348:1134–1149

Theurl M, Schgoer W, Albrecht-Schgoer K, Lener D, Wolf D, Wolf M, Demetz E, Tymoszuk P, Tancevski I, Fischer-Colbrie R, Franz WM, Marschang P, Kirchmair R (2015) Secretoneurin gene therapy improves hind limb and cardiac ischaemia in Apo E−/− mice without influencing systemic atherosclerosis. Cardiovasc Res 105:96–106

Tilemans D, Jacobs GFM, Andries M, Proost P, Devreese B, Van Damme J, Van Beeumen J, Denef C (1994) Isolation of two peptides from rat gonadotroph-conditioned medium displaying an amino acid sequence identical to fragments of secretogranin II. Peptides 15:537–545

Tooze SA (1998) Biogenesis of secretory granules in the trans-Golgi network of neuroendocrine and endocrine cells. Biochim Biophys Acta 1404:231–244

Troger J, Doblinger A, Leierer J, Laslop A, Schmid E, Teuchner B, Opatril M, Philipp W, Klimaschewski L, Pfaller K, Gottinger W, Fischer-Colbrie R (2005) Secretoneurin in the peripheral ocular innervation. Invest Ophthalmol Vis Sci 46:647–654

Troger J, Kirchmair R, Marksteiner J, Seidl CV, Fischer-Colbrie R, Saria A, Winkler H (1994) Release of secretoneurin and noradrenaline from hypothalamic slices and its differential inhibition by calcium channel blockers. Naunyn-Schmied Arch Pharmacol 349:565–569

Trudeau VL, Martyniuk CJ, Zhao E, Hu H, Volkoff H, Decatur WA, Basak A (2012) Is secretoneurin a new hormone? Gen Comp Endocrinol 175:10–18

Vaudry H, Conlon JM (1991) Identification of a peptide arising from the specific post-translation processing of secretogranin II. FEBS Lett 284:31–33

Wagner M, Stridsberg M, Peterson CG, Sangfelt P, Lampinen M, Carlson M (2013) Increased fecal levels of chromogranin a, chromogranin B, and secretoneurin in collagenous colitis. Inflammation 36:855–861

Watanabe Y, Kameoka S, Gopalakrishnan V, Aldape KD, Pan ZZ, Lang FF, Majumder S (2004) Conversion of myoblasts to physiologically active neuronal phenotype. Genes Dev 18:889–900

Wechselberger K, Schmid A, Posod A, Hock M, Neubauer V, Fischer-Colbrie R, Kiechl-Kohlendorfer U, Griesmaier E (2016) Secretoneurin serum levels in healthy term neonates and neonates with hypoxic-ischaemic encephalopathy. Neonatology 110:14–20

Wiedenmann B, Waldherr R, Buhr H, Hille A, Rosa P, Huttner WB (1988) Identification of gastroenteropancreatic neuroendocrine cells in normal and neoplastic human tissue with antibodies against synaptophysin, chromogranin a, secretogranin I (chromogranin B), and secretogranin II. Gastroenterology 95:1364–1374

Wolkersdorfer M, Laslop A, Lazure C, Fischer-Colbrie R, Winkler H (1996) Processing of chromogranins in chromaffin cell culture: effects of reserpine and a-methyl p-tyrosine. Biochem J 316:953–958

Xu R, Li Q, Zhou J, Zhou X, Perelman JM, Kolosov VP (2014) Secretoneurin induces airway mucus hypersecretion by enhancing the binding of EGF to NRP1. Cell Physiol Biochem 33:446–456

Yajima A, Narita N, Narita M (2008) Recently identified a novel neuropeptide manserin colocalize with the TUNEL-positive cells in the top villi of the rat duodenum. J Pept Sci 14:773–776

Yajima A, Ikeda M, Miyazaki K, Maeshima T, Narita N, Narita M (2004) Manserin, a novel peptide from secretogranin II in the neuroendocrine system. Neuroreport 15:1755–1759

Yoo SH, Chu SY, Kim KD, Huh YH (2007) Presence of secretogranin II and high-capacity, low-affinity Ca^{2+} storage role in nucleoplasmic Ca^{2+} store vesicles. Biochemistry 46:14663–14671

Chromogranins as Molecular Coordinators at the Crossroads between Hormone Aggregation and Secretory Granule Biogenesis

O. Carmon, F. Laguerre, L. Jeandel, Y. Anouar, and M. Montero-Hadjadje

Abstract Chromogranins are members of a family of soluble glycoproteins sharing common structural features and properties, known to be inducers of prohormone aggregation and sorting into secretory granules. There is now increasing evidence for a key role of chromogranins in hormone sorting to the regulated secretory pathway, resulting from the interaction of chromogranin-induced aggregates with the TGN membrane through either sorting receptors such as carboxypeptidase E, or lipids such as cholesterol. These molecular interactions would contribute to the TGN membrane remodeling, a prerequisite to the recruitment of cytosolic proteins inducing membrane curvature and consecutive secretory granule budding. The identification of the molecular cues involved in the biogenesis of secretory granules is currently under intense investigation. The diversity of chromogranins sharing common structural features but with possible non-redundant functions implies a variety of secretory granule populations whose existence and function remain to be established in a given neuroendocrine cell type.

The present chapter deals with the role of the different members of the chromogranin family in the processes of hormone aggregation, secretory granule biogenesis, and hormone sorting through their interaction with the TGN membrane. Finally, the alteration of chromogranin secretion is described in pathophysiological conditions linked to dysregulated hormone secretion.

1 Introduction

The production and release of neurohormones by neuroendocrine cells are crucial for the coordination of the physiological functions governing organisms. Neurohormones are molecular mediators that are stored in vesicular organelles, called secretory granules, generated by budding of the trans-Golgi network (TGN)

O. Carmon • F. Laguerre • L. Jeandel • Y. Anouar (✉) • M. Montero-Hadjadje
Inserm U1239, University of Rouen-Normandy, UNIROUEN, Institute for Research and Innovation in Biomedicine, Mont-Saint-Aignan, France
e-mail: youssef.anouar@univ-rouen.fr

membrane. The predominant components found in secretory granules are chromogranins including chromogranin A (CgA), chromogranin B (CgB) also named secretogranin I (SgI), secretogranin II (SgII), SgIII, 7B2 (SgV), NESP55 (SgVI), VGF (SgVII), and proSAAS (SgVIII), which constitute a family of soluble phospho-glycoproteins with similar structural features that confer to these proteins common functional attributes in the biogenesis of secretory granules and the sorting of prohormones. Indeed, chromogranins exhibit a high proportion of acidic amino acids and a high capacity to bind calcium, which support their ability to aggregate with each other and with neuropeptides in the low pH and high calcium conditions found in the TGN (Elias et al. 2010). Chromogranins also exhibit several dibasic sites which are the targets of processing enzymes, leading to the formation of biologically active peptides (Metz-Boutigue et al. 1993; Montero-Hadjadje et al. 2008) but also acting as sorting receptors for chromogranin-induced aggregates in order to direct them to secretory granules (Elias et al. 2010). These proteins are also able to concentrate catecholamines inside the secretory granules to control their release during the exocytotic process (Machado et al. 2010). In this chapter, we will provide a brief update on the central role of chromogranins in the concomitant molecular phenomena involved in the formation of secretory granules (molecular aggregation, sorting of hormone aggregates and budding of the TGN membrane) and leading to the establishment of the regulated secretory pathway in neuroendocrine cells.

2 I/ The Twenty-First Century, the Advent of Chromogranins as Critical Regulators of Secretory Granule Biogenesis

To date, several members of the chromogranin family are known to play an active role in the formation of secretory granules. The pioneer study was performed using an antisens RNA strategy to knockdown CgA expression in neuroendocrine PC12 cells, which showed a drastic reduction of secretory granule number (Kim et al. 2001). Following this initial observation, two strains of CgA knockout mice were generated which diverged for the outcome of secretory granules but which both exhibited hypertension due to an increase in circulating catecholamine levels (Mahapatra et al. 2005, Hendy et al. 2006). These data suggested therefore the existence of a link between CgA expression and the biogenesis of functional secretory granules. Interestingly, disruption of the CgB gene in mice revealed a similar phenotype in pancreatic and chromaffin cells regarding the secretory granule number, with a defective secretion of islet hormones and catecholamines (Obermüller et al. 2010, Díaz-Vera et al. 2010). Similarly, knockdown of SgII expression leads to a decrease in the number of secretory granules in PC12 cells (Courel et al. 2010).

Genetic ablation of SgIII in mice caused no apparent defects despite the ubiquitous expression of SgIII in neuroendocrine cells and tissues (Kingsley et al. 1990). In SgIII-deficient AtT20 cells, the intracellular retention of CgA and proopiomelanocortin (POMC) were impaired, but residual adrenocorticotropic hormone (ACTH)/POMC together with SgII were still localized to the remaining secretory granules, and were secreted in a regulated manner (Sun et al. 2013). The role of a less known member of the chromogranin family, VGF, in secretory granule biogenesis has been investigated more recently (Fargali et al. 2014). In this study, VGF-knockout mice exhibited decreased secretory granule size in noradrenergic chromaffin cells, decreased adrenal CgB and increased plasma epinephrine leading to hypertension, while knock-in of human VGF1–615 rescued the hypertensive knockout phenotype. Interestingly, knock-in of human VGF1–524, that lacks C-terminal peptide TLQP-21, resulted in a significant increase in blood pressure and infusion of TLQP-21 normalized hypertension. Together, these studies indicate that chromogranins and chromogranin-derived peptides may have redundant and non-redundant roles in the regulated secretory pathway, and in the regulation of catecholamine levels and blood pressure.

3 II/ Hormone Aggregation as the First Function Assigned to Chromogranins

After their biosynthesis in the rough endoplasmic reticulum, chromogranins are transported and stored as soluble glycoproteins in the TGN lumen which exhibits high calcium concentration (10–15 mM) and low pH (6–6.5) (Kim et al. 2006). Chromogranins share several structural properties that are conserved during evolution, such as global acidity that enables them to bind calcium with low affinity and high capacity (Taupenot et al. 2003). A structural analysis of CgA revealed that in the conditions of the TGN lumen, CgA can adopt a coiled-coil structure which promotes the initiation of a core for aggregate nucleation (Mosley et al. 2007). Besides, CgA and CgB exhibit a N-terminal disulfide-bonded loop which has been shown to contribute to the aggregation with prohormones at low pH (Chanat and Huttner 1991, Thiele and Huttner 1998). On the other hand, the C-terminal domain of CgA is involved in calcium/pH-dependent homodimerization/homotetramerization (Yoo and Lewis 1993), allowing the aggregation-mediated sorting of CgA to secretory granules (Cowley et al. 2000). These aggregates of high molecular weight sort away prohormones from constitutively secreted proteins (Dannies 2001). Homophilic or heterophilic aggregation is related to the tertiary conformation of the prohormones which direct specific interactions to give rise to various secretory granule populations in a same cell type (Kim et al. 2006).

4 III/ Hormone Sorting through Interaction of Chromogranins with the TGN Membrane

Concomitant to the process of hormone aggregation, chromogranin-induced aggregates interact with the TGN membrane through either sorting receptors or lipids (Fig. 1a). Indeed, secretogranin III (SgIII) interacts with cholesterol in lipid rafts (Hosaka et al. 2004) and with the modifying enzyme and membrane-associated carboxypeptidase E (CPE) (Hosaka et al. 2005). Since CgA has been shown to bind to SgIII (Hosaka et al. 2002), SgIII-cholesterol interaction results in the sorting of CgA-induced aggregates to the granules of the regulated secretory pathway. A recent work by Sun et al. (2013) has suggested a role for SgII as a protein with potential for sorting POMC, suggesting that endocrine cells may have developed a system of redundancy with multiple sorting systems that can partially or completely compensate for each other with the ultimate capacity of ensuring hormone delivery, as also proposed for CPE and SgIII in CgA and POMC sorting (Cawley et al. 2016).

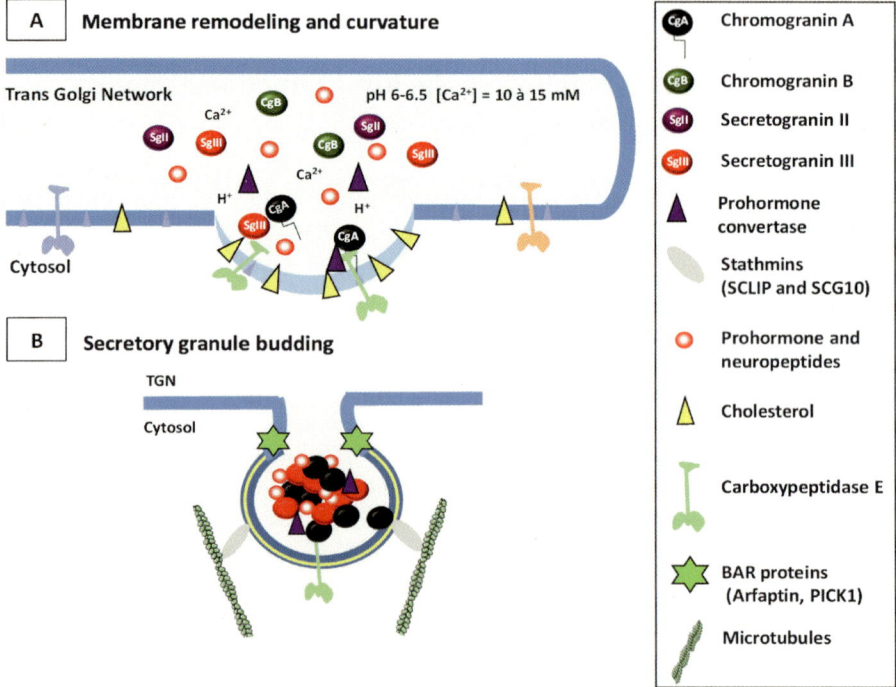

Fig. 1 Proposed model of chromogranin involvement in secretory granule biogenesis from TGN compartment. (**a**) Chromogranins induce the remodeling of the TGN membrane by interacting with distinct components: SgIII interacts with CgA, cholesterol and carboxypeptidase E. CgA also interacts with CPE. (**b**) At the cytosolic side of the TGN membrane, BAR proteins and stathmins are recruited to facilitate respectively the fission and the transfer to microtubules of nascent CgA-containing secretory granules for their transport towards the cell periphery

Their primary sequence exhibiting several dibasic sites, chromogranins are the substrates for intragranular prohormone-converting enzymes. Prohormone convertases are associated with cholesterol-rich lipid microdomains through α helices-rich hydrophobic domain(s) identified in their C-terminal part, thus acting as receptors that carry chromogranin-induced hormone aggregates to secretory granules (Dikeakos and Reudelhuber 2007).

Surprisingly, using a transcriptome/proteome-wide approach and co-immunoprecipitation experiments, SCLIP and SCG10 stathmins, cytosolic proteins known as regulators of microtubule polymerization, were identified as partners of CgA (Mahapatra et al. 2008). In this study, SCLIP and SCG10 were colocalized to the Golgi apparatus of chromaffin cells and their down-regulation altered chromaffin cell granules and abolished the level of secretion of endogenous CgA, SgII and CgB. Other cytosolic proteins are capable of sensing, inducing, and/or stabilizing membrane curvature through Bin/Amphiphysin/Rvs (BAR) domains which are crescent-shaped, dimeric α-helical modules and could thus participate to secretory granule formation (Frost et al. 2009, Rao and Haucke 2011). Recently, two members of this family, arfaptin 1 and PICK1, were related to the biogenesis of CgA-containing secretory granules through their binding to TGN membrane lipids (Fig. 1b) (Gehart et al. 2012, Cruz-Garcia et al. 2013, Holst et al. 2013, Pinheiro et al. 2014). Altogether, these data suggest that chromogranins influence the molecular organization of the TGN membrane to induce the recruitment of membrane and cytosolic proteins necessary for the budding of functional secretory granules, but the mechanism of action and the chronology of induced molecular events remain to be elucidated.

5 IV/ Are Chromogranins Linked to Hypersecretory Endocrine Pathologies?

Ubiquitous distribution of chromogranins in endocrine and neuroendocrine tissue from which they are secreted into the bloodstream, makes chromogranins useful markers of normal and tumoral neuroendocrine cells. In fact, multiple studies have documented the clinical value of detecting chromogranins and their derived peptides in tissues and measuring their circulating levels (Taupenot et al. 2003, Bartolomucci et al. 2011). Measurement of plasma chromogranins and derived peptide levels can be used to diagnose or monitor the progression of neuroendocrine tumors, the highest accuracy being observed in tumors characterized by an intense secretory activity (Guérin et al. 2010, Guillemot et al. 2006, 2014). This is the case of pheochromocytomas which are catecholamine-producing neoplasms arising from chromaffin cells of the adrenal medulla, leading to several clinical manifestations due to the actions of excess catecholamines, such as essential hypertension. Interestingly, the SgII-derived peptide EM66 has been established as a valuable marker to distinguish benign and malignant pheochromocytomas (Yon et al. 2003, Guillemot et al. 2006). Serum CgA has been also well documented as a marker for sympathoadrenal activity underlying cardiovascular regulation and essential hypertension (O'Connor and Bernstein 1984; Takiyyuddin et al. 1994) along with catecholamines, as well as severe inflammatory diseases such as sepsis and systemic inflammatory response syndrome (Zhang et al. 2008). Moreover, it has been shown that CgA genetic variants may cause profound changes in human autonomic activity, and may associate with the risk of developing hypertension (Rao et al. 2007, Chen et al. 2008), suggesting a link between CgA expression, catecholamine secretion and blood pressure regulation.

Furthermore, Colombo et al. (2002) have shown that CgA expression in neoplastic cells affects tumor growth and morphogenesis in mouse models, suggesting that abnormal secretion of CgA by neuroendocrine neoplastic cells could affect tumorigenic processes. A very interesting study based on the comparison of normotensive and hypertensive patients with pheochromocytoma revealed lower urinary catecholamines and a global decreased chromaffin gene expression, including SgII, in tumors from normotensive patients (Haissaguerre et al. 2013). In line with a potential link between SgII, hormone secretion and tumorigenesis, we have recently demonstrated that SgII is expressed in prostate cancer, that its increased levels correlate with high grade tumors and that its expression triggers a neuroendocrine differentiation of prostatic tumor cells as revealed by the appearance of secretory granules and a secretory activity. Because neuroendocrine differentiation is associated with a poor prognosis, these data suggest that SgII-induced secretion may play a pivotal role in prostate cancer progression (Courel et al. 2014).

Chromogranins seem to be also implicated in the alteration of hormone secretion observed in the context of metabolic disorders. Indeed, CgA knockout mice display increased adiposity and high levels of circulating leptin and catecholamines. As in diet-induced obese mice, desensitization of leptin receptors caused by hyperleptinemia is believed to contribute to the obese phenotype of these KO mice (Bandyopadhyay et al. 2012). In small bowel Crohn's disease, glucagon-like peptide 1 and CgA-immunopositive cells were significantly increased with appreciable mRNA increases for CgA, glucagon-like peptide 1 and Neurogenin 3, an enteroendocrine transcription factor, indicating an enhanced enteroendocrine cell activity (Moran et al. 2012). In accordance, modifications of CgA-immunoreactive cell density were also observed in the small intestine epithelium of patients with irritable bowel syndrome (IBS) after receiving dietary guidance, which may reflect a change in the densities of the small intestinal enteroendocrine cells, suggesting the contribution of their secretory activity to the improvement in the IBS symptoms (Mazzawi and El-Salhy 2016). Furthermore, the knockout of the CgB gene in mice provoked a reduction in stimulated secretion of insulin, glucagon and somatostatin in pancreatic islets, and consequently CgB-KO animals display some hallmarks of human type-2 diabetes (Obermüller et al. 2010).

6 Concluding Remarks

While aggregation ability of chromogranins allows the segregation of secretory granule cargo proteins, the anchor of the chromogranin-bound cargo proteins to the TGN membrane as the secretory granule forms or matures seems to result from a less universal mechanism. The study of chromogranin-expressing cells, which possess multiple types of secretory granules and thus the ability to selectively secrete different cocktails of biologically active components (Dannies 2001), may be informative on the variety of existing chromogranin-mediated sorting mechanisms. The identification of the membrane patches into the TGN recognized by the protein cargo, and the understanding of their interactions with cytosolic proteins would help to describe the transport, docking at the membrane, and exocytosis of secretory granules. What is already known is that chromogranins and their conserved domains are not only important for promoting the biogenesis of secretory granules, but also for the targeting and release of hormones through the regulated secretory pathway (Gondré-Lewis et al. 2012). These findings raise the interesting possibility that chromogranins could be relevant factors in the formation of secretory granules, probably by promoting the interactions between the secretory granule membrane and cytosolic proteins. Because several studies performed on neuroendocrine cells have provided clues regarding the redundant and non-redundant roles of chromogranins and chromogranin-derived peptides in the regulated secretory pathway, non-endocrine cell models or endocrine cell models devoid of a regulated secretory pathway expressing chromogranins could constitute useful and appropriate tools to further characterize the molecular mechanisms by which each chromogranin contributes to or establishes a regulated secretory pathway. The increasing evidence showing that chromogranins are crucial actors impacted in many diseases related to endocrine disruption indicates that chromogranins should be considered more than disease markers and warrants further studies to unravel their molecular mechanism of action and thereby to identify novel therapeutical strategies for the treatment of various pathologies including neuroendocrine tumors, essential hypertension and/or metabolic disorders which are increasingly affecting the populations.

References

Bandyopadhyay GK, Vu CU, Gentile S, Lee H, Biswas N, Chi N-W, O'Connor DT, Mahata SK (2012) Catestatin (chromogranin A(352-372)) and novel effects on mobilization of fat from adipose tissue through regulation of adrenergic and leptin signaling. J Biol Chem 287:23141–23151

Bartolomucci A, Possenti R, Mahata SK, Fischer-Colbrie R, Loh YP, Salton SRJ (2011) The extended granin family: structure, function, and biomedical implications. Endocr Rev 32:755–797

Cawley NX, Rathod T, Young S, Lou H, Birch N, Loh YP (2016) Carboxypeptidase E and secretogranin III coordinately facilitate efficient sorting of proopiomelanocortin to the regulated secretory pathway in AtT20 cells. Mol Endocrinol 30:37–47

Chanat E, Huttner WB (1991) Milieu-induced, selective aggregation of regulated secretory proteins in the trans-Golgi network. J Cell Biol 115:1505–1519

Chen Y, Rao F, Rodriguez-Flores JL et al (2008) Naturally occurring human genetic variation in the 3′-untranslated region of the secretory protein chromogranin A is associated with autonomic blood pressure regulation and hypertension in a sex-dependent fashion. J Am Coll Cardiol 52:1468–1481

Colombo B, Curnis F, Foglieni C, Monno A, Arrigoni G, Corti A (2002) Chromogranin A expression in neoplastic cells affects tumor growth and morphogenesis in mouse models. Cancer Res 62:941–946

Courel M, Soler-Jover A, Rodriguez-Flores JL, Mahata SK, Elias S, Montero-Hadjadje M, Anouar Y, Giuly RJ, O'Connor DT, Taupenot L (2010) Pro-hormone secretogranin II regulates dense core secretory granule biogenesis in catecholaminergic cells. J Biol Chem 285:10030–10043

Courel M, El Yamani F-Z, Alexandre D et al (2014) Secretogranin II is overexpressed in advanced prostate cancer and promotes the neuroendocrine differentiation of prostate cancer cells. Eur J Cancer 50:3039–3049

Cowley DJ, Moore YR, Darling DS, Joyce PB, Gorr SU (2000) N- and C-terminal domains direct cell type-specific sorting of chromogranin A to secretory granules. J Biol Chem 275:7743–7748

Cruz-Garcia D, Ortega-Bellido M, Scarpa M, Villeneuve J, Jovic M, Porzner M, Balla T, Seufferlein T, Malhotra V (2013) Recruitment of arfaptins to the trans-Golgi network by PI(4)P and their involvement in cargo export. EMBO J 32:1717–1729

Dannies PS (2001) Concentrating hormones into secretory granules: layers of control. Mol Cell Endocrinol 177:87–93

Díaz-Vera J, Morales YG, Hernández-Fernaud JR, Camacho M, Montesinos MS, Calegari F, Huttner WB, Borges R, Machado JD (2010) Chromogranin B gene ablation reduces the catecholamine cargo and decelerates exocytosis in chromaffin secretory vesicles. J Neurosci 30:950–957

Dikeakos JD, Reudelhuber TL (2007) Sending proteins to dense core secretory granules: still a lot to sort out: Figure 1. J Cell Biol 177(2):191–196

Elias S, Delestre C, Ory S, Marais S, Courel M, Vazquez-Martinez R, Bernard S, Coquet L, Malagon MM, Driouich A, Chan P, Gasman S, Anouar Y, Montero-Hadjadje M (2012) Chromogranin A Induces the Biogenesis of Granules with Calcium- and Actin-Dependent Dynamics and Exocytosis in Constitutively Secreting Cells. Endocrinol 153(9):4444–4456

Fargali S, Garcia AL, Sadahiro M et al (2014) The granin VGF promotes genesis of secretory vesicles, and regulates circulating catecholamine levels and blood pressure. FASEB J 28:2120–2133

Frost A, Unger VM, De Camilli P (2009) The BAR domain superfamily: membrane-molding macromolecules. Cell 137:191–196

Gehart H, Goginashvili A, Beck R, Morvan J, Erbs E, Formentini I, De Matteis MA, Schwab Y, Wieland FT, Ricci R (2012) The BAR domain protein Arfaptin-1 controls secretory granule biogenesis at the trans-Golgi network. Dev Cell 23:756–768

Gondré-Lewis MC, Park JJ, Loh YP (2012) Cellular mechanisms for the biogenesis and transport of synaptic and dense-core vesicles. Int Rev Cell Mol Biol 299:27–115

Guérin M, Guillemot J, Thouënnon E et al (2010) Granins and their derived peptides in normal and tumoral chromaffin tissue: implications for the diagnosis and prognosis of pheochromocytoma. Regul Pept 165:21–29

Guillemot J, Anouar Y, Montero-Hadjadje M et al (2006) Circulating EM66 is a highly sensitive marker for the diagnosis and follow-up of pheochromocytoma. Int J Cancer 118:2003–2012

Guillemot J, Guérin M, Thouënnon E, Montéro-Hadjadje M, Leprince J, Lefebvre H, Klein M, Muresan M, Anouar Y, Yon L (2014) Characterization and plasma measurement of the WE-14 peptide in patients with pheochromocytoma. PLoS One 9:e88698

Haissaguerre M, Courel M, Caron P et al (2013) Normotensive incidentally discovered pheochromocytomas display specific biochemical, cellular, and molecular characteristics. J Clin Endocrinol Metab 98:4346–4354

Hendy GN, Li T, Girard M, Feldstein RC, Mulay S, Desjardins R, Day R, Karaplis AC, Tremblay ML, Canaff L (2006) Targeted ablation of the chromogranin a (Chga) gene: normal neuroendocrine dense-core secretory granules and increased expression of other granins. Mol Endocrinol 20:1935–1947

Holst B, Madsen KL, Jansen AM et al (2013) PICK1 deficiency impairs secretory vesicle biogenesis and leads to growth retardation and decreased glucose tolerance. PLoS Biol 11:e1001542

Hosaka M, Watanabe T, Sakai Y, Uchiyama Y, Takeuchi T (2002) Identification of a chromogranin A domain that mediates binding to secretogranin III and targeting to secretory granules in pituitary cells and pancreatic beta-cells. Mol Biol Cell 13:3388–3399

Hosaka M, Suda M, Sakai Y, Izumi T, Watanabe T, Takeuchi T (2004) Secretogranin III binds to cholesterol in the secretory granule membrane as an adapter for chromogranin A. J Biol Chem 279:3627–3634

Hosaka M, Watanabe T, Sakai Y, Kato T, Takeuchi T (2005) Interaction between secretogranin III and carboxypeptidase E facilitates prohormone sorting within secretory granules. J Cell Sci 118:4785–4795

Kim T, Tao-Cheng JH, Eiden LE, Loh YP (2001) Chromogranin A, an "on/off" switch controlling dense-core secretory granule biogenesis. Cell 106:499–509

Kim T, Gondré-Lewis MC, Arnaoutova I, Loh YP (2006) Dense-core secretory granule biogenesis. Physiology (Bethesda) 21:124–133

Kingsley DM, Rinchik EM, Russell LB, Ottiger HP, Sutcliffe JG, Copeland NG, Jenkins NA (1990) Genetic ablation of a mouse gene expressed specifically in brain. EMBO J 9:395–399

Machado JD, Díaz-Vera J, Domínguez N, Alvarez CM, Pardo MR, Borges R (2010) Chromogranins A and B as regulators of vesicle cargo and exocytosis. Cell Mol Neurobiol 30:1181–1187

Mahapatra NR, O'Connor DT, Vaingankar SM et al (2005) Hypertension from targeted ablation of chromogranin A can be rescued by the human ortholog. J Clin Invest 115:1942–1952

Mahapatra NR, Taupenot L, Courel M, Mahata SK, O'Connor DT (2008) The trans-Golgi proteins SCLIP and SCG10 interact with chromogranin A to regulate neuroendocrine secretion. Biochemistry 47:7167–7178

Mazzawi T, El-Salhy M (2016) Changes in small intestinal chromogranin A-immunoreactive cell densities in patients with irritable bowel syndrome after receiving dietary guidance. Int J Mol Med 37:1247–1253

Metz-Boutigue MH, Garcia-Sablone P, Hogue-Angeletti R, Aunis D (1993) Intracellular and extracellular processing of chromogranin A. Determination of cleavage sites. Eur J Biochem 217:247–257

Montero-Hadjadje M, Vaingankar S, Elias S, Tostivint H, Mahata SK, Anouar Y (2008) Chromogranins A and B and secretogranin II: evolutionary and functional aspects. Acta Physiol (Oxf) 192:309–324

Moran GW, Pennock J, McLaughlin JT (2012) Enteroendocrine cells in terminal ileal Crohn's disease. J Crohns Colitis 6:871–880

Mosley CA, Taupenot L, Biswas N et al (2007) Biogenesis of the secretory granule: chromogranin A coiled-coil structure results in unusual physical properties and suggests a mechanism for granule core condensation. Biochemistry 46:10999–11012

O'Connor DT, Bernstein KN (1984) Radioimmunoassay of chromogranin A in plasma as a measure of exocytotic sympathoadrenal activity in normal subjects and patients with pheochromocytoma. N Engl J Med 311:764–770

Obermüller S, Calegari F, King A et al (2010) Defective secretion of islet hormones in chromogranin-B deficient mice. PLoS One 5:e8936

Pinheiro PS, Jansen AM, de Wit H, Tawfik B, Madsen KL, Verhage M, Gether U, Sørensen JB (2014) The BAR domain protein PICK1 controls vesicle number and size in adrenal chromaffin cells. J Neurosci 34:10688–10700

Rao Y, Haucke V (2011) Membrane shaping by the Bin/amphiphysin/Rvs (BAR) domain protein superfamily. Cell Mol Life Sci 68:3983–3993

Rao F, Wen G, Gayen JR et al (2007) Catecholamine release-inhibitory peptide catestatin (chromogranin A(352-372)): naturally occurring amino acid variant Gly364Ser causes profound changes in human autonomic activity and alters risk for hypertension. Circulation 115:2271–2281

Sun M, Watanabe T, Bochimoto H, Sakai Y, Torii S, Takeuchi T, Hosaka M (2013) Multiple sorting systems for secretory granules ensure the regulated secretion of peptide hormones. Traffic 14:205–218

Takiyyuddin MA, Brown MR, Dinh TQ, Cervenka JH, Braun SD, Parmer RJ, Kennedy B, O'Connor DT (1994) Sympatho-adrenal secretion in humans: factors governing catecholamine and storage vesicle peptide co-release. J Auton Pharmacol 14:187–200

Taupenot L, Harper KL, O'Connor DT (2003) The chromogranin-secretogranin family. N Engl J Med 348:1134–1149

Thiele C, Huttner WB (1998) The disulfide-bonded loop of chromogranins, which is essential for sorting to secretory granules, mediates homodimerization. J Biol Chem 273:1223–1231

Yon L, Guillemot J, Montero-Hadjadje M, Grumolato L, Leprince J, Lefebvre H, Contesse V, Plouin P-F, Vaudry H, Anouar Y (2003) Identification of the secretogranin II-derived peptide EM66 in pheochromocytomas as a potential marker for discriminating benign versus malignant tumors. J Clin Endocrinol Metab 88:2579–2585

Yoo SH, Lewis MS (1993) Dimerization and tetramerization properties of the C-terminal region of chromogranin A: a thermodynamic analysis. Biochemistry 32:8816–8822

Zhang D, Lavaux T, Voegeli AC, Lavigne T, Castelain V, Meyer N, Sapin R, Aunis D, Metz-Boutigue MH, Schneider F (2008) Prognostic value of chromogranin A at admission in critically ill patients: a cohort study in a medical intensive care unit. Clin Chem 54:1497–1503

Involvement of Chromogranin A and Its Derived Peptides to Fight Infections

Marie-Hélène Metz-Boutigue and Francis Schneider

Abstract In this chapter, after the presentation of the expression of Chromogranin A and its antimicrobial derived peptides in circulation, we report a clinical analysis for Vasostatin-I (VS-I) and the preparation of antimicrobial coating with Cateslytin (CTL). In a clinical multicenter analysis of VS-I in the plasma of critically ill patients (481 patients and 13 healthy staff), we have shown for the first time, that an increase in the VS-I concentration was present on admission in non-selected critically ill patients and that it was associated with poor outcome at day 28 after admission. By using a combination of VS-I, lactate and age values, our assessment of prognosis was better than taking the parameters alone. Concerning the preparation of antimicrobial biomaterial we decided to use polysaccharide multilayer film based on CTL functionalized hyaluronic acid as polyanion and chitosan as polycation, (HA-CTL-C/CHI), with the aim of designing a self-defensive coating against both bacteria (*Staphylococcus aureus*) and yeast (*Candida albicans*). The ability of *S. aureus* and *Candida* species to degrade HA, by producing hyaluronidase, allows the active peptide to be released from film only in the presence of the pathogens. We have demonstrated the antimicrobial activities of HA-CTL in solution and HA-CTL-C/CHI films against *S. aureus* and *C. albicans*. We have shown the penetration of the fluorescently labeled HA FITC -CTL-C into the cell membrane of *C. albicans* and the cytotoxicity of HA-CTL-C/CHI films was tested through human gingival fibroblasts (HGFs) viability.

M.-H. Metz-Boutigue (✉)
Inserm U1121, Université de Strasbourg, FMTS, Faculté de Chirurgie Dentaire Hôpital Civil, Porte de l'hôpital, 67000 Strasbourg, France
e-mail: marie-helene.metz@inserm.fr

F. Schneider
Inserm U1121, Université de Strasbourg, FMTS, Service de Réanimation Médicale, Hôpital Hautepierre, Avenue Molière, 67000 Strasbourg, France
e-mail: Francis.Schneider@chru-strasbourg.fr

1 Innate Immunity and Antimicrobial Peptides

Multidrug antibiotic resistance is a worldwide crucial health problem. A major factor in the emergence of antibiotic resistant organisms is the overuse of antibiotics in the hospital or the community. To overcome this abuse, numerous efforts are undertaken to reduce antibiotics prescription and/or promote synergistic effects by others molecules. The production of new potent antibiotics, acting alone or in combination is urgent.

The innate immune system is, since 2 billion years, the primary defense in most living organisms and antimicrobial peptides (AMPs) are fundamental components of the innate immune defense of multicellular organisms, either animal or plant [for review, (Bulet et al. 2004; Aerts et al. 2008; Manners 2007)]. The antimicrobial peptides (AMPs) have been well conserved throughout the evolution and they ensure the organism's defense against a large number of pathogens. They serve as endogenous antibiotics that are able to rapidly kill bacteria, fungi and viruses. In addition to their direct antimicrobial activity, they also have a wide range of functions in modulating both innate and adaptive immunity and acting on inflammation (Fig. 1) (Hilchie et al. 2013). To date more than 2619 AMPs have been identified, (antimicrobial peptides database http://aps.unmc.edu/AP/main.php), including peptides from several tissues and cell types from, bacteria, invertebrates, plants and mammals (Wang et al. 2016). The cationic character of AMPs induces an electrostatic attraction to the negatively-charged phospholipids of microbial membranes and their hydrophobicity aids the integration into the microbial cell membrane, leading to membrane disruption. Furthermore, the amphipathic structure also allows the peptides to be soluble both in aqueous environments and in lipid membranes (Yeaman and Yount 2003).

2 Occurrence of Chromogranin A in Circulation

Chromogranins/secretogranins (CGs/SGs) constitute the granin family of genetically distinct acidic proteins present in secretory vesicles of nervous, endocrine and immune cells (Helle 2010). CGs are widespread among mammals and also in vertebrates (Montero-Hadjadje et al. 2008). Plasma CGA is by now a commonly used diagnostic and prognostic marker for tumors of neuroendocrine origin, using antibodies raised to a range of epitopes along the CGA molecule (Guérin et al. 2010; Stridsberg et al. 2004). There is a relatively constant background of granins in the peripheral circulation and in the cerebrospinal fluid. Normal human serum contains 40 µg/L; close to 1 nM of CGA (O'Connor et al. 1993). A vast number of reports on pathologically high plasma CGA have accumulated since the first documentation of increased levels in patients with neuroendocrine tumors (O'Connor and Bernstein 1984). Plasma CGA is elevated in patients with a range of systemic diseases including renal and hepatic failure, cardiac arrest, and essential hypertension (Taupenot

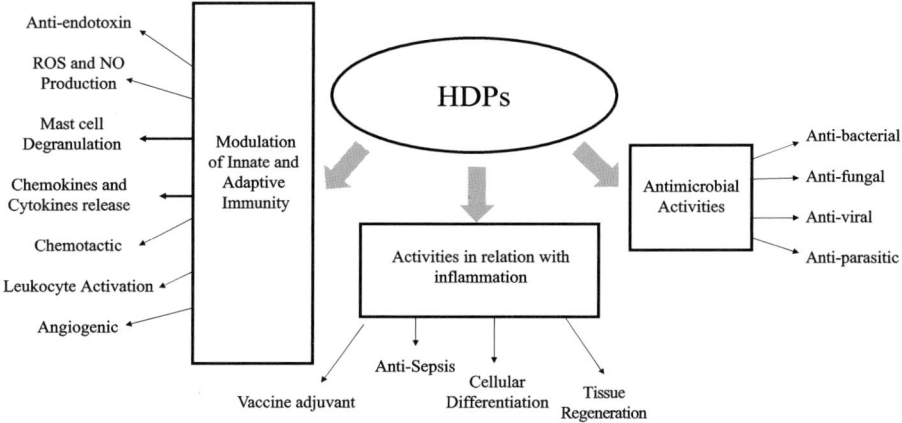

Fig. 1 Biological activities of Host Defense Peptides (HDPs)

et al. 2003) as well as in inflammatory conditions such as heart failure (Corti et al. 2000; Ceconi et al. 2002), acute coronary syndromes (Jansson et al. 2009), rheumatoid arthritis (Di Comite et al. 2006, 2009), systemic lupus erythematosis (Di Comite et al. 2006) and acute systemic inflammatory response syndrome (Zhang et al. 2009a). It seems well established that increased plasma CGA is predictive of shorter survival, not only in patients with metastatic neuroendocrine tumors (Arnold et al. 2008; Nikou et al. 2008), but also in chronic heart failure (Corti et al. 2000) and in the critically ill, nonsurgical patients (Zhang et al. 2008, 2009a). Whether plasma CGA, serves solely as passive marker of the secretory state of the various elements of the diffuse endocrine system or, in addition, as active and functional contributors to homeostatic regulations of normal and clinical conditions, would depend on the ability of the prohormone and/or its derived peptides to activate or modulate relevant cellular functions.

3 Chromogranin A Processing in Different Tissues

The natural processing of bovine CGs is well described in granules of sympathoadrenal medullary chromaffin cells, where the resulting peptides are co-secreted with catecholamines (Metz-Boutigue et al. 1993). For the bovine sequence numerous cleavage sites were identified. They correspond to positions 3–4, 64–65, 76–77, 78–79, 115–116, 247–248, 291–292, 315–316, 331–332, 350–351, 353–354, 358–359, 386–387. The major generated fragments are CgA1-76 (Vasostatin-I; VS-I) and CgA79-431 (pro-chromacin; ProChrom) (Strub et al. 1996). The N-terminal domain of CGA corresponding to vasostatin-I (VS-I) CGA1-76 is highly conserved across vertebrates from fish to man (Turquier et al. 1999). Furthermore, CGA and derived peptides were also identified in neutrophils (PMNs), in heart extracts and in vitreous. It has been demonstrated that processing of CGs is tissue- and cell-specific (Yoshida et al. 2009). The numerous cleavage sites are consistent with the specificity of prohormone convertases (PC1/3 and PC2) and carboxypeptidase E (CPE), that reside within chromaffin granules (Metz-Boutigue et al. 1993; Seidah and Chrétien 1999). In addition, cathepsin L (Biswas et al. 2009), carboxypeptidases, aminopeptidases (Hook et al. 1982), and trypsin-like proteases (Evangelista et al. 1982) are also involved in the proteolytic processing of CGA. These are typically regulated by endogenous protease inhibitors such as endopins 1 and 2, which possess high homology to alpha 1 antitrypsin A1AT (Hook and Metz-Boutigue 2002).

The discovery that pancreastatin, a CGA-derived peptide inhibits insulin secretion from pancreatic beta-cells, initiated the concept of prohormone (Eiden 1987; Nakano et al. 1989). The release of these CGs-derived peptides from chromaffin cells results from the nicotinic cholinergic stimulation and regulates several neuroendocrine functions (Helle and Serck-Hanssen 1975).

The natural processing of CGA in PMNs after stimulation by the leukocidin Panton-Valentin (Lugardon et al. 2000; Briolat et al. 2005) generates a range of CGA-derived fragments containing VS-I and catestatin CAT. Wherever PMNs accumulate in response to invading microorganisms, tissue inflammation, and sites of mechanical injury, these released peptides may affect a wide range of cells involved in inflammatory responses, e.g. endothelial, endocardial and epithelial cells, other leucocytes, fibroblasts, cardiomyocytes, and vascular and intestinal smooth muscle.

Furthermore, Leukotoxins E/D (LukE/D) from *Staphylococcus aureus* were also used to induce PMNs secretions. Full-length CGA (1–439) and several CGA-derived fragments were identified including CGA1–291, CgA1–249, CgA1–209 and CgA1–115 (Aslam et al. 2013a).

CGA-immunoreactive fractions were also detected in rat heart extracts (Glattard et al. 2006) by using western blot and TOF/TOF mass spectrometry. Four endogenous N-terminal CGA-derived peptides containing the vasostatin sequence were characterized as CGA4-113, CGA1-124, CGA1-135 and CGA1-199. Post-translational modifications of these fragments were identified: phosphorylation at S96 and O-glycosylation (trisaccharide, NAcGal-Gal-NeuAc) at T126. This first identification of CGA-derived peptides containing the VS-I motif in rat heart supports their role in cardiac physiology by an autocrine/paracrine mechanism.

We evaluated the presence and processing of CGA in the vitreous of diabetic patients (DV) compared with non-diabetic vitreous (NDV) (Fournier et al. 2011) because inflammation has been linked to the development of diabetic retinopathy. ELISA, Western blot, RP-HPLC, dot blot, protein sequencing and mass spectrometry were used to study the quantitative expression and the processing of CGA. Expression was higher in DV than in NDV. Mean concentration of CGA evaluated by ELISA was 90.8 ng/L (±90.1) in DV *vs.* 29.7 ng/L (±20.9) in NDV (p = 0.039). In DV, proteomic analyses showed that long CGA-derived fragments and A1AT were overexpressed, suggesting possible inhibition of the proteolytic process. In DV, the increase of the presence of complete CGA and the attenuation of its endogenous proteolytic processing could participate in diabetic retinopathy progression by reducing the presence of

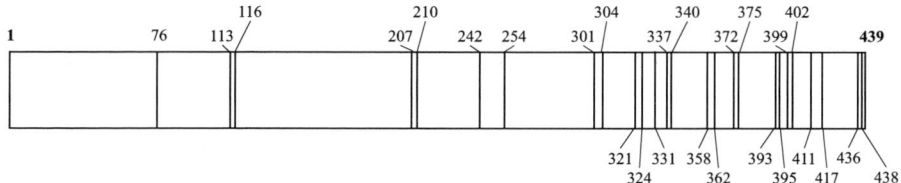

Fig. 2 Representation of the different endogenous cleavage sites of human CGA

regulatory peptides (VS-I), important for the pro–/anti-angiogenic balance in the eye. The alteration of the processing may be related to high levels of A1AT in DV compared with NDV, in accordance with the results of a recent study reporting high rates of A1AT in diabetic patients with proliferative diabetic retinopathy compared with patients without diabetic retinopathy (Gao et al. 2008). In addition, the reduction of the proteolytic activity observed in DV could also be due to protein modifications resulting from posttranslational formation of advanced glycation end products. Finally, taking into account all these studies a large number of endogenous fragments derived from human CGA are generated (Fig. 2).

4 The Antimicrobial Chromogranin A Derived-Peptides

During the two past decades, our laboratory has characterized new antimicrobial CGA-derived peptides (Strub et al. 1996; Metz-Boutigue et al. 1998; Lugardon et al. 2000, 2001; Briolat et al. 2005; Helle et al. 2007; Shooshtarizadeh et al. 2010) (Fig. 3). The corresponding sequences are highly conserved in human along evolution. Interestingly, the main cleavage site in position 78–79 of bCGA and the subsequent remove of the two basic residues K77 and K78 by the carboxypeptidase H (Metz-Boutigue et al. 1993) produces two antimicrobial fragments: VS-I (bCGA1-76) (Lugardon et al. 2000) and prochromacin (Prochrom; bCGA79-431) (Strub et al. 1996). For these N- and C-terminal domains with antimicrobial activities several shorter active fragments were identified: for VS-I, bCGA1-40 (N CGA; NCA) (Helle et al. 2007), bCGA47-66 (chromofungin; CHR) and for ProChrom, bCGA173–194 (Chromacin; Chrom) (Strub et al. 1996) and bCGA344-364 (Catestatin; CAT) (Briolat et al. 2005). The unique disulfide bridge of bCGA is present in VS-I and NCA sequences. Two post-translational modifications are important for the expression of the antibacterial activity of Chrom: the phosphorylation of Y173 and the O-glycosylation of S186 (Strub et al. 1996). Furthermore, it is important to point out that a dimerization motif GXXXG similar to that reported for Glycophorin A (Brosig and Langosch 1998) is present in the Chrom sequence (G184-G188).

These peptides act at micromolar range against bacteria, fungi, yeasts and are non-toxic for mammalian cells (Shooshtarizadeh et al. 2010). They are recovered in biological fluids involved in defense mechanisms (serum, saliva) and in secretions of stimulated human neutrophils (Lugardon et al. 2000; Briolat et al. 2005).

Peptide	Location	Sequence	Net charge
CGA			
VS-1	1-76	LPVNSPMNKGDTEVMKC*IVEVISDTLSKPSPMPVSKEC*FETLRGDERILSILRHQNLLKELQDLALQGAKERTHQQ	+3
NCA	4-40	LPVNSPMNKGDTEVMKC*IVEVISDTLSKPSPMPVSKEC*FE	-1
CHR	47-66	RILSILRHQNLLKELQDLAL	+1,5
Chrom	173-194	YPGPQAKEDSEGPSQGPASREK	-1
CAT	344-364	RSMRLSFRARGYGFRGPGLQL	+5
CCA	418-427	LEKVAHQLEE	-2
ProChrom	79-431	HSSYEDELSEVLEK.....	-37

Fig. 3 Location, sequence and net charge of the antimicrobial peptides derived from CgA; *C** cysteine residues involved in disulfide bridges, *Y* phosphorylated residue, *S* O-glycosylated residue (Helle et al. 2007)

These new AMPs are integrated in the concept that highlights the key role of the adrenal medulla in the immunity (Sternberg 2006) as previously reported for adrenaline and neuropeptide Y that regulate immunity systemically once released from the adrenal medulla. Furthermore, the adrenal medulla contains and releases large amounts of interleukin 6 (IL-6) and tumor necrosis factor-alpha (TNF-alpha) in response to pro-inflammatory stimuli such as lipopolysaccharide (LPS), IL-1 alpha and IL-1 beta. The discovery of the presence of Toll like receptors (TLRs) on the adrenal cortex cells raises the interesting possibility that the adrenal gland might have a direct role in the response to pathogens, activation of innate immune response and clearing of infectious agents (Sternberg 2006). We have more particularly studied VS-I and CAT.

VS-I displays antimicrobial activity against (i) Gram-positive bacteria (*Micrococcus luteus* and *Bacillus megaterium*) with a minimal inhibitory concentration (MIC) in the range 0.1–1 μM; (ii) filamentous fungi (*Neurospora crassa, Aspergillus fumigatus, Alternaria brassicola, Nectria haematococca, Fusarium culmorum, Fusarium oxysporum*) with a MIC of 0.5–3 μM and (iii) yeast cells (*Saccharomyces cerevisiae, Candida albicans*) with a MIC of 2 μM (Lugardon et al. 2000). VS-I possesses structural features specific for antimicrobial peptides, such as a global positive charge (+3), an equilibrated number of polar and hydrophobic residues (20:23) and the presence of a helical region CGA40-66 characterized to be a calmodulin-binding sequence (Lugardon et al. 2001; Yoo 1992; Zhang et al. 2009b) (Fig. 4a). The loss of the antibacterial activity of CGA7-57 suggests that the N- and C-terminal sequences are essential and CGA7-57 is less efficient than VS-I against fungi (Lugardon et al. 2000). Besides, the disulfide bridge is essential for the antibacterial, but not the antifungal property. Altogether, these data suggest that antibacterial and antifungal activities of VS-I have different structural requirements (Lugardon et al. 2000). Interestingly, two helix-helix dimerization motifs important for the interaction with membranes such as LXXXXXXL, present in dopamine transporter sequences (Torres et al. 2003) are present in the bovine and human VS-I

A

```
             1                                                                           78
b CGA        LPVNSPMNKGDTEVMKCIVEVISDTLSKPSPMPVSKECFETLRGDERILSILRHQNLLKELQDLALQGAKERTHQQKK
hrVS-1       STA                          Q                                       A
p CGA                                     Q                                       S
r CGA                 T   K    VL    S    P  L Q   V                              AQ   QQ
m CGA                 T   K    VL    S    P  L Q                                  A    PLK
o CGA              TNN    K             N L IIE  L         I              EI V  N Q    RR
e CGA              DT                     V  Q                                    AP   KH
f CGA              SSLEGEDNTK           N V ITQD L         I              E  A  M LQKAKK
```

B

```
             344                   364
b CAT        RSMRLSFRARGYGFRGPGLQL
h CAT        S  K    A           P
h P370L      S  K    A           L
h G364S      S  K    A S         P
p CAT           K    PA S
r CAT           K    A         D P
m CAT           K    T A       D
e CAT           K    A
f CAT           KIPTKDQK       E ASEE
```

Fig. 4 Sequence alignment of VS-I and CAT sequences of different species. (**a**) Sequence alignment of bovine CGA1–78 with hrVS-1 and corresponding fragments from several species. pCGA (porcine, p), rCGA (rat, r), mCGA (mouse, m), oCGA (ostrich, o), eCGA (equine, e), fCGA (frog, f). (**b**) Sequence alignment of bovine CGA344-364 with the corresponding fragments from several species hCGA (human, h and mutations G364S, P370L), pCGA (porcine, p), rCGA (rat, r), mCGA (mouse, m), eCGA (equine, e), fCGA (frog, f)

sequences (L42-L49; L57-L64). Surface interaction of rhodamine-labelled bCGA1-40 was demonstrated using confocal microscopy after incubation of the labeled peptide with *A. fumigatus*, *A. brassicola* and *N. haematococca* (Blois et al. 2006).

In addition, the interaction of bCGA1-40 with monolayers of phospholipids and sterols, as models for the interaction with mammalian and fungal membranes was investigated by the surface tension technique (Blois et al. 2006; Maget-Dana et al. 2002). These studies demonstrated that the N-terminal bCGA1-40 fragment interacts with model membrane phospholipids in a manner consistent with an amphiphilic penetration into membranes in a concentration range relevant for biological activity in mammalian tissue (Blois et al. 2006).

When VS-I was treated with the protease Glu-C from *S. aureus V8*, one of the generated peptide, chromofungin (CHR) (Lugardon et al. 2001), is the shortest active VS-I-derived peptide with antimicrobial activities (Fig. 3). It is well conserved during evolution (Fig. 4a) and displays antifungal activity at 2–15 µM against filamentous fungi (*N. crassa*, *A. fumigatus*, *A. brassicola*, *N. haematococca*, *F. culmorum*, *F. oxysporum*) and yeast cells (*C. albicans*, *C. tropicalis*, *C. neoformans*) (Lugardon et al. 2001). Since this peptide was generated after digestion of the material present in chromaffin secretory vesicles by the protease Glu-C from *S. aureus*, it may be hypothesized that it is produced during infections by this pathogen. The 3-D structure of CHR has been determined in water-trifluoroethanol (50:50) by using ^{1}H-NMR spectroscopy. This analysis revealed the amphipathic helical structure of the sequence 53–56, whereas the segment 48–52 confers hydrophobic character

(Lugardon et al. 2001). The importance of the amphipathic sequence for antifungal activity was demonstrated by the loss of such activity against *N. crassa* when two proline residues were substituted for L61 and L64, disrupting the helical structure, the amphipathic character and the dimerization motif helix-helix L57-L64 (Lugardon et al. 2001).

Concerning CAT several studies reported that it may be produced after the extensive processing of CGA. Two CGA-derived fragments bCGA333-364 and bCGA343-362 were characterized after the extensive processing by PC 1/3 or 2 in chromaffin granules (Taylor et al. 2000). In vitro cathepsin L is able to generate after digestion of recombinant hCGA, a catestatin (CAT)–derived fragment hCGA360-373 (Biswas et al. 2009). In addition to the inhibitory effect of CAT on catecholamine release from chromaffin cells (Mahata et al. 1997), we have shown for this peptide and its shorter active sequence bCGA344-358 (cateslytin, CTL), a potent antimicrobial activity with a MIC in the low micromolar range against Gram-positive bacteria (*M. luteus, B. megaterium* at concentration of 0.8 µM), Gram-negative bacteria (*E. coli* D22 at concentration of 8 µM), filamentous fungi (*N. crassa, A. fumigatus, N. haematococca* at concentration of 0.2–10 µM) and yeasts (*C. albicans, C. tropicalis, C. glabrata, C. neoformans* at concentration of 1.2–8 µM). The sequence of CAT (Fig. 4b) is highly conserved during evolution (Briolat et al. 2005). The two human variants P370L and G364S display antibacterial activity against *M. luteus* with a MIC of 2 and 1 µM, respectively, and against *E. coli* with a MIC of 20 and 10 µM, respectively (Briolat et al. 2005). However, the most active peptide corresponds to the bovine sequence. Bovine CTL, a cationic sequence with a global net charge of +5 (R344, R347, R351, R353, R358) and five hydrophobic residues (M346, L348, F360, Y355, F357) (Fig. 4b), is able to completely kill bacteria at concentration lower than 10 µM even in the presence of NaCl (0–150 mM) (Briolat et al. 2005). The C-terminal sequence bCGA352-358 is inactive, whereas the N-terminal sequences bCGA344-351 and bCGA348-358 are antibacterial at 20 µM.

5 Degradation of Antimicrobial Chromogranin A-Derived Peptides with Bacterial Proteases

The AMPs avoidance mechanisms deployed by bacteria include the proteolytic degradation of the active forms by the bacterial proteases. In order to examine the effects of bacterial proteases on the isolated AMPs derived from CGs, we have tested the effects of protease Glu-C from *S. aureus* V8 and several supernatants from *S. aureus, Salmonella enterica, Klebsiella oxytoca, Shigella sonnei* and *Vibrio cholera* on AMPs integrity (Thesis of Rizwan Aslam). The different strains were isolated from patients of the Strasbourg Civil Hospital by the Institute of Bacteriology. After incubation with *S. aureus* V8 protease Glu-C of the proteic intragranular material of chromaffin cells present in the adrenal medulla, 21 new peptides were isolated by HPLC and analysed by sequencing and mass spectrometry. These peptides were tested against Gram positive bacteria (*M. luteus* and *S. aureus*), Gram negative bacteria (*E. coli*), fungi (*N. crassa*) and yeast (*C. albicans*). They are not antibacterial, but 5 peptides corresponding to CGA47-60, CGA418-426 and CGB279-291, CGB450-464 and CGB470-486 display antifungal activity at the micromolar range against *N. crassa*. Thus, *S. aureus* subverts innate immunity to degrade the antibacterial CGA-derived peptides and produce new antifungal peptides (Thesis of Rizwan Aslam).

Three antimicrobial CGA-derived peptides (bCAT, hCAT and bCTL were incubated in presence of staphylococcal supernatants from strains ATCC 25923 and ATCC 49775 (Aslam et al. 2013b). CTL, the active domain of CAT, is able to completely kill *S. aureus* at 30 µM, but the two other peptides are inactive. By using a proteomic analysis (HPLC, sequencing and mass spectrometry) we demonstrated that CTL was not degraded by supernatants, whereas CHR, bovine and human CAT are processed to produce several fragments (Fig. 5).

Proteases from diarrheogenic bacteria (*K. oxytoca, S. enterica, S. sonnei* and *V. cholera*) were also tested with different peptides. The four strains have a clinical interest because apart from inducing diarrhea, they may cause other infections. We have tested bovine, rat and human CAT corresponding to the sequences bCgA344-364, rCgA6344-364 and hCgA352-372, bovine CTL located at bCgA344-358, two short fragments hCgA360-372 and the conserved tetrapeptide LSFR (bCgA348-351). In addition, we have tested a scrambled peptide relative to the sequence of bCAT and the procatestatin fragment bCgA332-364. By using HPLC we have compared the profiles of the peptide alone and the peptide with the inoculated medium. After sequencing and MALDI-TOF analysis of the different fractions we demonstrated that except CTL all the peptides are completely degraded by the different bacterial supernatants.

6 Synergy of the Combination of Antimicrobial Peptides with Antibiotics

Synergy is the combined activity of two antimicrobial agents that can never be attained by any one of them singly (Serra et al. 1977). The emergence of multi-drug resistant bacteria, with therapeutic failure against *S. aureus, Klebsiella pneumonia, Acinebacter baumannii* and *Pseudomonas aeruginosa* have paved the way to develop new therapeutic agents acting by synergism.

In our group, we have examined the synergical effects of three bovine CGA-derived peptides (CAT, CTL and CHR) with Minocycline, Amoxicillin and Linezolide against *S. aureus* and Voriconazole against *C. albicans* (Thesis of Menonve Atindehou). To demonstrate that antimicrobial peptides are able to reduce the doses of antibiotics used and to potentiate the activity of antibiotics, antimicrobial tests were carried out by combining the antibiotic peptides at doses below the

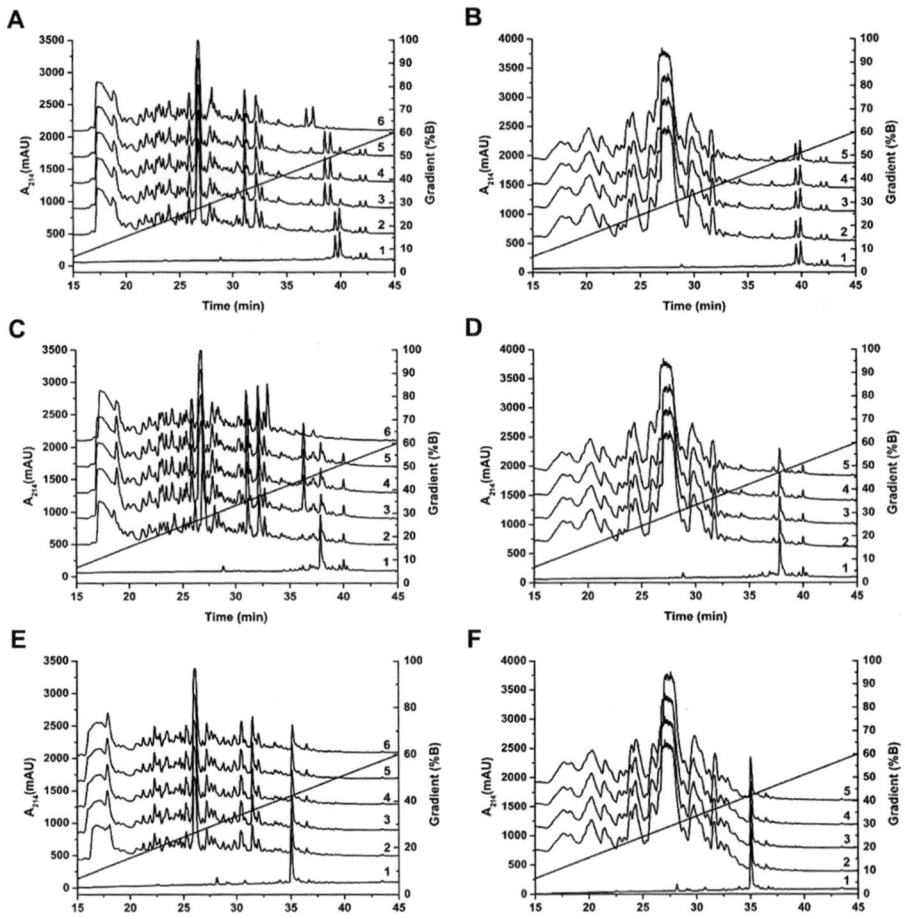

Fig. 5 HPLC chromatograms of bCAT, hCAT and CTL alone or with different bacterial strain supernatants (ATCC 49775 and ATCC 25923) with (**b, d, f**) or not (**a, c, e**) protease inhibitors (Aslam et al. 2013b)

MIC. The comparison was made with the antibiotic or peptide separately at the same doses. Minocycline has a MIC of 2 μg/ml alone against the *S. aureus* ATCC 49775, but when it was combined with CTL at a concentration corresponding to 75 % of the MIC, the concentration of Minocycline was lowered to 0.5 μg/ml. Similar data were obtained by the use of the amidated bCTL and Minocycline against *S. aureus* and CTL with Voriconazole against *C. albicans* and *C. tropicalis* (Thesis of Menonve Atindehou).

To conclude, these different assays show that bCTL facilitates the effect of antibiotic drugs and might allow the decrease of the administered concentrations and prevent the antibiotic resistance process.

7 Chromogranin A and VS-1 in Plasma of Critically ill Patients from Intensive Care Unit

In critically ill patients admitted to intensive care units (ICUs), severe diseases include multiple organ failure with clinical and biochemical signs of systemic inflammation: this corresponds to the systemic inflammatory response syndrome (SIRS) which is typical of the early phase response to a stress of unspecified etiology. During this response, major cardiovascular changes occur, and an activation of the adrenomedullary response as well (Wortsman et al. 1984). The initial triggering factor in such patients is frequently severe infection (Brun-Buisson 2000). Physicians have been used to measure the time-dependent changes in plasma concentrations of C-reactive protein and pro-calcitonin as sensitive and specific biomarkers to evaluate the level of inflammation related to these pathophysiological conditions. However, both of these parameters are unable to produce significant information as to outcome of such patients (Harbarth et al. 2001). Therefore, clinicians are seeking for new biomarkers that may reflect a link between initial SIRS, acute inflammation and outcome.

CGA is a marker of stress that exerts numerous autocrine, paracrine and endocrine functions (Taupenot et al. 2003); thus, it is capable of influencing the vascular tone (Aardal and Helle 1992), the activity of the myocardium (Glattard et al. 2006), and to reflect the level of activation of the sympathetic tone (Cryer et al. 1991). Vasostatin-I (VS-I; CGA1–76), the highly conserved N-terminal fragment of CGA (Figs. 3 and 4a), is predominantly produced and released in the circulation (Metz-Boutigue et al. 1993; Aardal and Helle 1992). Its initial property was considered to be its ability to relax contracted vessels (Aardal et al. 1993), but later it was shown to stabilize heart functioning by counterbalancing the adrenergic-dependent stress (Tota et al. 2007). The mechanistic actions of VS-I are still unclear in the absence of classic receptors for this molecule, but it is involved in negative inotropism through different biological pathways that could be integrated in a "whip-brake" conception of cardiovascular homoeostasis (Tota et al. 2010). Furthermore, we have reported that VS-1 is produced by PMNs and that it is involved in innate immunity by displaying antimicrobial properties (Lugardon et al. 2000). In addition, CHR the active sequence from VS-I activates PMNs (Zhang et al. 2009b).

In order to understand the role of CGA as a marker of stress during the time of SIRS, in a first step we have analyzed the possibility of a correlation between CGA concentrations, the inflammation biomarkers and the outcome in non-surgical critically ill patients with multiple organ failure (Zhang et al. 2008, 2009a). In a second step, we decided to evaluate the plasma VS-I levels in critically ill patients in order to learn whether their changes would correlate with the severity of disease (defined as death 28 days after admission) (Schneider et al. 2012).

Concerning the analysis of CGA in the plasma of critically ill patients, the population consisted of 67 participants (Zhang et al. 2008) (53 participants and 14 healthy controls). In septic patients (n = 31), the infection focus was the respiratory apparatus, the urinary or digestive tracts (n = 17, 8 and 6 respectively). On admission, acutely stressed patients with a SIRS (n = 44) had significantly higher plasma concentrations of CGA than controls (115 µg/L (68–202.8) *vs* 40 µg/L (35–52.5), $p < 0.001$). In addition, the septic group (sepsis, severe sepsis and septic shock) had significantly higher CGA levels than patients with SIRS but without infections (138.5 µg/L (65–222.3) *vs* 110 µg/L (81–143), $p < 0.01$). Interestingly, CGA plasma levels positively correlated with biomarkers of inflammation (C-reactive protein, procalcitonin and white cell counts) (Zhang et al. 2008).

In a multicentre analysis of VS-I in the plasma of critically ill patients (Tota et al. 2010), the population includes 481 patients and 13 healthy staff. Plasma VS-I concentrations were systematically higher in patients than in controls ($p < 0.01$). In univariate analysis, only severe SIRS patients [5.77 ng/ml], severe sepsis [6.07 ng/ml] and septic shock patients [6.21 ng/ml] displayed significantly higher plasma VS-I concentrations when compared to no SIRS, no infection patients [3.56 ng/ml, $p < 0.01$] and sepsis patients [3.6 ng/ml, $p < 0.01$]. Finally, there was a significant increase in all patients demonstrating circulatory failure when compared to those without circulatory failure [5.93 *vs*. 3.58 ng/ml, $p < 0.001$] and also in non-survivors *vs*. survivors [5.75 *vs*. 3.70 ng/ml, $p < 0.001$], but there was no significant difference in patients whether infection was present or not [4.41 *vs*. 3.88 ng/ml, $p = 0.547$]. As shown in Fig. 6, from admission to the 60th hour of admission within the ICU, plasma VS-I concentrations were systematically higher in patients than in controls ($p < 0.01$). We did not find differences of kinetics when comparing survivors and non-survivors, or infected patients and non-infected patients.

The results of this study show for the first time that an increase in the VS-I concentration was present on admission in non-selected critically ill patients and that a single measurement of this protein was associated with poor outcome at day 28 after admission, without having to specify a primary diagnosis.

Studying the VS-I/CGA ratios, we have established that in physiological conditions (in healthy controls), an average of 8% of the CGA is processed into VS-I, whereas in critically ill patients this ratio was lower (around 5%). Despite this absence of statistical difference among healthy controls and the entire cohort of patients, surviving patients had in average a significantly lower VS-I/CGA ratio than healthy controls ($p < 0.05$). Despite increased CGA levels after acute illness, we only measured half of such a ratio in patients. This suggests that the cleavage of CGA is limited under the pathological conditions of multiple organ failure. Different hypothetical mechanisms can be proposed to explain this phenomenon. First, CGA may aggregate (Yoo 1992) and this may hide the cleavage site located at position 76–77 on CGA. Then, many enzymes (both from host and from pathogens), including metallo-proteases are overexpressed in sepsis (Yazdan-Ashoori et al. 2011) and are candidates to degrade CGA and VS-I (Lauhio et al. 2011). In addition, after acute stress, disseminated intravascular coagulation often occurs (Lee and Downey 2000), which is a source of plasmin release that may interfere with the cleavage of CGA (Parmer et al. 2000). Furthermore, VS-I accumulation could result from an increase in the production of CGA and VS-I by nervous, endocrine and also immune cells activated as a result of the initial stressing injury (Wortsman et al. 1984).

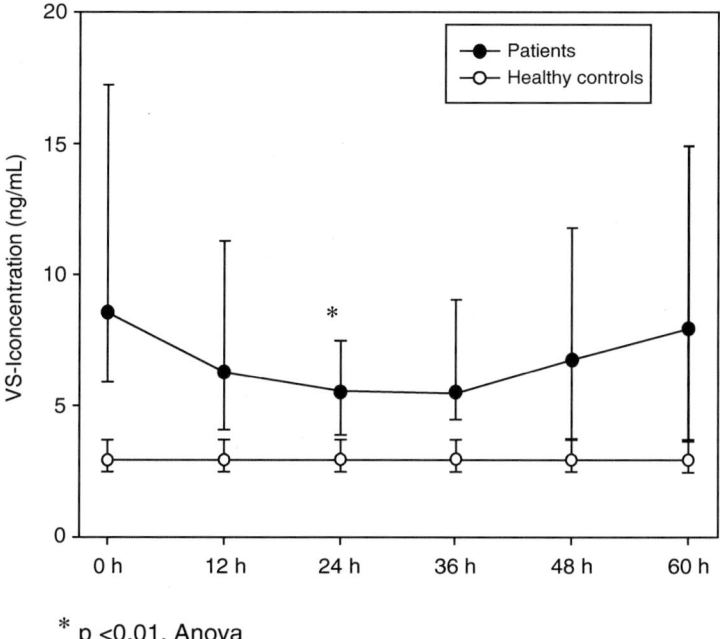

Fig. 6 Serial twice-a-day VS-I sampling in patients with simultaneous SIRS and acute circulatory failure criteria (Schneider et al. 2012)

The third piece of information available is that a single VS-I plasma concentration evaluation on admission can affect the assessment of 28-day outcome within 5 h after admission (time required for the performance of the test). We have gathered three data suggesting that patients who present increased plasma VS-I concentrations on admission have a higher risk of death within 28 days: first, patients that die have on average higher VS-I concentrations on admission, whatever the disease on admission; second, death rates are significantly higher when patients have an admission VS-I level beyond a median value of 3.97 ng/ml; third, the VS-I concentration in plasma is an independent risk factor for survival. Why the increased VS-I circulating concentration is an ominous sign remains to be determined since protective effects have been reported in animals (Roatta et al. 2011); one possible explanation is that protective effects (at high concentrations) are related to intracellular concentrations of VS-I and not circulating ones.

Lactate has often been considered as the best biomarker of severity early after onset of a severe disease and has therefore gained considerable acceptance (Trzeciak et al. 2007). Finally, using a combination of VS-I, lactate and age values, which are all available early after admission, our assessment of prognosis was better than taking the parameters alone (Schneider et al. 2012). Our findings enable us to hope for better routine decision-making for critically ill patients and also to improve emergency triage in patients who display increased lactate levels without obvious clinical signs of severity.

8 New Biomaterials with Antimicrobial Chromogranin Derived Peptides

Implantable medical devices are widely used in surgery not only to replace altered or lost tissues but also in critical care for fluid or gas administration using catheters or tracheal tube, respectively. These devices constitute an open gate for pathogens invasion and nosocomial infections (Von Eiff et al. 2005). Prevention of pathogen colonization of medical implants constitutes a major medical and financial issue. Indeed each year in Europe, 5% of patients admitted to hospitals suffer from hospital-acquired infections leading to a mortality of 10% (Guggenbichler et al. 2011). *S. aureus*, a Gram-positive bacterium, is responsible for hospital acquired infections especially in immunocompromized patients and 82% of the strains are methicillin resistant (Paniagua-Contreras et al. 2012). *C. albicans*, the most common human yeast pathogen, possesses the ability to form biofilms that are sources of local and systemic infection. Moreover, *C. albicans* biofilms allow the formation of *S. aureus* microcolonies on their surface and even enhanced *S. aureus* resistance to antibiotics (Harriott and Noverr 2009). The recent resistance of *C. albicans* to antifungal therapies (Ramage et al. 2006) and of *S. aureus* to antibiotics points out the need of multifunctional coatings that prevent infections of both yeast and bacteria.

Polyelectrolyte multilayer (PEM) films, based on an alternated deposition of polycations and polyanions onto a solid surface, emerged as a simple and efficient approach to functionalize surfaces in a controlled way (Decher 1997; Gribova et al. 2012). Natural AMPs, secreted by numerous living organisms against pathogens, gain increased attention due to their broad spectrum of antimicrobial activity and their low cytotoxicity (Glinel et al. 2012). They predominantly cause disruption of the membrane integrity of pathogen agents and thus unlikely initiate the development of resistance (Glinel et al. 2012). We decided to use bovine CTL, corresponding to bovine CgA344-358 and acting in the innate immunity system (Zhang et al. 2009b). We used polysaccharide multilayer films based on CTL-C-functionalized hyaluronic acid (HA) as polyanion and chitosan (CHI) as polycation, (HA-CTL-C/CHI), that were deposited on a planar surface with the aim of designing a self-defensive coating against both bacteria and yeasts (Fig. 7). A cysteine residue (C) was added at the C-terminal end of the CTL sequence to allow its grafting to HA. HA and CHI are biodegradable by enzymatic hydrolysis with hyaluronidase. Both polysaccharides are already widely used in biomedical applications due to their interesting intrinsic properties (Volpi et al. 2009; Riva et al. 2011). The ability of *S. aureus* (Larkin et al. 2009), *Candida* species (Shimizu et al. 1996) and *M. luteus* to degrade HA, by producing hyaluronidase, allows the CTL–C to be released from PEM films only in the presence of the pathogens. In this context we developed a new self-defensive coating (Cado et al. 2013) where the release of the antimicrobial peptide is triggered by enzymatic degradation of the film due to the pathogens

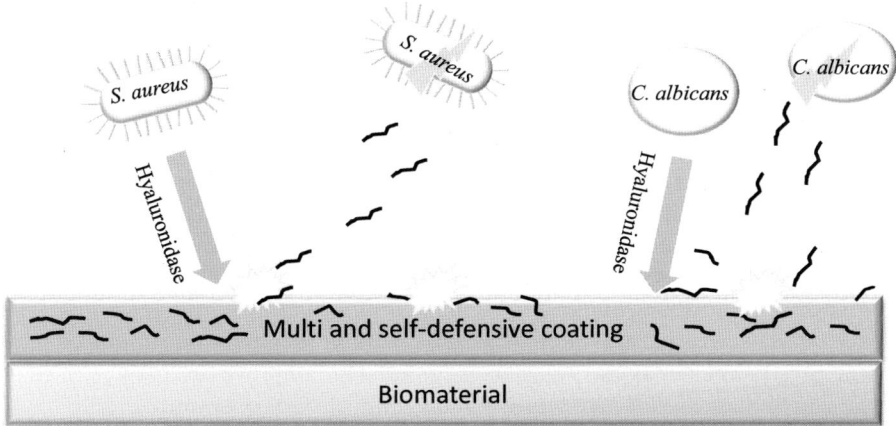

Fig. 7 Schematic representation of the antimicrobial activity of the CHI-HA coating of biomaterial. By action of hyaluronidase the active peptide is released to kill *S. aureus* and *C. albicans* (Cado et al. 2013)

themselves (Fig. 7). Polysaccharides adsorbed mass was determined by surface plasmon resonance, the buildup and the topography of the films were characterized by atomic force microscopy. Antibacterial and antifungal activities of HA-CTL-C in solution and HA-CTL-C/CHI films were tested against two strains Gram-positive bacteria, (*S. aureus* and *M. luteus*) and one strain of yeast strain (*C. albicans*), respectively, by using microdilution assays. Confocal laser scanning microscopy allowed following the penetration of the fluorescently labeled HA FITC -CTL-C, diluted in solution or embedded in a PEM film, into the cell membrane of *C. albicans*. Finally, the cytotoxicity of HA-CTL-C/CHI films was tested through human gingival fibroblasts (HGFs) viability.

In conclusion, we designed a new surface coating based on polysaccharide multilayer films containing a functionalized HA with 5% of CTL-C, a peptide possessing both antibacterial and antifungal properties. Antimicrobial properties of CTL-C were preserved when grafted on HA either in solution or when embedded into PEM films. After 24 h of incubation, HA-CTL-C/CHI films fully inhibit the development of *S. aureus* and *C. albicans,* which are common and virulent pathogens agents encountered in care-associated diseases. The presence of CTL-C peptides on HA allows the penetration of the modified polysaccharide inside *C. albicans* after 45 min of contact. The secretion of hyaluronidase by all tested pathogens seems to be responsible for HA-CTL-C release from the film and for its activity. The film can keep its activity during 3 cycles of use against fresh incubated *C. albicans* suspension. Furthermore, the limited fibroblasts adhesion, without cytotoxicity, on HA-CTL-C/CHI films highlights a medically relevant application to prevent infections on catheters or tracheal tubes where fibrous tissue encapsulation is undesirable.

9 General Conclusion

This review summarizes data concerning the processing of CGA and the biological roles of the antimicrobial CGA-derived peptides. For two peptides VS-I and CAT, we have characterized the active sequences (CHR, CTL) and their effects in presence of administered antibiotics. We demonstrated that CTL resists to bacterial proteases. Several peptides might allow the decrease of the antibiotics concentrations prescribed by clinicians.

The clinical studies show that high concentration of CGA in plasma with sepsis and SIRS correlated with markers of inflammation. In addition, by using a combination of VS-I, lactate and age values, the assessment of prognosis was improved than taking each parameter alone. This data is important to improve emergency triage in patients with increased lactate levels without clinical signs of severity.

CTL, the active domain of CAT, was used to elaborate new self-defensive material against *S. aureus* and *C. albicans*, two pathogens responsible of nosocomial infections. Interestingly, the limited fibroblast adhesion, without cytotoxicity highlights a medically relevant application to prevent infections on catheters or tracheal prosthesis.

References

Aardal S, Helle KB (1992) The vasoinhibitory activity of bovine chromogranin A fragment (vasostatin) and its independence of extracellular calcium in isolated segments of human blood vessels. Regul Pept 41(1):9–18

Aardal S, Helle KB, Elsayed S, Reed RK, Serck-Hanssen G (1993) Vasostatins, comprising the N-terminal domain of chromogranin A, suppress tension in isolated human blood vessel segments. J Neuroendocrinol 5(4):405–412

Aerts AM, François IE, Cammue BP, Thevissen K (2008) The mode of antifungal action of plant, insect and human defensins. Cell Mol Life Sci 65(13):2069–2079

Arnold R, Wilke A, Rinke A, Mayer C, Kann PH, Klose KJ et al (2008) Plasma chromogranin A as marker for survival in patients with metastatic endocrine gastroenteropancreatic tumors. Clin Gastroenterol Hepatol 6(7):820–827

Aslam R, Laventie BJ, Marban C, Prévost G, Keller D, Strub JM et al (2013a) Activation of neutrophils by the two-component leukotoxin LukE/D from *Staphylococcus aureus*: proteomic analysis of the secretions. J Proteome Res 12(8):3667–3678

Aslam R, Marban C, Corazzol C, Jehl F, Delalande F, Van Dorsselaer A et al (2013b) Cateslytin, a chromogranin A derived peptide is active against Staphylococcus aureus and resistant to degradation by its proteases. PLoS One 8(7):e68993

Biswas N, Rodriguez-Flores JL, Courel M, Gayen JR, Vaingankar SM, Mahata M et al (2009) Cathepsin L colocalizes with chromogranin A in chromaffin vesicles to generate active peptides. Endocrinology 150(8):3547–3557

Blois A, Holmsen H, Martino G, Corti A, Metz-Boutigue MH, Helle KB (2006) Interactions of chromogranin A-derived vasostatins and monolayers of phosphatidylserine, phosphatidylcholine and phosphatidylethanolamine. Regul Pept 134(1):30–37

Briolat J, Wu SD, Mahata SK, Gonthier B, Bagnard D, Chasserot-Golaz S et al (2005) New antimicrobial activity for the catecholamine release-inhibitory peptide from chromogranin A. Cell Mol Life Sci 62(3):377–385

Brosig B, Langosch D (1998) The dimerization motif of the glycophorin A transmembrane segment in membranes: importance of glycine residues. Protein Sci 7(4):1052–1056

Brun-Buisson C (2000) The epidemiology of the systemic inflammatory response. Intensive Care Med 26(Suppl 1):S64–S74

Bulet P, Stöcklin R, Menin L (2004) Anti-microbial peptides: from invertebrates to vertebrates. Immunol Rev 198(1):169–184

Cado G, Aslam R, Séon L, Garnier T, Fabre R, Parat A, Chassepot A, Voegel JC, Senger V, Schneider F, Frère Y, Jierry L, Schaaf P, Kerdjoudj H, Metz-Boutigue MH, Boulmedais F (2013) Self-defensive biomaterial coating against bacteria and yeasts: polysaccharide multilayer film with embedded antimicrobial peptide. Adv Funct Mater 23:4801–4809

Ceconi C, Ferrari R, Bachetti T, Opasich C, Volterrani M, Colombo B et al (2002) Chromogranin A in heart failure; a novel neurohumoral factor and a predictor for mortality. Eur Heart J 23(12):967–974

Corti A, Ferrari R, Ceconi C (2000) Chromogranin A and tumor necrosis factor-alpha (TNF) in chronic heart failure. Adv Exp Med Biol 482:351–359

Cryer PE, Wortsman J, Shah SD, Nowak RM, Deftos LJ (1991) Plasma chromogranin A as a marker of sympathochromaffin activity in humans. Am J Phys 260(2 Pt 1):E243–E246

Decher G (1997) Fuzzy nanoassemblies: toward layered polymeric multicomposites. Science 277(5330):1232–1237

Di Comite G, Marinosci A, Di Matteo P, Manfredi A, Rovere-Querini P, Baldissera E et al (2006) Neuroendocrine modulation induced by selective blockade of TNF-alpha in rheumatoid arthritis. Ann N Y Acad Sci 1069:428–437

Di Comite G, Rossi CM, Marinosci A, Lolmede K, Baldissera E, Aiello P et al (2009) Circulating chromogranin A reveals extra-articular involvement in patients with rheumatoid arthritis and curbs TNF-alpha-elicited endothelial activation. J Leukoc Biol 85(1):81–87

Eiden LE (1987) Is chromogranin a prohormone? Nature 325(6102):301

Evangelista R, Ray P, Lewis RV (1982) A "trypsin-like" enzyme in adrenal chromaffin granules: a proenkephalin processing enzyme. Biochem Biophys Res Commun 106(3):895–902

Fournier I, Gaucher D, Chich JF, Bach C, Shooshtarizadeh P, Picaud S et al (2011 Feb) Processing of chromogranins/secretogranins in patients with diabetic retinopathy. Regul Pept 167(1):118–124

Gao BB, Chen X, Timothy N, Aiello LP, Feener EP (2008) Characterization of the vitreous proteome in diabetes without diabetic retinopathy and diabetes with proliferative diabetic retinopathy. J Proteome Res 7(6):2516–2525

Glattard E, Angelone T, Strub JM, Corti A, Aunis D, Tota B et al (2006) Characterization of natural vasostatin-containing peptides in rat heart. FEBS J 273(14):3311–3321

Glinel K, Thebault P, Humblot V, Pradier CM, Jouenne T (2012) Antibacterial surfaces developed from bio-inspired approaches. Acta Biomater 8(5):1670–1684

Gribova V, Auzely-Velty R, Picart C (2012) Polyelectrolyte multilayer assemblies on materials surfaces: from cell adhesion to tissue engineering. Chem Mater 24(5):854–869

Guérin M, Guillemot J, Thouënnon E, Pierre A, El-Yamani FZ, Montero-Hadjadje M, Dubessy C, Magoul R, Lihrmann I, Anouar Y, Yon L (2010) Granins and their derived peptides in normal and tumoral chromaffin tissue: implications for the diagnosis and prognosis of pheochromocytoma. Regul Pept 165(1):21–29

Guggenbichler JP, Assadian O, Boeswald M, Kramer A (2011) Incidence and clinical implication of nosocomial infections associated with implantable biomaterials - catheters, ventilator-associated pneumonia, urinary tract infections. GMS Krankenhhyg Interdiszip 6(1):Doc18

Harbarth S, Holeckova K, Froidevaux C, Pittet D, Ricou B, Grau GE et al (2001) Geneva Sepsis Network. Diagnostic value of procalcitonin, interleukin-6, and interleukin-8 in critically ill patients admitted with suspected sepsis. Am J Respir Crit Care Med 164(3):396–402

Harriott MM, Noverr MC (2009) *Candida albicans* and *Staphylococcus aureus* form polymicrobial biofilms: effects on antimicrobial resistance. Antimicrob Agents Chemother 53(9):3914–3922

Helle KB (2010) Chromogranins A and B and secretogranin II as prohormones for regulatory peptides from the diffuse neuroendocrine system. Results Probl Cell Differ 50:21–44

Helle KB, Serck-Hanssen G (1975) The adrenal medulla: a model for studies of hormonal and neuronal storage and release mechanisms. Mol Cell Biochem 6(2):127–146

Helle KB, Corti A, Metz-Boutigue MH, Tota B (2007) The endocrine role for chromogranin A: a prohormone for peptides with regulatory properties. Cell Mol Life Sci 64(22):2863–2886

Hilchie AL, Wuerth K, Hancock RE (2013) Immune modulation by multifaceted cationic host defense (antimicrobial) peptides. Nat Chem Biol 9(2):761–768

Hook VY, Metz-Boutigue MH (2002) Protein trafficking to chromaffin granules and proteolytic processing within regulated secretory vesicles of neuroendocrine chromaffin cells. Ann N Y Acad Sci 971:397–405

Hook VY, Eiden LE, Brownstein MJ (1982) A carboxypeptidase processing enzyme for enkephalin precursors. Nature 295(5847):341–342

Jansson AM, Røsjø H, Omland T, Karlsson T, Hartford M, Flyvbjerg A et al (2009) Prognostic value of circulating chromogranin A levels in acute coronary syndromes. Eur Heart J 30(1):25–32

Larkin EA, Carman RJ, Krakauer T, Stiles BG (2009) Staphylococcus aureus: the toxic presence of a pathogen extraordinaire. Curr Med Chem 16(30):4003–4019

Lauhio A, Hastbacka J, Pettila V, Tervahartiala T, Karlsson S, Varpula T et al (2011) Serum MMP-8, −9 and TIMP- 1 in sepsis: high serum levels of MMP- 8 and TIMP-1 are associated with fatal outcome in a multicentre, prospective cohort study. Hypothetical impact of tetracyclines. Pharmacol Res 64(6):590–594

Lee WL, Downey GP (2000) Coagulation inhibitors in sepsis and disseminated intravascular coagulation. Intensive Care Med 26(11):1701–1706

Lugardon K, Raffner R, Goumon Y, Corti A, Delmas A, Bulet P et al (2000) Antibacterial and antifungal activities of vasostatin-1, the N-terminal fragment of chromogranin a. J Biol Chem 275(15):10745–10753

Lugardon K, Chasserot-Golaz S, Kieffer AE, Maget-Dana R, Nullans G, Kieffer B et al (2001) Structural and biological characterization of chromofungin, the antifungal chromogranin A-(47-66)-derived peptide. J Biol Chem 276(38):35875–35882

Maget-Dana R, Metz-Boutigue MH, Helle KB (2002) The N-terminal domain of chromogranin A (CgA1-40) interacts with monolayers of membrane lipids of fungal and mammalian compositions. Ann N Y Acad Sci 971:352–354

Mahata SK, O'Connor DT, Mahata M, Yoo SH, Taupenot L, Wu H et al (1997) Novel autocrine feedback control of catecholamine release. A discrete chromogranin A fragment is a noncompetitive nicotinic cholinergic antagonist. J Clin Invest 100(6):1623–1633

Manners JM (2007) Hidden weapons of microbial destruction in plant genomes. Genome Biol 8(9):225

Metz-Boutigue MH, Garcia-Sablone P, Hogue-Angeletti R, Aunis D (1993) Intracellular and extracellular processing of chromogranin A. Determination of cleavage sites. Eur J Biochem 217(1):247–257

Metz-Boutigue MH, Goumon Y, Lugardon K, Strub JM, Aunis D (1998) Antibacterial peptides are present in chromaffin cell secretory granules. Cell Mol Neurobiol 18(2):249–266

Montero-Hadjadje M, Vaingankar S, Elias S, Tostivint H, Mahata SK, Anouar Y (2008) Chromogranins A and B and secretogranin II: evolutionary and functional aspects. Acta Physiol (Oxf) 192(2):309–324

Nakano I, Funakoshi A, Miyasaka K, Ishida K, Makk G, Angwin P et al (1989) Isolation and characterization of bovine pancreastatin. Regul Pept 25(2):207–213

Nikou GC, Marinou K, Thomakos P, Papageorgiou D, Sanzanidis V, Nikolaou P et al (2008) Chromogranin A levels in diagnosis, treatment and follow-up of 42 patients with non-functioning pancreatic endocrine tumours. Pancreatology 8(4–5):510–519

O'Connor DT, Bernstein KN (1984) Radioimmunoassay of chromogranin A in plasma as a measure of exocytotic sympathoadrenal activity in normal subjects and patients with pheochromocytoma. N Engl J Med 311(12):764–770

O'Connor DT, Cervenka JH, Stone RA, Parmer RJ, Franco-Bourland RE, Madrazo I et al (1993) Chromogranin A immunoreactivity in human cerebrospinal fluid: properties, relationship to noradrenergic neuronal activity and variation in neurological disease. Neuroscience 56(4):999–1007

Paniagua-Contreras G, Sáinz-Espuñes T, Monroy-Pérez E, Rodríguez-Moctezuma JR, Arenas-Aranda D, Negrete-Abascal E et al (2012) Virulence markers in *Staphylococcus aureus* strains isolated from hemodialysis catheters of Mexican patients. Adv Microbiol 2(4):476–487

Parmer RJ, Mahata SK, Jiang O, Taupenot L, Gong Y, Mahata M et al (2000) Tissue plasminogen activator and chromaffin cell function. Adv Exp Med Biol 482:179–192

Ramage G, Martínez JP, López-Ribot JL (2006) *Candida* biofilms on implanted biomaterials: a clinically significant problem. FEMS Yeast Res 6(7):979–986

Riva R, Ragelle H, des Rieux A, Duhem N, Jerome C, Preat V (2011) Chitosan and chitosan derivatives in drug delivery and tissue engineering. Chitosan for biomaterials II. Springer, Berlin, Heidelberg, pp 19–44

Roatta S, Passatore M, Novello M, Colombo B, Dondossola E, Mohammed M et al (2011) The chromogranin A-derived N-terminal peptide vasostatin-I: in vivo effects on cardiovascular variables in the rabbit. Regul Pept 168(1–3):10–20

Schneider F, Bach C, Chung H, Crippa L, Lavaux T, Bollaert PE et al (2012) Vasostatin-I, a chromogranin A-derived peptide, in non-selected critically ill patients: distribution, kinetics, and prognostic significance. Vasostatin-I, a chromogranin A-derived peptide, in non-selected critically ill patients: distribution, kinetics, and prognostic significance. Intensive Care Med 38(9):1514–1522

Seidah NG, Chrétien M (1999) Proprotein and prohormone convertases: a family of subtilases generating diverse bioactive polypeptides. Brain Res 848(1–2):45–62

Serra P, Brandimarte C, Martino P, Carlone S, Giunchi G (1977) Synergistic treatment of enterococcal endocarditis: in vitro and in vivo studies. Arch Intern Med 137(11):1562–1567

Shimizu MT, Almeida NQ, Fantinato V, Unterkircher CS (1996) Studies on hyaluronidase, chondroitin sulphatase, proteinase and phospholipase secreted by Candida species. Mycoses 39(5–6):161–167

Shooshtarizadeh P, Zhang D, Chich JF, Gasnier C, Schneider F, Haïkel Y et al (2010) The antimicrobial peptides derived from chromogranin/secretogranin family, new actors of innate immunity. Regul Pept 165(1):102–110

Sternberg EM (2006) Neural regulation of innate immunity: a coordinated nonspecific host response to pathogens. Nat Rev Immunol 6(4):318–328

Stridsberg M, Eriksson B, Oberg K, Janson ET (2004) A panel of 11 region-specific radioimmunoassays for measurements of human chromogranin A. Regul Pept 117(3):219–227

Strub JM, Goumon Y, Lugardon K, Capon C, Lopez M, Moniatte M et al (1996) Antibacterial activity of glycosylated and phosphorylated chromogranin A-derived peptide 173-194 from bovine adrenal medullary chromaffin granules. J Biol Chem 271(45):28533–28540

Taupenot L, Harper KL, O'Connor DT (2003) The chromogranin-secretogranin family. N Engl J Med 348(12):1134–1149

Taylor CV, Taupenot L, Mahata SK, Mahata M, Wu H, Yasothornsrikul S et al (2000) Formation of the catecholamine release-inhibitory peptide catestatin from chromogranin A. Determination of proteolytic cleavage sites in hormone storage granules. J Biol Chem 275(30):22905–22915

Torres GE, Carneiro A, Seamans K, Fiorentini C, Sweeney A, Yao WD et al (2003) Oligomerization and trafficking of the human dopamine transporter. Mutational analysis identifies critical domains important for the functional expression of the transporter. J Biol Chem 278(4):2731–2739

Tota B, Quintieri AM, Di Felice V, Cerra MC (2007) New biological aspects of chromogranin A-derived peptides: focus on vasostatins. Comp Biochem Physiol A Mol Integr Physiol 147(1):11–18

Tota B, Cerra MC, Gattuso A (2010) Catecholamines, cardiac natriuretic peptides and chromogranin A: evolution and physiopathology of a 'whip-brake' system of the endocrine heart. J Exp Biol 213(Pt 18):3081–3103

Trzeciak S, Dellinger RP, Chansky ME, Arnold RC, Schorr C, Milcarek B et al (2007) Serum lactate as a predictor of mortality in patients with infection. Intensive Care Med 33(6):970–977

Turquier V, Vaudry H, Jégou S, Anouar Y (1999) Frog chromogranin A messenger ribonucleic acid encodes three highly conserved peptides. Coordinate regulation of proopiomelanocortin and chromogranin A gene expression in the pars intermedia of the pituitary during background color adaptation. Endocrinology 140(9):4104–4112

Volpi N, Schiller J, Stern R, Soltés L (2009) Role, metabolism, chemical modifications and applications of hyaluronan. Curr Med Chem 16(14):1718–1745

Von Eiff C, Jansen B, Kohnen W, Becker K (2005) Infections associated with medical devices: pathogenesis, management and prophylaxis. Drugs 65(2):179–214

Wang G, Li X, Wang Z (2016) APD3: the antimicrobial peptide database as a tool for research and education. Nucleic Acids Res 44(D1):D1087–D1093

Wortsman J, Frank S, Cryer PE (1984) Adrenomedullary response to maximal stress in humans. Am J Med 77(5):779–784

Yazdan-Ashoori P, Liaw P, Toltl L, Webb B, Kilmer G, Carter DE et al (2011) Elevated plasma matrix metalloproteinases and their tissue inhibitors in patients with severe sepsis. J Crit Care 26(6):556–565

Yeaman MR, Yount NY (2003) Mechanisms of antimicrobial peptide action and resistance. Pharmacol Rev 55(1):27–55

Yoo SH (1992) Identification of the Ca(2+)-dependent calmodulin-binding region of chromogranin A. Biochemistry 31(26):6134–6140

Yoshida C, Ishikawa T, Michiue T, Zhao D, Komatsu A, Quan L et al (2009) Immunohistochemical distribution of chromogranin A in medicolegal autopsy materials. Leg Med (Tokyo) 11(Suppl 1):S231–S233

Zhang D, Lavaux T, Voegeli AC, Lavigne T, Castelain V, Meyer N et al (2008) Prognostic value of chromogranin A at admission in critically ill patients: a cohort study in a medical intensive care unit. Clin Chem 54(9):1497–1503

Zhang D, Lavaux T, Sapin R, Lavigne T, Castelain V, Aunis D et al (2009a) Serum concentration of chromogranin A at admission: an early biomarker of severity in critically ill patients. Ann Med 41(1):38–44

Zhang D, Shooshtarizadeh P, Laventie BJ, Colin DA, Chich JF, Vidic J et al (2009b) Two chromogranin A-derived peptides induce calcium entry in human neutrophils by calmodulin-regulated calcium independent phospholipase A2. PLoS One 4(2):e4501

Conserved Nature of the Inositol 1,4,5-Trisphosphate Receptor and Chromogranin Coupling and Its Universal Importance in Ca^{2+} Signaling of Secretory Cells

Coupling of Inositol 1,4,5-Trisphosphate Receptor/Ca^{2+} Channel and Chromogranin

Seung Hyun Yoo

Abstract Secretory granules have been demonstrated to be the major inositol 1,4,5-trisphosphate (IP$_3$)-sensitive intracellular Ca^{2+} stores in secretory cells, primarily due to the presence of high concentrations of the Ca^{2+} storage proteins chromogranins and the IP$_3$ receptor (IP$_3$R)/Ca^{2+} channels in addition to 40 mM Ca^{2+}. Among the many extraordinary features of secretory granules is the coupling to the IP$_3$R/Ca^{2+} channels and the channel-modulatory roles of chromogranins, increasing the open probability and the mean open time of the channels multiple-fold. Particularly noteworthy is that most conserved regions of chromogranins A (CGA) and B (CGB) and of the IP$_3$Rs, i.e., the near amino (N)-terminal region of chromogranins A and B and the last intraluminal loop (L3-2 loop) of the IP$_3$Rs, participate in the coupling. Given that the two regions which participate in the coupling are highly conserved in the biokingdom, the coupling between chromogranins and the IP$_3$R/Ca^{2+} channels appears to be a very ancient molecular design forming the IP$_3$-dependent Ca^{2+} signaling systems of organisms. Although the coupling is uncovered in secretory cells, similar molecular arrangements which involve the IP$_3$R/Ca^{2+} channels and some type of Ca^{2+} storage and/or Ca^{2+} channel-modulatory proteins that are functionally equivalent to chromogranins in secretory cells will also exist in nonsecretory cells.

S.H. Yoo (✉)
Gran Med Inc., Get Pearl Tower 304, Get Pearl Ro 12, Yeonsu Gu,
Incheon 21999, South Korea
e-mail: shyoo@granmedical.com

© Springer International Publishing AG 2017
T. Angelone et al. (eds.), *Chromogranins: from Cell Biology to Physiology and Biomedicine*, UNIPA Springer Series, DOI 10.1007/978-3-319-58338-9_5

1 Introduction

Secretory granules from a wide variety of secretory cells, ranging from primitive unicellular organisms to humans, serve as the major IP$_3$-sensitive intracellular Ca^{2+} store (Yoo 2010, 2011). The main reason secretory granules can function as the major IP$_3$-sensitive intracellular Ca^{2+} store is the abundant presence of the IP$_3$R/Ca^{2+} channels and the high capacity, low affinity Ca^{2+} storage proteins chromogranins in secretory granules in addition to 40 mM Ca^{2+} (Haigh et al. 1989; Winkler and Westhead 1980). Secretory granules contain the highest concentrations and the largest amounts of all three IP$_3$R types and chromogranins A (CGA) and B (CGB) in secretory cells (Huh et al. 2005b, 2005c). A significant portion of cellular contents of these molecules is also contained in the endoplasmic reticulum (ER) and the nucleus with the exception of CGA which is absent in the nucleus (Yoo et al. 2002).

As if to highlight the physiological significance of colocalization of these molecules in the same organelles, chromogranins A and B directly interact with the IP$_3$R/Ca^{2+} channels (Yoo et al. 2001) and activate the channels; increasing the open probability 8- to 16-fold and the mean open time 9- to 42-fold, respectively (Thrower et al. 2002, 2003; Yoo and Jeon 2000). Yet, the interaction properties of chromogranins with the IP$_3$Rs differ depending on the pH milieu: chromogranin A binds and activates the IP$_3$R/Ca^{2+} channels only at intragranular pH 5.5, dissociating from each other and becoming ineffective at near physiological pH 7.5, in contrast to chromogranin B that stays coupled to and activates the IP$_3$R/Ca^{2+} channels at both the intragranular pH 5.5 and the near physiological pH 7.5 (Thrower et al. 2003; Yoo et al. 2000).

Moreover, being the first example of granin proteins existing outside of secretory granules chromogranin B has further been shown to localize in the nucleus, along with all three IP$_3$R types (Yoo et al. 2005; Huh and Yoo 2003); chromogranin B and the IP$_3$Rs colocalize in numerous small nucleoplasmic Ca^{2+} store vesicles (Huh et al. 2006b; Yoo et al. 2005, 2014). Although the normal physiological pH of cells is ~7.4, including that of the nucleus and endoplasmic reticulum (ER), secretory granules maintain an acidic pH 5.5, thus providing a different environment for the chromogranin-IP$_3$R/Ca^{2+} channel interaction and function. This pH-dependence of chromogranin-IP$_3$R interaction and different localization of chromogranins A and B in subcellular organelles appear to reflect the differences each chromogranin contributes to the overall Ca^{2+} homeostasis in secretory cells.

In spite of the apparent differences in each chromogranin's role in intracellular Ca^{2+} homeostasis, there lies an invariable common molecular organization in the cooperative works between the IP$_3$Rs and chromogranins that the IP$_3$R/Ca^{2+} channel modulatory role of chromogranins is through the interaction between the highly conserved near amino (N)-terminal regions of chromogranins A and B (Table 1) (Helle 2000; Montero-Hadjadje et al. 2008; Taupenot et al. 2003; Winkler and Fischer-Colbrie 1992; Bartolomucci et al. 2011) and the equally well conserved intraluminal L3-2 loops of the IP$_3$Rs (Fig. 1 and Table 2). The near N-terminal region of chromogranins is conserved not only between chromogranins A and B but also across different species. Similarly, the L3-2 loop of the IP$_3$Rs is also well conserved not only between the three types but also across different species, including

Table 1 Conserved amino acid sequences near the amino terminal regions of chromogranins A and B across species

Protein		*Amino Acid Sequence	Genbank Accession Code
CGA	human	^{35}CIVEVISDTLSKPSPMPVSQEC^{56}	AAB53685.1
	bovine	^{35}CIVEVISDTLSKPSPMPVSKEC^{56}	AAC48700.1
	mouse	^{35}CVLEVISDSLSKPSPMPVSPEC^{56}	AAA37457.1
	rat	^{35}CVLEVISDSLSKPSPMPVSPEC^{56}	AEB41036.1
	pig	^{35}CIVEVISDTLSKPSPMPVSQEC^{56}	NP_001157477.2
CGB	human	^{36}CIIEVLSNALSKSSAPPITPEC^{57}	AAB53685.1
	bovine	^{36}CIIEVLSNALLKSSAPPITPEC^{57}	AAC48700.1
	mouse	^{36}CIIEVLSNALSKSSVPTITPEC^{57}	AAA37457.1
	rat	^{36}CIIEVLSNALSKSSAPTITPEC^{57}	AEB41036.1
	pig	^{36}CIIEVLSNALSKSNAPPITPEC^{57}	NP_001157477.2

ᵃThe numbers shown indicate the position of the amino acids that were numbered from the signal peptides of each chromogranin and the colors denote identical (*red*), conserved (*blue*), and different (*black*) residues

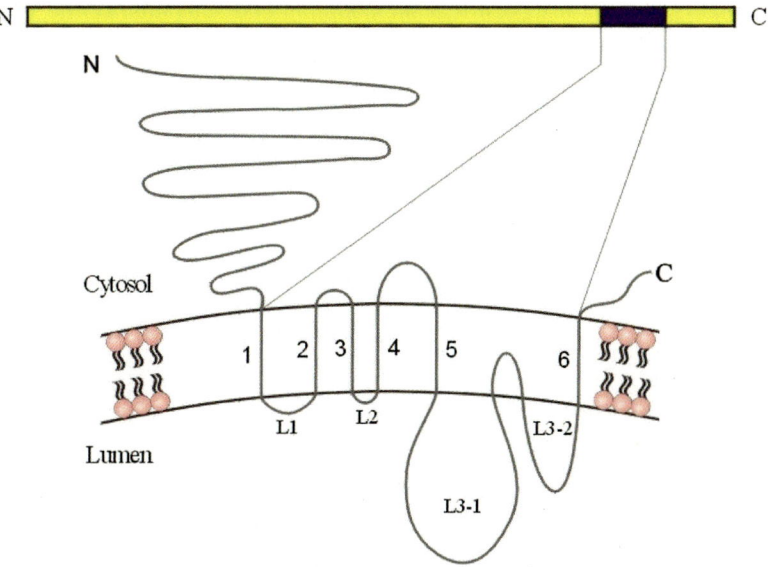

Fig. 1 Topology of the IP$_3$R showing the intraluminal loops. There are six transmembrane regions near the C-terminal end in the IP$_3$Rs. The intraluminal loops formed are shown as L1, L2, L3-1, and L3-2

Table 2 Conserved amino acid sequences of the chromogranin-interacting intraluminal loop L3-2 of the IP$_3$ receptors across species

Protein		*Amino Acid Sequence	Genbank Accession Code
IP$_3$ Receptor type 1	bovine	^{2510}DVLRKPSKEEPLFAARVIYD^{2529}	NP_777266.1
	human isoform 1	^{2511}DVLRKPSKEEPLFAARVIYD^{2530}	NP_001093422.2
	human isoform 2	^{2496}DVLRKPSKEEPLFAARVIYD^{2515}	NP_002213.5
	human isoform 3	^{2544}DVLRKPSKEEPLFAARVIYD^{2563}	NP_001161744.1
IP$_3$ Receptor type 2	bovine	^{2502}DVLRRPSKDEPLFAARVVYD^{2521}	NP_776794.1
	human	^{2502}DVLRRPSKDEPLFAARVVYD^{2521}	NP_002214.2
IP$_3$ Receptor type 3	bovine	^{2471}DILRKPSKDESLFPARVVYD^{2490}	NP_776795.1
	human	^{2478}DILRKPSKDESLFPARVVYD^{2497}	NP_002215.2
IP$_3$ Receptor	nematode (*Trichinella britovi*)	^{2492}DVLRKPHPQEPLFYMRILYD^{2511}	KRY53103.1

^{a}The numbers shown indicate the position of the amino acids in each IP$_3$ receptor and the colors denote identical (*red*), conserved (*blue*), and different (*black*) residues

primitive species such as protozoa and nematodes that contain only a single IP$_3$R type. These facts amply suggest that the coupling and activation of the IP$_3$R/Ca^{2+} channels by chromogranins has a very fundamental significance and a deep-rooted origin in the intracellular Ca^{2+} signaling systems in the biokingdom.

In particular, the ability of CGB to bind and activate the IP$_3$R/Ca^{2+} channels at both acidic and physiological pH appears to underscore the critical importance of the IP$_3$R/Ca^{2+} channel modulatory role of CGB in secretory cells, encompassing both the cytoplasm in form of secretory granules and the ER, and the nucleus in the form of IP$_3$-sensitive nucleoplasmic Ca^{2+} store vesicles. Hence, the molecular basis of the coupling between the IP$_3$Rs and chromogranins, and its role in intracellular Ca^{2+} homeostasis of secretory cells are discussed here.

2 IP$_3$R Topology

The IP$_3$R monomers are composed of ~2800 amino acids with six transmembrane domains near the C-terminal end (Fig. 1). There are three IP$_3$R types, types 1, 2, and 3, and a complex of homo- and/or heterotetramers of these isoforms forms a Ca^{2+} channel (Foskett et al. 2007; Fan et al. 2015). The six transmembrane domains of the IP$_3$Rs are located in amino acid residues number 2500–2700, which form three intraluminal loops, but the third loop is made of two half-loops due to a partial burial of the middle region of this loop in the membrane (Fig. 1). As a result, the loops are termed L1, L2, and L3-1 and L3-2.

Despite the seemingly equal contribution of these IP$_3$R isoforms to form the IP$_3$R/Ca^{2+} channels, there is relatively little sequence homology between the isoforms with the exception of the L3-2 loop that maintains well conserved amino acid sequences throughout the different isoforms across species. The sequence comparison of the L3-2 regions of the IP$_3$Rs among the isoforms and different species is shown in Table 2. Assuming the substitutions in amino acids of similar property as conserved, the degree of conservation reaches ~75%, even including the nematode. In line with its potential importance in the function of IP$_3$Rs, the L3-2 loop of the IP$_3$R has been shown to participate in the interaction with chromogranins (Yoo and Lewis 1994, 1995, 1998, 2000). Underscoring such a structural fit between chromogranins and the IP$_3$Rs, a near-atomic level structure of the IP$_3$R1 that was resolved by cryo-electron microscopy has indeed revealed the protruding intraluminal L3-2 loops in its tetrameric form (Fan et al. 2015).

3 CGA and CGB Sequence Comparison

There are two conserved regions not only among the same type of chromogranins from different species but also among the different types of chromogranins from a wide variety of species; one being the near N-terminal region, usually amino acid residues number 20–40, and the other being the 23–25 amino acids from the C-terminal end (Bartolomucci et al. 2011; Helle 2000; Montero-Hadjadje et al. 2008; Taupenot et al. 2003; Winkler and Fischer-Colbrie 1992). As shown in sequence comparison (Table 1), the near N-terminal region of chromogranins A and B is highly conserved, strongly implicating potentially important roles of this region in the function of chromogranins. Interestingly, the conserved near N-terminal region is boxed between two cysteine residues that are invariably present in this region of chromogranins A and B across species.

As if to explain the reasons for the high degree of sequence conservation, the conserved near N-terminal region of chromogranins has been shown to specifically interact with the L3-2 loop of the IP$_3$Rs (see below). This fact further underscored the possibility of the critical role of the near N-terminal region in the function of chromogranins. Accordingly, the conserved near N-terminal region of chromogranins has indeed been shown to be indispensable in targeting chromogranins to secretory granules (Courel et al. 2006; Glombik et al. 1999; Kang and Yoo 1997; Taupenot et al. 2002; Yoo and Kang 1997; Thiele and Huttner 1998; Montero-Hadjadje et al. 2009; Yoo et al. 2014), i.e., chromogranin fragments without the near N-terminal region failed to localize in secretory granules and were shown to be constitutively secreted. Moreover, the conserved C-terminal regions of chromogranins A and B participate in dimerization and tetramerization of chromogranins (Thiele and Huttner 1998; Yoo and Lewis 1992, 1993, 1996): CGA forms homodimer at pH 7.5 and homotetramer at pH 5.5, but in the presence of CGB, CGA and CGB form CGA-CGB heterodimer at pH 7.5 and CGA$_2$-CGB$_2$ heterotetramer at pH 5.5 (Yoo and Lewis 1996).

4 Interaction Between the L3-2 Loop and the Near N-Terminal Region

Reflecting the pH dependency of the interaction between chromogranins and the IP$_3$Rs, chromogranin A bound the L3-2 loop of the IP$_3$R only at the intragranular pH 5.5, but not at pH 7.5. However chromogranin B bound the L3-2 loop of the IP$_3$R at both pH 5.5 and 7.5, although the interaction at pH 7.5 was considerably weaker than that at pH 5.5. Moreover, it was further shown that tetrameric chromogranins interact with four molecules of L3-2 at pH 5.5 (Yoo and Lewis 1995) (Fig. 2). The chromogranin region that interacted with the L3-2 loop of the IP$_3$R was shown to be the conserved near N-terminal regions of both CGA and CGB. As was the case with intact CGA, the near N-terminal region of CGA also interacted with the L3-2 loop of the IP$_3$R only at pH 5.5, but dissociated completely from each other at near physiological pH 7.5. However, the near N-terminal region of CGB still interacted with the L3-2 loop even at pH 7.5, as was the case with intact CGB.

Although the conserved near N-terminal region of chromogranins is shown to interact with the L3-2 loop of the IP$_3$Rs, the interaction strength of whole chromogranins for the L3-2 loop is at least two orders of magnitude higher than that of the near N-terminal region for the L3-2 loop (Table 3). The interaction strength between

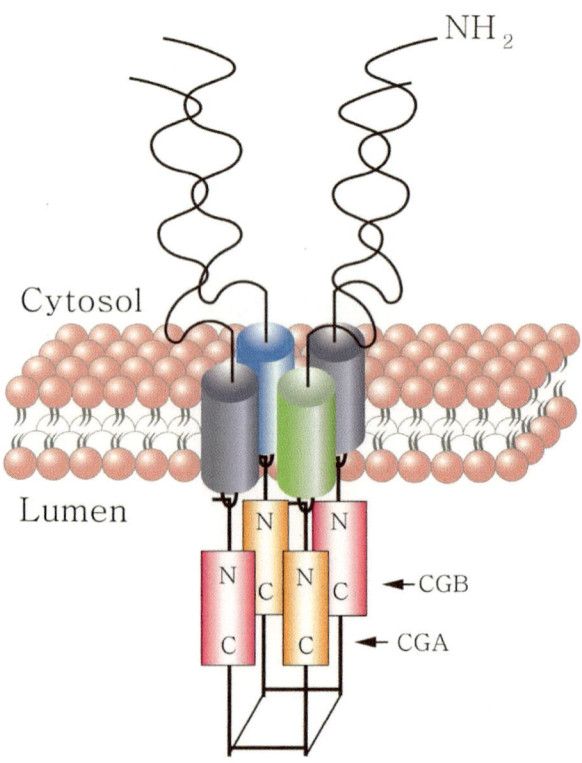

Fig. 2 Schematic model showing the coupling between the tetrameric IP$_3$Rs and tetrameric chromogranins. Heterotetrameric IP$_3$R/Ca^{2+} channel and CGA$_2$CGB$_2$ heterotetrameric chromogranins are coupled through the L3-2 loops of IP$_3$Rs and the near N-terminal regions of chromogranins A and B

Table 3 The interaction strength of the intraluminal loop (L3-2) of the IP$_3$R with intact chromogranins A (CGA) and B (CGB), and with the conserved near N-terminal regions of each chromogranin at different pH as expressed in the changes of standard free energy ($\Delta G°$) at 37 °C (310.15 K)

	Chromogranin A		Chromogranin B	
	Intact CGA-IP$_3$R L3-2	N-terminal CGA fragment-IP$_3$R L3-2	Intact CGB-IP$_3$R L3-2	N-terminal CGB fragment-IP$_3$R L3-2
$\Delta G°$ (kcal/mol), pH 5.5	$-6.0 \sim -6.9$ (Kd: 11 ~ 17 µM)	-3.57 (Kd: 3.04 mM)	CGB or the N-terminal fragment strongly self-associates at pH 5.5, leading to aggregation (not measurable)	
$\Delta G°$ (kcal/mol), pH 7.5	CGA and the L3–2 dissociate at pH 7.5 (no interaction).		-8.10 (Kd: 1.9 µM)	-4.70 (Kd: 0.48 mM)

the intact CGA and the L3-2 loop peptide is $-6.0 \sim -6.9$ kcal/mol (*Kd*, 11–17 µM) at pH 5.5 whereas that between the near N-terminal peptide of CGA and the L3-2 loop peptide is -3.57 kcal/mol (*Kd*, 3.04 mM) (Yoo and Lewis 1995, 1998), indicating approximately two orders of magnitude stronger interaction by intact CGA. This result further suggests that the interaction strength between intact CGA and intact IP$_3$Rs in secretory granules will be several orders of magnitude higher than that shown between intact CGA and the L3-2 loop at pH 5.5.

Moreover, in line with the more prominent role of CGB in activating the IP$_3$R/Ca^{2+} channels compared to CGA, the interaction of CGB with the IP$_3$Rs is substantially stronger than that of CGA. Highlighting such strong binding, the interaction strength between intact CGB and the L3-2 loop even at pH 7.5, which is known to be significantly weaker than that at pH 5.5, is -8.10 kcal/mol (*Kd*, 1.9 µM) (Table 3) (Yoo and Lewis 2000). This value indicates that the interaction between intact CGB and the L3-2 loop at pH 7.5 is still an order of magnitude stronger than that shown between intact CGA and the L3-2 loop at pH 5.5, whereby amply underscoring the extraordinarily strong binding of CGB for the IP$_3$R. Yet, reflecting a weaker interaction than that shown between intact CGB and the L3-2 loop of the IP$_3$Rs, the interaction strength between the conserved near N-terminal region of CGB and the L3-2 loop at pH 7.5 is -4.70 kcal/mol (*Kd*, 0.48 mM). Nonetheless, this interaction is still significantly stronger than that shown with the near N-terminal region of CGA and the L3-2 loop at pH 5.5, i.e., -3.57 kcal/mol (*Kd*, 3.04 mM).

Given that the interaction of the near N-terminal region of chromogranins with the IP$_3$R loop is significantly weaker than that of intact chromogranins, the actual interaction between intact CGB and the IP$_3$R/Ca^{2+} channels in cells will be far stronger than the estimated values. Furthermore, considering that chromogranin interaction with the IP$_3$R at pH 5.5 is substantially stronger than at pH 7.5, the interaction strength of CGB for the IP$_3$Rs at pH 5.5 is certain to be several orders of magnitude higher than that between CGA and the IP$_3$Rs.

5 Physiological Significance of the Chromogranin-IP$_3$R Interaction

Despite the many similarities in physicochemical properties between CGA and CGB and their near universal presence in a myriad of secretory cells (Helle 2000; Montero-Hadjadje et al. 2008; Taupenot et al. 2003; Winkler and Fischer-Colbrie 1992; Bartolomucci et al. 2011), the subcellular localization of CGA is limited to the cytoplasm, i.e., secretory granules, the ER, and the Golgi albeit most of it is in acidic secretory granules. Hence, the interaction between CGA and the IP$_3$Rs will be the strongest and most stable inside secretory granules which maintain the IP$_3$R/Ca^{2+} channels in ordered and release-ready state (Yoo and Jeon 2000). However, unlike the interaction of CGA with the IP$_3$R/Ca^{2+} channels in secretory granules, CGA in the ER is incapable of interacting with the IP$_3$Rs due to its inability to bind the IP$_3$Rs at the physiological pH of the ER.

The failure of CGA to bind the IP$_3$Rs directly at near physiological pH 7.5 is shown with all three types of the IP$_3$R (Yoo et al. 2001). As shown in Fig. 3, CGA binds directly with the IP$_3$R at the intragranular pH 5.5, but it dissociates completely from the IP$_3$R at pH 7.5. Yet CGA interacts with CGB at pH 7.5, forming CGA-CGB heterodimer, thereby suggesting the presence of CGA-CGB heterodimers in the lumen of the ER. Considering the fact that CGB interacts with all three IP$_3$R types even at pH 7.5, it is expected that the CGA-CGB heterodimers in the lumen of the ER stay coupled to the IP$_3$R/Ca^{2+} channels of the ER, and this is indeed the case as shown in Fig. 3. As a result, the effect of CGA on the IP$_3$R/Ca^{2+} channels will be exerted mainly in secretory granules, and that in the ER will be only through its interaction with CGB, which in turn interacts with the IP$_3$R/Ca^{2+} channels (Fig. 3). Differing from CGA that exists in the cytoplasm only, CGB localizes inside the nucleus as well. Moreover, CGB interacts with the IP$_3$Rs and activates the IP$_3$R/Ca^{2+} channels regardless of the different pH environment. Therefore, the IP$_3$R/Ca^{2+} channel-activating effect of CGB will be evident in both the cytoplasm and the nucleus (Huh et al. 2006a).

One of the major physiological effects of the strong interaction of CGB with the IP$_3$Rs is the prominent IP$_3$R/Ca^{2+} channel-activating role of CGB, increasing the

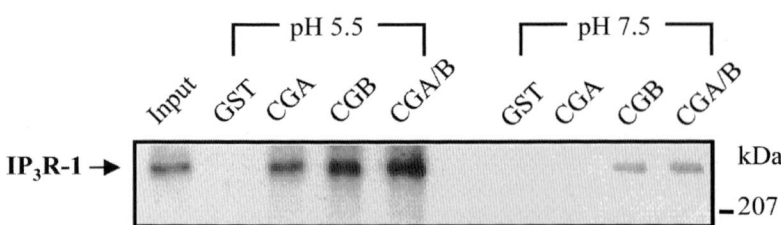

Fig. 3 pH-dependent interaction of the IP$_3$R with chromogranins. Purified IP$_3$R1 (0.7–1.0 µg) was reacted with GST-CGA, GST-CGB fusion proteins, and an equimolar mixture of the two (CGA/B) at pH 5.5 and 7.5 in the presence of 2 mM Ca^{2+}. The bound IP$_3$R1 was separated on a 7.5% SDS gel and analyzed by immunoblot

open probability 8- to 16-fold and the mean open time of the channel 24- to 42-fold (Thrower et al. 2003). This IP$_3$R/Ca^{2+} channel-activating role is also shared with CGA to a certain extent, but the magnitude of CGA effect on the Ca^{2+} channel is markedly smaller than that exerted by CGB, consistent with the weaker interaction strength of CGA for the IP$_3$Rs. As a result, the IP$_3$-dependent Ca^{2+} mobilization in the cell is heavily affected by the subcellular localization of CGB inside the cell.

6 Concentration of CGB and the IP$_3$Rs in Secretory Granules, ER, and Nucleus

The CGB concentrations in secretory granules and the ER in the typical neuroendocrine adrenal chromaffin cells are ~200 μM and 120 μM, respectively (Huh et al. 2005a), while the relative IP$_3$R concentrations in secretory granules and the ER are ~3.5 and ~1.0, respectively, when that of the nucleus is taken as 1.0 (Huh et al. 2005c). That secretory granules contain the highest concentrations of CGB and the IP$_3$Rs, and the largest amounts of CGB and the IP$_3$Rs, than any other organelle in the cytoplasm naturally points to the more prominent role of secretory granules in IP$_3$-dependent Ca^{2+} signaling, i.e., higher IP$_3$ sensitivity of the granule IP$_3$R/Ca^{2+} channels than that of the ER and the markedly larger amounts of Ca^{2+} released from secretory granules than from the ER (Huh et al. 2007).

Unlike the more easily identifiable molecular organization of secretory granules and the ER, the situation in the nucleus is a bit more complicated. The CGB concentration in the nucleus is estimated to be ~80 μM (Huh et al. 2005a), a figure that shows average CGB concentration in the whole nucleus although CGB localization is limited only to the small IP$_3$-sensitive nucleoplasmic Ca^{2+} store vesicles (Huh et al. 2006b; Yoo et al. 2005, 2014). Hence, the CGB concentration in the small nucleoplasmic vesicles will be at least several-fold higher than the estimated 80 μM, the value given for the whole nucleus, and is most likely substantially higher than 200 μM CGB present in secretory granules.

In addition to the concentrations of CGB and the IP$_3$Rs contained in the organelles, the molecular contents and organization inside the organelles will also be of crucial importance in the formation of the IP$_3$-sensitive Ca^{2+} store properties of respective Ca^{2+} store organelles. Although both secretory granules and the nucleoplasmic Ca^{2+} store vesicles contain CGB and the IP$_3$Rs and release Ca^{2+} in response to IP$_3$, there appears to be a fundamental difference in the intrinsic functions between the two organelles; while secretory granules store hormones and other essential bioactive molecules in addition to high concentrations of Ca^{2+}, the small nucleoplasmic Ca^{2+} store vesicles do not appear to store high concentrations of other bioactive proteins, but rather appears to store primarily large amounts of Ca^{2+}. In this respect, the primary function of the nucleoplasmic Ca^{2+} store vesicles is thought to be Ca^{2+} storage/control in the nucleoplasm, which is consistent with the high IP$_3$ sensitivity of the IP$_3$R/Ca^{2+} channels in the nucleoplasm (Huh et al. 2007).

7 Effects of CGB Presence in the Cytoplasm and the Nucleus

The difference in the interaction property of CGA and CGB appeared to directly influence the IP_3-dependent Ca^{2+} mobilization effects of CGA and CGB; the effect of CGB on the magnitude of IP_3-dependent Ca^{2+} releases is substantially greater than that of CGA even in the cytoplasm. The magnitude of IP_3-mediated Ca^{2+} releases attributed to CGB in the cytoplasm is approximately twice as big as that attributed to CGA, underscoring the critical importance of CGB in the IP_3-dependent Ca^{2+} signaling in the cytoplasm.

Differences in such effects have clearly been demonstrated by expressing chromogranins in nonsecretory NIH3T3 cells that do not contain intrinsic chromogranins (Fig. 4): CGA expression in NIH3T3 cells increased the IP_3-induced Ca^{2+} release in the cytoplasm ~100% whereas CGB expression increased it ~200% (Fig. 4A) (Huh et al. 2006a). On the other hand, suppression of CGA expression in secretory PC12 cells that contain intrinsic chromogranins decreased the IP_3-induced Ca^{2+} release in the cytoplasm ~50% whereas suppression of CGB expression in PC12 cells decreased the IP_3-induced Ca^{2+} release ~75% (Huh et al. 2006a). Given that chromogranin-containing cytoplasmic organelles include secretory granules and the ER, the IP_3-dependent Ca^{2+} mobilization effect of each chromogranin will be through its IP_3R/Ca^{2+} channel-modulatory role in these organelles.

In contrast, since there is no CGA in the nucleus, only CGB participates in the IP_3-dependent nucleoplasmic Ca^{2+} signaling. In line with the IP_3R/Ca^{2+} channel-activating role of CGB, the presence of CGB in the nucleus markedly enhanced the

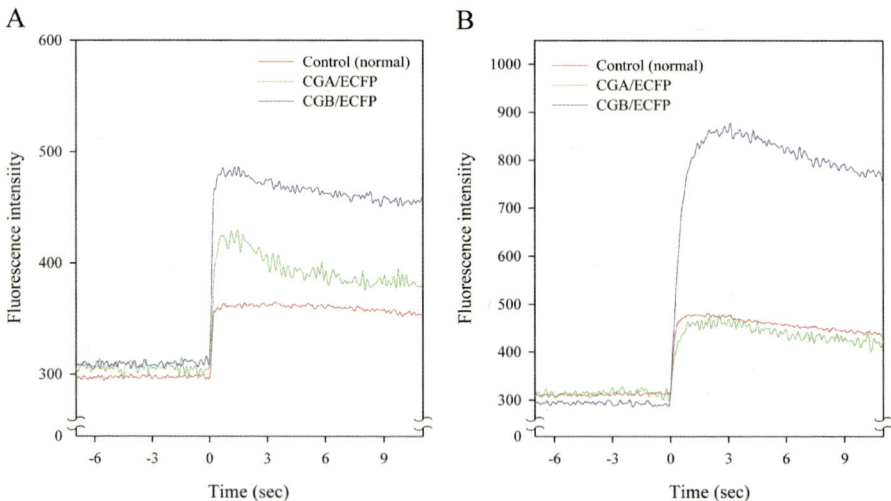

Fig. 4 Effects of chromogranins on the IP_3-induced Ca^{2+} release in the cytoplasm and nucleus of NIH3T3 cells. The IP_3-induced Ca^{2+} releases in the cytoplasm (**A**) and nucleus (**B**) of NIH3T3 cells after transfection with CGA (CGA/ECFP) and CGB (CGB/ECFP) (Figure is modified from Fig. 2 of Ref. (Huh et al. 2006a))

magnitude of IP_3-mediated nucleoplasmic Ca^{2+} releases (Fig. 4B). Although CGA expression in nonsecretory NIH3T3 cells was without any effect on the IP_3-induced Ca^{2+} release in the nucleus, CGB expression in NIH3T3 cells increased the IP_3-induced Ca^{2+} release in the nucleus ~315% (Fig. 4B) (Huh et al. 2006a), thereby demonstrating the crucial role of CGB in the nuclear Ca^{2+} control.

Nonsecretory cells that do not contain chromogranins would not need as subtle and diverse Ca^{2+} control machineries as secretory cells, and for that matter, the nonsecretory cells would be able to meet their cellular Ca^{2+} control needs without the service of multi-functional molecules such as chromogranins. Instead, the roles played by chromogranins in secretory cells will probably be met by other molecules that are functionally equivalent to chromogranins, and these molecules might as well exist in the IP_3-sensitive nucleoplasmic Ca^{2+} store vesicles of nonsecretory cells as CGB exists inside the small IP_3-sensitive nucleoplasmic Ca^{2+} store vesicles of secretory cells.

8 Conclusion

Direct coupling between the two ancient family of proteins chromogranins and the IP_3Rs, via the interaction between the near N-terminal regions of chromogranins and the L3-2 loops of the IP_3Rs, is likely to have served as the beginning of IP_3-dependent intracellular Ca^{2+} signaling systems in the biokingdom. In view of the requirements of the existence of signaling molecules, i.e., Ca^{2+} and IP_3, and of the molecules that interact with them in order for the Ca^{2+} signaling systems to operate, the conserved nature of the coupling between the Ca^{2+} storage proteins chromogranins and the IP_3R/Ca^{2+} channels appears to testify a very ancient origin of the emergence of the IP_3-dependent Ca^{2+} signaling systems in organisms. That chromogranins store Ca^{2+} and modulate the Ca^{2+} channels while the IP_3Rs sense the Ca^{2+} mobilization signal IP_3 and double as the Ca^{2+} channels in the same organelles seems to lend a strong support to such a notion. Further, given the highly conserved nature of the L3-2 loops of the IP_3Rs even in protozoa and nematodes that have only one type of IP_3R, it appears inevitable that some type of Ca^{2+} storage and Ca^{2+} channel-modulatory proteins that are functionally equivalent to chromogranins of secretory cells also widely exist in a variety of nonsecretory cells across the biokingdom.

Acknowledgement The author thanks Yong Suk Hur for help in preparing the manuscript.

Reference

Bartolomucci A, Possenti R, Mahata SK, Fischer-Colbrie R, Loh YP, Salton SR (2011) The extended granin family: structure, function, and biomedical implications. Endocr Rev 32:755–797

Courel M, Rodemer C, Nguyen ST, Pance A, Jackson AP, O'Connor DT, Taupenot L (2006) Secretory granule biogenesis in sympathoadrenal cells: identification of a granulogenic determinant in the secretory prohormone chromogranin A. J Biol Chem 281:38038–38051

Fan G, Baker ML, Wang Z, Baker MR, Sinyagovskiy PA, Chiu W, Ludtke SJ, Serysheva II (2015) Gating machinery of InsP3R channels revealed by electron cryomicroscopy. Nature 527:336–341

Foskett JK, White C, Cheung KH, Mak DO (2007) Inositol trisphosphate receptor Ca2+ release channels. Physiol Rev 87:593–658

Glombik MM, Kromer A, Salm T, Huttner WB, Gerdes HH (1999) The disulfide-bonded loop of chromogranin B mediates membrane binding and directs sorting from the trans-Golgi network to secretory granules. EMBO J 18:1059–1070

Haigh JR, Parris R, Phillips JH (1989) Free concentrations of sodium, potassium and calcium in chromaffin granules. Biochem J 259:485–491

Helle KB (2000) The chromogranins. Historical perspectives. Adv Exp Med Biol 482:3–20

Huh YH, Bahk SJ, Ghee JY, Yoo SH (2005a) Subcellular distribution of chromogranins A and B in bovine adrenal chromaffin cells. FEBS Lett 579:5145–5151

Huh YH, Chu SY, Park SY, Huh SK, Yoo SH (2006a) Role of nuclear chromogranin B in inositol 1,4,5-trisphosphate-mediated nuclear Ca^{2+} mobilization. Biochemistry 45:1212–1226

Huh YH, Huh SK, Chu SY, Kweon HS, Yoo SH (2006b) Presence of a putative vesicular inositol 1,4,5-trisphosphate-sensitive nucleoplasmic Ca^{2+} store. Biochemistry 45:1362–1373

Huh YH, Jeon SH, Yoo JA, Park SY, Yoo SH (2005b) Effects of chromogranin expression on inositol 1,4,5-trisphosphate-induced intracellular Ca^{2+} mobilization. Biochemistry 44:6122–6132

Huh YH, Kim KD, Yoo SH (2007) Comparison of and chromogranin effect on inositol 1,4,5-trisphosphate sensitivity of cytoplasmic and nucleoplasmic inositol 1,4,5-trisphosphate receptor/Ca2+ channels. Biochemistry 46:14032–14043

Huh YH, Yoo JA, Bahk SJ, Yoo SH (2005c) Distribution profile of inositol 1,4,5-trisphosphate receptor isoforms in adrenal chromaffin cells. FEBS Lett 579:2597–2603

Huh YH, Yoo SH (2003) Presence of the inositol 1,4,5-triphosphate receptor isoforms in the nucleoplasm. FEBS Lett 555:411–418

Kang YK, Yoo SH (1997) Identification of the secretory vesicle membrane binding region of chromogranin a. FEBS Lett 404:87–90

Montero-Hadjadje M, Elias S, Chevalier L, Benard M, Tanguy Y, Turquier V, Galas L, Yon L, Malagon MM, Driouich A, Gasman S, Anouar Y (2009) Chromogranin A promotes peptide hormone sorting to mobile granules in constitutively and regulated secreting cells: role of conserved N- and C-terminal peptides. J Biol Chem 284:12420–12431

Montero-Hadjadje M, Vaingankar S, Elias S, Tostivint H, Mahata SK, Anouar Y (2008) Chromogranins A and B and secretogranin II: evolutionary and functional aspects. Acta Physiol (Oxf) 192:309–324

Taupenot L, Harper KL, Mahapatra NR, Parmer RJ, Mahata SK, O'Connor DT (2002) Identification of a novel sorting determinant for the regulated pathway in the secretory protein chromogranin A. J Cell Sci 115:4827–4841

Taupenot L, Harper KL, O'Connor DT (2003) The chromogranin-secretogranin family. N Engl J Med 348:1134–1149

Thiele C, Huttner WB (1998) The disulfide-bonded loop of chromogranins, which is essential for sorting to secretory granules, mediates homodimerization. J Biol Chem 273:1223–1231

Thrower EC, Choe CU, So SH, Jeon SH, Ehrlich BE, Yoo SH (2003) A functional interaction between chromogranin B and the inositol 1,4,5-trisphosphate receptor/Ca^{2+} channel. J Biol Chem 278:49699–49706

Thrower EC, Park HY, So SH, Yoo SH, Ehrlich BE (2002) Activation of the inositol 1,4,5-trisphosphate receptor by the calcium storage protein chromogranin A. J Biol Chem 277:15801–15806

Winkler H, Fischer-Colbrie R (1992) The chromogranins A and B: the first 25 years and future perspectives. Neuroscience 49:497–528

Winkler H, Westhead E (1980) The molecular organization of adrenal chromaffin granules. Neuroscience 5:1803–1823

Yoo SH (2010) Secretory granules in inositol 1,4,5-trisphosphate-dependent Ca2+ signaling in the cytoplasm of neuroendocrine cells. FASEB J 24:653–664

Yoo SH (2011) Role of secretory granules in inositol 1,4,5-trisphosphate-dependent Ca(2+) signaling: from phytoplankton to mammals. Cell Calcium 50:175–183

Yoo SH, Huh YH, Huh SK, Chu SY, Kim KD, Hur YS (2014) Localization and projected role of phosphatidylinositol 4-kinases IIalpha and IIbeta in inositol 1,4,5-trisphosphate-sensitive nucleoplasmic Ca store vesicles. Nucleus 5:341–351

Yoo SH, Jeon CJ (2000) Inositol 1,4,5-trisphosphate receptor/Ca^{2+} channel modulatory role of chromogranin A, a Ca^{2+} storage protein of secretory granules. J Biol Chem 275:15067–15073

Yoo SH, Kang YK (1997) Identification of the secretory vesicle membrane binding region of chromogranin B. FEBS Lett 406:259–262

Yoo SH, Lewis MS (1992) Effects of pH and Ca^{2+} on monomer-dimer and monomer-tetramer equilibria of chromogranin A. J Biol Chem 267:11236–11241

Yoo SH, Lewis MS (1993) Dimerization and tetramerization properties of the C-terminal region of chromogranin A: a thermodynamic analysis. Biochemistry 32:8816–8822

Yoo SH, Lewis MS (1994) pH-dependent interaction of an intraluminal loop of inositol 1,4,5-trisphosphate receptor with chromogranin A. FEBS Lett 341:28–32

Yoo SH, Lewis MS (1995) Thermodynamic study of the pH-dependent interaction of chromogranin A with an intraluminal loop peptide of the inositol 1,4,5-trisphosphate receptor. Biochemistry 34:632–638

Yoo SH, Lewis MS (1996) Effects of pH and Ca^{2+} on heterodimer and heterotetramer formation by chromogranin A and chromogranin B. J Biol Chem 271:17041–17046

Yoo SH, Lewis MS (1998) Interaction between an intraluminal loop peptide of the inositol 1,4,5-trisphosphate receptor and the near N-terminal peptide of chromogranin A. FEBS Lett 427:55–58

Yoo SH, Lewis MS (2000) Interaction of chromogranin B and the near N-terminal region of chromogranin B with an intraluminal loop peptide of the inositol 1,4, 5-trisphosphate receptor. J Biol Chem 275:30293–30300

Yoo SH, Nam SW, Huh SK, Park SY, Huh YH (2005) Presence of a nucleoplasmic complex composed of the inositol 1,4,5-trisphosphate receptor/Ca^{2+} channel, chromogranin B, and phospholipids. Biochemistry 44:9246–9254

Yoo SH, Oh YS, Kang MK, Huh YH, So SH, Park HS, Park HY (2001) Localization of three types of the inositol 1,4,5-trisphosphate receptor/Ca^{2+} channel in the secretory granules and coupling with the Ca^{2+} storage proteins chromogranins A and B. J Biol Chem 276:45806–45812

Yoo SH, So SH, Kweon HS, Lee JS, Kang MK, Jeon CJ (2000) Coupling of the inositol 1,4,5-trisphosphate receptor and chromogranins A and B in secretory granules. J Biol Chem 275:12553–12559

Yoo SH, You SH, Kang MK, Huh YH, Lee CS, Shim CS (2002) Localization of the secretory granule marker protein chromogranin B in the nucleus. Potential role in transcription control J Biol Chem 277:16011–16021

Chromogranin A in Endothelial Homeostasis and Angiogenesis

Flavio Curnis, Fabrizio Marcucci, Elisabetta Ferrero, and Angelo Corti

Abstract The unbalanced production of factors capable of regulating vascular homeostasis and angiogenesis is a common denominator of many pathological conditions, including cardiovascular diseases, macular degeneration, rheumatoid arthritis, neoplastic diseases and many others. Among the various regulatory factors so far identified, chromogranin A (CgA), a protein released in circulation by many normal and neoplastic cells of the diffuse neuroendocrine system, is emerging as an important player. Indeed, a growing body of evidence suggests that circulating CgA and its fragments contribute to the homeostatic regulation of the endothelial barrier function and angiogenesis in normal conditions, and that alteration of their relative levels, either by changes in their secretion or by extracellular proteolytic processing, might represent important mechanisms that contribute to regulate angiogenesis. Here, we review these studies and discuss the potential role of CgA and its fragments as regulators of vascular physiology in cancer.

1 Introduction

It is well known that *endothelial cells*, either in physiological or pathological conditions, can sense a number of factors released by other cells, locally or in circulation, to regulate blood flow and tissue perfusion, its *barrier function* and transport of plasma components to tissues, leucocyte adhesion and extravasation, *angiogenesis* and tissue remodelling. For example, the formation of new blood vessels from pre-existing vessels is tightly regulated by the coordinated action of anti- and pro-angiogenic factors that are released by the endothelium or other cells (Ribatti 2009). The unbalanced production of these factors is a common denominator of many pathological conditions, including cardiovascular diseases, macular degeneration, skin diseases, diabetic ulcers, stroke, rheumatoid arthritis, neoplastic diseases and

F. Curnis • E. Ferrero (✉) • A. Corti (✉)
Division of Experimental Oncology, San Raffaele Scientific Institute,
via Olgettina, 58 Milan, Italy
e-mail: ferreo.elisabetta@hsr.it; corti.angelo@hsr.it

F. Marcucci
Department of Pharmacological and Biomolecular Sciences, University of Milan, Milan, Italy

many others. Thus, the identification of factors that regulate vascular homeostasis and angiogenesis, and the elucidation of their mechanisms of action, is of great experimental and clinical interest.

Among the various vasoregulatory factors so far identified, *chromogranin A* (CgA), a protein secreted by many normal and neoplastic cells of the diffuse neuroendocrine system, is emerging as an important player. Indeed, a growing body of evidence suggests that this protein and its fragments can regulate the endothelial barrier function and can also work as a proteolytic-dependent angiogenic switch. These studies and the potential role of CgA and its fragments as regulators of vascular physiology in cancer and other pathophysiological conditions are reviewed and discussed in the following.

2 CgA Structure and Expression

Human CgA is a glycosylated, sulphated and phosphorylated protein, 439-residue long, stored in the secretory granules of many normal and neoplastic cells of the diffuse neuroendocrine system (Portela-Gomes et al. 2010; Helle et al. 2007; Taupenot et al. 2003; Deftos 1991; Janson et al. 1997). CgA is also expressed by human polymorphonuclear neutrophils, by wound keratinocytes, and by myocardiocytes (Helle et al. 2007; Taupenot et al. 2003; Lugardon et al. 2000; Pieroni et al. 2007; Steiner et al. 1990; Biswas et al. 2010; Glattard et al. 2006; Radek et al. 2008; Stridsberg et al. 2004). In addition, certain *tumors*, such as prostate cancer, non-small cell lung cancer, breast cancer, gastric and colonic adenocarcinomas, may undergo neuroendocrine differentiation and present focal expression of CgA (Portel-Gomes et al. 2001, 2010; Taupenot et al. 2003).

3 Physiological Functions of CgA

Within cells CgA has an important biological function in secretory granule biogenesis and control of secretion (Kim et al. 2001; Mosley et al. 2007; Courel et al. 2010; Montesinos et al. 2008). These intracellular functions are described with more details in other chapters of this book. CgA has also extracellular functions as precursor of various bioactive peptides (Helle et al. 2007). *Proteolytic processing* of CgA by intra-granular and/or extracellular proteases, such as prohormone convertase 1 and 2, furin, cathepsin L, plasmin, and thrombin, is important for the regulation of its biological activity (Eskeland et al. 1996; Doblinger et al. 2003; Colombo et al. 2002a; Biswas et al. 2008; Biswas et al. 2009; Crippa et al. 2013; Bianco et al. 2016). CgA-derived fragments can exert a variety of biological effects: for example, a fragment corresponding to residues 1–76, named vasostatin I (VS-I), inhibits vasoconstriction in isolated blood vessels (Helle et al. 2007; Aardal and Helle 1992), induces negative inotropism in isolated working frog and rat heart (Tota et al. 2008;

Imbrogno et al. 2004; Corti et al. 2004), inhibits parathyroid hormone secretion (Russell et al. 1994), is neurotoxic in neuronal/microglial cell cultures (Ciesielski-Treska et al. 1998) and induces antibacterial/antifungal effects (Helle et al. 2007; Lugardon et al. 2000). Another fragment corresponding to residues 248–293, called pancreastatin, regulates glucose and lipid metabolism (Zhang et al. 2006; O'Connor et al. 2005; Sanchez-Margalet et al. 2010; Gayen et al. 2009), whereas a peptide corresponding to residues 352–372, called catestatin, is a potent inhibitor of nicotinic-cholinergic-stimulated catecholamine secretion, acts as a vasodilator in rats and humans, inhibits the inotropic and lusitropic properties of the rodent heart, induces chemotaxis, stimulates rat mast cells to release histamine, and acts as an antimicrobial and antimalarial peptide (Mahata et al. 2010). Finally, peptides derived from proteolytic cleavage of the penultimate and the last pair of basic residues at the C-terminus of CgA, called serpinins, have protective effects against oxidative stress on neurons and pituitary cells and enhance both myocardial contractility and relaxation (Loh et al. 2012).

CgA may also work as a negative modulator of *cell adhesion* and as a precursor of positive modulators, depending on proteolytic processing. For instance, CgA, isolated from human pheochromocytomas, inhibits the adhesion of human and mouse fibroblasts to plates coated with collagen I, collagen IV, or with fibronectin (Colombo et al. 2002a; Gasparri et al. 1997; Ratti et al. 2000a), whereas the recombinant CgA fragment 1–78 promotes *fibroblast* adhesion (Lugardon et al. 2000; Gasparri et al. 1997; Dondossola et al. 2010). CgA1-78-dependent cytoskeletal rearrangements and changes in cell adhesion have been observed also with smooth muscle cells, cardiomyocytes, keratinocytes and endothelial cells (Colombo et al. 2002a; Gasparri et al. 1997; Curnis et al. 2012; Angelone et al. 2010; Ratti et al. 2000b). Of note, the region 45–78 of this fragment is crucial for this activity: this region contains an *RGD* site capable of interacting with $\alpha v \beta 6$ *integrin* on keratinocytes, and is followed by an amphipathic α-helix (residues 47–66), 100%-conserved in human, porcine, bovine, equine and mouse CgA (Simon and Aunis 1989; Turquier et al. 1999). In addition, this region contains the highly conserved hydrophilic residues 67–78, which share strong sequence and structural similarity with ezrin-radixin-moesin binding phosphoprotein 50 (EBP50), a protein that works as a molecular linkage between membrane-cytoskeleton adapter proteins and the cytoplasmic domain of various membrane proteins and receptors (Dondossola et al. 2010; Bretscher et al. 2002).

4 Role of CgA and Its Fragments in Preserving the Endothelial Barrier Integrity

The endothelium forms a semi-permeable barrier that regulates blood-tissue exchange of plasma fluid and proteins by trans-cellular and para-cellular transport mechanisms (Komarova and Malik 2010). While the trans-endothelial pathway represents the main mechanism for the transport of molecules in physiological conditions, the paracellular transport becomes a prominent way for plasma leakage in inflammation, ischemia-reperfusion injury, adult respiratory distress syndrome, tumor growth and metastatization, and several other pathological conditions (Komarova and Malik 2010; Hu et al. 2008; Mehta and Malik 2006; Wallez and Huber 2008). The *endothelial barrier* can be altered by various inflammatory and vasoactive agents, which, consequently, increase endothelial permeability. Experimental evidence suggests that CgA, at low concentrations, can preserve the integrity of the endothelial barrier function when these cells are exposed to inflammatory stimuli, such as tumor necrosis factor-α (TNF) or vascular endothelial growth factor (VEGF) (Ferrero et al. 2004). Structure-function studies showed that an active site is located in the N-terminal domain. Indeed, the CgA1-78 fragment is sufficient to protect vessels from *TNF*-induced *vascular leakage in vivo* and to inhibit the paracellular flux of radio-labeled albumin through endothelial cell monolayers in vitro (Ferrero et al. 2004). Mechanistic studies showed that CgA1-78 can inhibit the TNF-induced disassembly of VE-cadherin-dependent adherence junctions and the conversion of peripheral actin bundles into stress fibers in endothelial cell monolayers (Ferrero et al. 2004). The inhibitory effect of CgA and CgA1-78 on TNF-induced alteration of endothelial barrier integrity is unlikely related to inhibition of TNF/TNF-receptor interactions, as CgA does not bind these molecules (Ferrero et al. 2004). Experimental data obtained with bovine pulmonary arterial endothelial cells showed that CgA1-78 can inhibit TNF-induced phosphorylation of p38MAPK by a pertussis toxin-sensitive mechanism, implicating a role for CgA1-78 in protecting the endothelial integrity via regulation of the stress-activated MAPK pathway in a G-protein-dependent manner (Blois et al. 2006a). CgA1-78 can also inhibit TNF-induced intercellular cell adhesion (ICAM)-1 expression, monocyte chemoactractant protein (MCP)-1 release and relocation of high-mobility group box (HMGB)-1 in human microvascular endothelial cells, further supporting an anti-inflammatory role of this CgA fragment (Di Comite et al. 2009a).

The *receptors* responsible for these biological effects are still unknown. It has been reported that bovine aorta endothelial cells bind and internalize 1 nM ^{125}I-labeled CgA (Mandalà et al. 2000). Furthermore, fluorescein-isotiocyanate (FITC)-labeled CgA1-78 is bound and internalized in endocytotic vesicles by human umbilical vein endothelial cells (HUVECs) in culture (Ferrero et al. 2004). In vitro binding studies performed with the Langmuir apparatus showed that CgA1-78 can interact with membrane phospholipids, particularly with phosphatidylserine (Dondossola et al. 2010; Blois et al. 2006b). The specific, enhanced fluidity exerted by low nanomolar concentrations of CgA1-78 in monolayers of phosphatidylserine may suggest that CgA1-78 can indirectly interact with, or perturb, specific receptors in phosphatidylserine-rich microdomains in the outer leaflet of the plasma membrane (Blois et al. 2006b). CgA1-78 can also induce a marked increase of caveolae-dependent endocytosis in bovine aortic endothelial cells, which is significantly reduced by heparinase III and by wortmannin, a specific phosphoinositide 3-kinase (PI3K) inhibitor (Ramella et al. 2010). As these compounds also abolished the CgA1-78-dependent phosphorylation of endothelial nitric oxide synthase (eNOS), a mechanism based on heparan sulfate proteoglycans interaction, caveolae endocytosis and a PI3K-dependent eNOS phosphorylation has been proposed (Ramella et al. 2010). Although the data so far generated provide some hints, further studies are necessary to have a more complete picture on the mechanisms of the interaction of CgA and CgA1-78 with endothelial cells.

5 The CgA-Dependent Angiogenic Switch

A growing body of evidence suggests that CgA and some of its fragments may have a regulatory role in angiogenesis. For example, it has been shown that recombinant full-length CgA can inhibit angiogenesis in the chick embryo chorioallantoic membrane model (Crippa et al. 2013). Furthermore, CgA (0.1–0.2 nM) can inhibit capillary sprouting induced by fibroblast growth factor *(FGF)-2* or vascular endothelial growth factor *(VEGF)*, two potent pro-angiogenic factors, from rat aortic rings (RAR) cultured in 3D-collagen gels (Crippa et al. 2013). Structure-function studies showed that a functional site of CgA is located in the C-terminal region. However, also CgA1-76, but not CgA1-373, 1–400 and 1–409, can inhibit VEGF- and FGF-2-induced angiogenesis at 0.1–1 nM concentrations, suggesting that an additional anti-angiogenic site is present in the N-terminal region, albeit in a latent (or less active) form (Crippa et al. 2013). Interestingly, CgA1-78 was 30-fold less potent than CgA1-76 (VS-I) in the RAR angiogenesis assay. It would appear, therefore, that the anti-angiogenic site in the N-terminal region of CgA requires cleavage of the first dibasic pair residues ($K_{77}K_{78}$) and removal of C-terminal lysines for full activation. Considering that biologically relevant levels of both CgA1-439 and VS-I are present in circulation in normal subjects (see below), these findings suggest that both molecules, but not large fragments of CgA lacking the C-terminal region, contribute to the homeostatic inhibition of angiogenesis in normal conditions. The concept that the N-terminal fragment of CgA is an anti-angiogenic molecule is also supported by the observation that 0.3 μM CgA_{1-78} can inhibit in endothelial cells hypoxia-driven morphological changes, vascular-endothelial *(VE)-cadherin* redistribution, intercellular gap formation, tube morphogenesis, and hypoxia inducible factor (HIF)-1α nuclear translocation (Veschini et al. 2011) as well as cell proliferation, migration and invasion induced by VEGF (Ferrero et al. 2004; Blois et al. 2006a; Di Comite et al. 2009a). Furthermore, a recent study has shown that CgA1-78 can prevent choroidal neovascularization and vascular leakage in an established mouse model of laser-induced ocular neovascularization (Maestroni et al. 2015), suggesting that this polypeptide might have therapeutic activity in human ocular diseases involving neovascularization or excessive vascular permeability.

Interestingly, while full-length CgA1-439 inhibit FGF-2–induced angiogenesis, the fragments CgA1-373 and CgA352-372 (catestatin) can induce the release of FGF-2 from endothelial cells and exert pro-angiogenic effects, pointing to CgA as a paradoxical player in the regulation of angiogenesis. Indeed, CgA1-439, CgA1-373 and CgA1-76 can counterbalance the pro-/anti-angiogenic activity of each other in the rat aortic ring (RAR) assay (Bianco et al. 2016) suggesting that CgA and its fragments may form a balance of anti- and pro-angiogenic factors tightly regulated by proteolysis.

6 Proteases Involved in CgA Fragmentation and Angiogenesis Regulation

Considering that 50% of CgA is proteolytically processed in bovine chromaffin cells (Metz-Boutigue et al. 1993), prohormone convertases and furin, capable of cleaving dibasic residues, might be involved in the intra-granular processing of CgA N-terminal and C-terminal regions (Metz-Boutigue et al. 1993; Koshimizu et al. 2010). However, also extracellular proteases might come into play for local angiogenesis regulation. Interestingly, recent studies have shown that the R_{373}-R_{374} dibasic site of CgA is efficiently cleaved by limited digestion with *thrombin* and *plasmin* (Crippa et al. 2013; Bianco et al. 2016), which may tip the balance toward a pro-angiogenic state in pathophysiological conditions characterized by their activation, e.g. in wound healing and cancer. However, extensive digestion of CgA with thrombin may cause additional cleavage at R362 and R394 (Crippa et al. 2013). Furthermore, prolonged digestion with plasmin may generate even more fragments with unknown functions (Bianco et al. 2016). Thus, extensive proteolysis of CgA and its fragments may represent other important regulatory mechanisms that deserve further investigation. Cleavage by carboxypeptidases might also represent additional mechanisms for CgA regulation: interestingly, CgA1-78 is rapidly converted to CgA1-76 when injected in mice (Crippa et al. 2013), suggesting that a *carboxypeptidase* B-like enzyme capable of removing the C-terminal dibasic residues of CgA1-78 is likely present in circulation. As CgA1-76 is more potent than CgA1-78 in the RAR assay, this enzyme might represent an important element for the homeostatic regulation of blood vessels at the systemic level. All these enzymes may contribute, therefore, to the generation and processing of the circulating fragments of CgA in normal and pathological conditions.

Finally, it is important to mention that the dose-response curve of CgA and its fragments in angiogenesis assays are U-shaped (Crippa et al. 2013). The mechanism underlying this behavior is unknown.

7 Circulating CgA and Fragments in Health and Disease

Given the different, and sometimes opposite, functions of CgA and its fragments in the regulation of vascular physiology, quantification of the circulating levels of these molecules in health and disease represents an important issue. The normal values of *circulating CgA* reported by different laboratories, as measured with immunoassays unable to discriminate between full-length CgA and fragments, vary within the 0.5–2 nM range, depending on the different antibodies and reagents used (Helle et al. 2007). Increased levels of circulating CgA have been detected in patients with *carcinoids* or with other *neuroendocrine tumors* (O'Connor and Bernstein 1984; O'Connor and Deftos 1986). Elevated serum levels have been detected also in subpopulations of patients with breast, prostate or non-small cell lung cancer, or in patients with renal failure, hypertension, atrophic gastritis, sepsis and other inflammatory diseases (Helle et al. 2007; Taupenot et al. 2003; O'Connor and Bernstein 1984; Ligumsky et al. 2001; Waldum and Brenna 2000; O'Connor et al. 2000; Syversen et al. 2004; Borch et al. 1997; Zhang et al. 2009; Di Comite et al. 2009b; Castoldi et al. 2010). Circulating CgA is markedly increased also in patients with heart failure, depending on the severity of the disease, or with rheumatoid arthritis (Pieroni et al. 2007; Di Comite et al. 2009a; Ceconi et al. 2002). In patients with heart failure and rheumatoid arthritis CgA correlates with soluble tumor necrosis factor receptors (sTNF-Rs), which are sensitive markers of inflammation (Corti et al. 2000). Furthermore, increased levels of serum CgA have been detected in subjects treated with *proton pump inhibitors*, a class of drugs commonly used to treat acid-peptic disorders (Giusti et al. 2004; Sanduleanu et al. 1999).

Thus, several pathological conditions are characterized by variable and abnormal plasma levels of CgA, for a variety of reasons. This implies that endothelial cells, either close to CgA-secreting cells or distant, are exposed to variable levels of this protein.

Hematological studies performed with a series of enzyme-linked immunosorbent assays (*ELISA*) capable of discriminating between intact and cleaved molecules showed that various CgA-derived polypeptides are present in the blood of healthy subjects, consisting of full-length CgA (about 0.1 nM) and a larger proportion of fragments lacking part of, or the entire, C-terminal region (Crippa et al. 2013). In addition, normal plasma contains a considerable amount of vasostatin-1 (about 0.3–0.4 nM) (Crippa et al. 2013). Other fragments, not detected by these assays, might be also present. For example, catestatin has been reported to be present in the blood of normal subjects at concentrations ranging from 0.03 nM (Ji et al. 2012) to 0.33 nM (Meng et al. 2013). The discrepancies between these values might be related to the use of different antibodies that detect catestatin as well as uncleaved precursors with different efficiency.

Although many reports describe the levels of circulating CgA in different pathological conditions, it is important to remark that most studies have been performed with immunoassays unable to discriminate between full-length CgA and fragments. Thus, only few studies have addressed the circulating levels of CgA fragments in pathological conditions. One of these studies has shown that VS-I plasma levels are increased in critically ill patients and that VS-I concentrations >0.44 nM (i.e. >1.4-fold higher than normal values), are associated with poor outcome (Schneider et al. 2012). Other investigators showed a reduction of CgA processing in patients with diabetic retinopathy (Fournier et al. 2011). Studies on the catestatin region, performed with an anti-catestatin antibody capable of cross-reacting with free peptide and full-length CgA, showed a reduction of immunoreactivity in patients with essential hypertension or in normotensive subjects with a family history of hypertension and increased epinephrine secretion (O'Connor et al. 2002). On the other hand, plasma catestatin levels, measured with a commercial assay kit, were found increased in patients with coronary heart disease and after acute myocardial infarction (Meng et al. 2013; Liu et al. 2013). However, also in this case, the catestatin-containing molecular entities detected by these assays in plasma were not characterized.

Regarding cancer, different proportions of full-length CgA and fragments have been detected in healthy subjects and *multiple myeloma* patients, including *CgA1-439, CgA1-373, CgA1-76* and other fragments (Bianco et al. 2016). The relative levels of circulating pro- and anti-angiogenic molecules (e.g. CgA1-373/CgA1-439 and CgA1-373/CgA1-76) were higher in these patients at diagnosis, compared to healthy subjects, suggesting that a proteolytic mechanism capable of "turning-on" the CgA-angiogenic switch was active in multiple myeloma. Furthermore, the bone marrow plasma of patients contained higher levels of CgA1-373 and lower levels of CgA1-439 than peripheral-blood plasma, suggesting that cleavage at R373 occurred in the bone marrow (Bianco et al. 2016). Relevant to the proposed role of CgA in angiogenesis and vascular homeostasis, the ratio between pro- and anti-angiogenic forms correlates with the bone marrow plasma levels of VEGF and FGF-2, two pro-angiogenic factors known to have key roles in the cross-talk between myeloma cells, stromal cells and endothelial cells in the bone marrow (Bianco et al. 2016; Ribatti et al. 2006; Jakob et al. 2006). Mechanistic studies showed that multiple myeloma and proliferating endothelial cells can promote CgA C-terminal cleavage by activating the plasminogen activator/plasmin system (Bianco et al. 2016). These findings suggest that CgA and its fragments may represent new players in the regulation of angiogenesis in the bone marrow microenvironment of multiple myeloma patients and, possibly, also in other diseases.

8 CgA and Tumor Vascular Biology

Aberrant angiogenesis in tumors can lead to the formation of blood vessels typically characterized by highly disorganized and dilated architecture, excessive branching, shunts and fenestrations, when compared to normal vessels (Marcucci and Corti 2011). Alteration of endothelial cell-cell adhesion and barrier function can lead to heterogeneous permeability and increased interstitial fluid pressure (Marcucci and Corti 2011). These structural and functional abnormalities of the vasculature may contribute to tumor cell proliferation, migration, invasion, and metastasis formation (Corti et al. 2011), and may also cause uneven penetration of drugs in tumors and poor therapeutic responses (Marcucci and Corti 2011; Marcucci and Corti 2012).

Considering the vasoregulatory functions of CgA (discussed above) and the fact that the tumor vasculature is exposed to CgA released in circulation by the neuroendocrine system and, in the case of neuroendocrine tumors, by the tumor cells themselves, it reasonable to hypothesize that CgA and its fragments may have a regulatory role on tumor vascular physiology. This may have a potential impact on angiogenesis, as well as on tumor cell trafficking and *metastasis* formation, drug transport, and tumor progression.

This hypothesis is supported by the observation that lymphoma and a mammary adenocarcinoma cell lines genetically engineered to secrete CgA are characterized by reduced growth rate, when implanted subcutaneously in mice, compared to non-secreting parental cells (Veschini et al. 2011; Colombo et al. 2002b). Systemic administration of CgA (1 μg) to lymphoma-bearing mice reduced TNF-induced penetration of patent blue, a synthetic dye, in tumor tissue, pointing to a protective effect on the endothelial barrier in tumors (Dondossola et al. 2011). CgA can also inhibit the transport of chemotherapeutic drugs in tumor tissues induced by *NGR-TNF*, a tumor necrosis factor-α (TNF)-based vascular targeting agent originally developed by our group and currently tested in phase II and III clinical studies (Dondossola et al. 2011; Curnis et al. 2000, 2002; Corti et al. 2013). In particular, studies performed in murine lymphoma and melanoma models have shown that patho-physiologically relevant levels of circulating CgA can inhibit the NGR-TNF−induced penetration of drugs in tumor tissues and inhibit its synergism with doxorubicin and melphalan (Dondossola et al. 2011). Notably, two-fold enhancement of endogenous circulating CgA, e.g. obtained by pharmacological treatment with omeprazole, significantly reduced the NGR-TNF−induced penetration of *doxorubicin* in tumors. Similar effects were obtained also by administration of CgA1-78 to tumor-bearing mice. Interestingly, mouse mammary adenocarcinomas genetically engineered to secrete CgA1-78 and implanted subcutaneously in mice are characterized by reduced vascular density and more regular vessels, compared with parental cells (Veschini et al. 2011). Considering that CgA1-78 can inhibit the nuclear translocation of *HIF-1α* (Veschini et al. 2011) it is possible that this fragment can inhibit hypoxia-driven endothelial cell activation and abnormal vascularization, and, consequently, lead to the formation of more regular vessels.

Besides affecting the transport of drugs in tumors, CgA can also regulate the trafficking of tumor cells through the endothelial barrier and, consequently, the tumor metastatization and *self-seeding* processes (Dondossola et al. 2012). This is an important effect, as cancer progression typically involves the seeding of malignant cells in circulation and the colonization of distant organs, as well as the tumor reinfiltration by aggressive circulating tumor cells. Studies in a murine model of mammary adenocarcinoma showed that CgA can inhibit (i) the shedding of cancer cells in circulation by primary tumors, (ii) the homing of circulating tumor cells to primary tumors (necessary for the self-seeding process), and (iii) the engraftment in lungs by circulating tumor cells (another important step of the metastatic cascade) (Dondossola et al. 2012). Mechanistic studies showed that CgA reduced gap formation induced by tumor cell–derived factors in endothelial cells, decreased vascular leakage in tumors, and inhibited the transendothelial migration of cancer cells (Dondossola et al. 2012). These findings point to a role for circulating CgA in the regulation of tumor cell trafficking from tumor-to-blood and from blood-to-tumor/normal tissues. The capability of CgA to strengthen the endothelial barrier function in tumors and to reduce vascular leakage is also suggested by the observation that exogenous CgA can inhibit the in vivo penetration of an anti-Thy1.1 antibody into lymphomas genetically engineered to express the Thy1.1 antigen, whereas the neutralization of endogenous CgA with an anti-CgA antibody (mAb 5A8) promotes the penetration of the anti-Thy1.1 antibody in the neoplastic tissues (Dondossola et al. 2012).

These observations rise the question as to whether the increased production of CgA observed in certain cancer patients is good or bad. Increased expression of CgA is associated with decreasing malignancy in *neuroendocrine tumors*, being higher in well-differentiated carcinomas (low grade) and lower in poorly differentiated (high grade) carcinomas (Helpap and Kollermann 2001). However, neuroendocrine differentiation in *prostate tumors* is associated with poorer prognosis (Young et al. 2000) and large cell carcinomas of the lung with neuroendocrine features are more clinically aggressive than classic large cell carcinomas (Iyoda et al. 2001). Furthermore, CgA correlates with worsening conditions, extension of the disease, and is an independent negative prognostic indicator of mortality in patients with *non small cell lung cancer* (Gregorc et al. 2007). It is, therefore, very difficult to speculate on whether CgA is good or bad in patients based on these associations. The overall biological effects of CgA likely dependent on its cellular source, its local concentration, its proteolytic processing and its post-translational modifications, which may vary from patient to patient. Possibly, while a regulated production and processing of CgA is likely crucial for maintaining the vascular homeostasis in physiological conditions, the unbalanced production of pro-/anti-angiogenic CgA polypeptides observed in certain cancer patients may contribute to sustain angiogenesis and tumor growth.

9 Conclusions

The experimental evidence accumulated so far suggests that circulating CgA and its fragments contribute to the homeostatic regulation of blood vessels in normal conditions and that alteration of their relative levels, either by changes in their secretion or by proteolytic processing, might represent important mechanisms for angiogenesis activation and regulation. Considering that CgA is widely used as a serum marker for various neoplastic and non-neoplastic diseases, selective quantification of anti- and pro-angiogenic CgA-related polypeptides in plasma samples of patients could represent an important step ahead for understanding the pathophysiological significance of this protein and could provide important prognostic information in cancer and other diseases with an angiogenesis component.

Acknowledgments This work was supported by Associazione Italiana per la Ricerca sul Cancro (AIRC 14338 and 9965) of Italy.

References

Aardal S, Helle KB (1992) The vasoinhibitory activity of bovine chromogranin A fragment (vasostatin) and its independence of extracellular calcium in isolated segments of human blood vessels. Regul Pept 41(1):9–18

Angelone T, Quintieri AM, Goumon Y, Di Felice V, Filice E, Gattuso A, Mazza R, Corti A, Tota B, Metz-Boutigue MH, Cerra MC (2010) Cytoskeleton mediates negative inotropism and lusitropism of chromogranin A-derived peptides (human vasostatin1-78 and rat CgA(1-64)) in the rat heart. Regul Pept 165(1):78–85

Bianco M, Gasparri AM, Colombo B, Curnis F, Girlanda S, Ponzoni M, Bertilaccio MT, Calcinotto A, Sacchi A, Ferrero E, Ferrarini M, Chesi M, Bergsagel PL, Bellone M, Tonon G, Ciceri F, Marcatti M, Caligaris Cappio F, Corti A (2016) Chromogranin A is preferentially cleaved into pro-angiogenic peptides in the bone marrow of multiple myeloma patients. Cancer Res. doi:10.1158/0008-5472.CAN-15-1637

Biswas N, Vaingankar SM, Mahata M, Das M, Gayen JR, Taupenot L, Torpey JW, O'Connor DT, Mahata SK (2008) Proteolytic cleavage of human chromogranin a containing naturally occurring catestatin variants: differential processing at catestatin region by plasmin. Endocrinology 149(2):749–757

Biswas N, Rodriguez-Flores JL, Courel M, Gayen JR, Vaingankar SM, Mahata M, Torpey JW, Taupenot L, O'Connor DT, Mahata SK (2009) Cathepsin L colocalizes with chromogranin a in chromaffin vesicles to generate active peptides. Endocrinology 150(8):3547–3557

Biswas N, Curello E, O'Connor DT, Mahata SK (2010) Chromogranin/secretogranin proteins in murine heart: myocardial production of chromogranin A fragment catestatin (Chga(364-384)). Cell Tissue Res 342(3):353–361. doi:10.1007/s00441-010-1059-4

Blois A, Srebro B, Mandalà M, Corti A, Helle KB, Serck-Hanssen G (2006a) The chromogranin A peptide vasostatin-I inhibits gap formation and signal transduction mediated by inflammatory agents in cultured bovine pulmonary and coronary arterial endothelial cells. Regul Pept 135:78–84

Blois A, Holmsen H, Martino G, Corti A, Metz-Boutigue MH, Helle KB (2006b) Interactions of chromogranin A-derived vasostatins and monolayers of phosphatidylserine, phosphatidylcholine and phosphatidylethanolamine. Regul Pept 134(1):30–37

Borch K, Stridsberg M, Burman P, Rehfeld JF (1997) Basal chromogranin A and gastrin concentrations in circulation correlate to endocrine cell proliferation in type-A gastritis. Scand J Gastroenterol 32(3):198–202

Bretscher A, Edwards K, Fehon RG (2002) ERM proteins and merlin: integrators at the cell cortex. Nature Reviews in Molecular and Cellular Biology 3(8):586–599

Castoldi G, Antolini L, Bombardi C, Perego L, Mariani P, Vigano MR, Torti G, Casati M, Corti A, Zerbini G, Valsecchi MG, Stella A (2010) Oxidative stress biomarkers and chromogranin A in uremic patients: effects of dialytic treatment. Clin Biochem 43(18):1387–1392. doi:10.1016/j.clinbiochem.2010.08.028

Ceconi C, Ferrari R, Bachetti T, Opasich C, Volterrani M, Colombo B, Parrinello G, Corti A (2002) Chromogranin A in heart failure; a novel neurohumoral factor and a predictor for mortality. Eur Heart J 23(12):967–974

Ciesielski-Treska J, Ulrich G, Taupenot L, Chasserot-Golaz S, Corti A, Aunis D, Bader MF (1998) Chromogranin A induces a neurotoxic phenotype in brain microglial cells. J Biol Chem 273(23):14339–14346

Colombo B, Longhi R, Marinzi C, Magni F, Cattaneo A, Yoo SH, Curnis F, Corti A (2002a) Cleavage of chromogranin A N-terminal domain by plasmin provides a new mechanism for regulating cell adhesion. J Biol Chem 277(48):45911–45919

Colombo B, Curnis F, Foglieni C, Monno A, Arrigoni G, Corti A (2002b) Chromogranin a expression in neoplastic cells affects tumor growth and morphogenesis in mouse models. Cancer Res 62(3):941–946

Corti A, Ferrari R, Ceconi C (2000) Chromogranin A and tumor necrosis factor alpha in heart failure. Chromogranins: Functional and Clinical Aspects (Advances in Experimental Medicine and Biology) 482:351–359

Corti A, Mannarino C, Mazza R, Angelone T, Longhi R, Tota B (2004) Chromogranin A N-terminal fragments vasostatin-1 and the synthetic CGA 7-57 peptide act as cardiostatins on the isolated working frog heart. Gen Comp Endocrinol 136(2):217–224

Corti A, Pastorino F, Curnis F, Arap W, Ponzoni M, Pasqualini R (2011) Targeted drug delivery and penetration into solid tumors. Med Res Rev. doi:10.1002/med.20238

Corti A, Curnis F, Rossoni G, Marcucci F, Gregorc V (2013) Peptide-mediated targeting of cytokines to tumor vasculature: the NGR-hTNF example. BioDrugs 27(6):591–603. doi:10.1007/s40259-013-0048-z

Courel M, Soler-Jover A, Rodriguez-Flores JL, Mahata SK, Elias S, Montero-Hadjadje M, Anouar Y, Giuly RJ, O'Connor DT, Taupenot L (2010) Pro-hormone secretogranin II regulates dense core secretory granule biogenesis in catecholaminergic cells. J Biol Chem 285(13):10030–10043. doi:10.1074/jbc.M109.064196

Crippa L, Bianco M, Colombo B, Gasparri AM, Ferrero E, Loh YP, Curnis F, Corti A (2013) A new chromogranin A-dependent angiogenic switch activated by thrombin. Blood 121(2):392–402. doi:10.1182/blood-2012-05-430314

Curnis F, Sacchi A, Borgna L, Magni F, Gasparri A, Corti A (2000) Enhancement of tumor necrosis factor alpha antitumor immunotherapeutic properties by targeted delivery to aminopeptidase N (CD13). Nat Biotechnol 18(11):1185–1190

Curnis F, Sacchi A, Corti A (2002) Improving chemotherapeutic drug penetration in tumors by vascular targeting and barrier alteration. J Clin Invest 110(4):475–482

Curnis F, Gasparri A, Longhi R, Colombo B, D'Alessio S, Pastorino F, Ponzoni M, Corti A (2012) Chromogranin A binds to αvβ6-integrin and promotes wound healing in mice. Cell Mol Life Sci 69(16):2791–2803

Deftos LJ (1991) Chromogranin A: its role in endocrine function and as an endocrine and neuroendocrine tumor marker. Endocr Rev 12(2):181–187

Di Comite G, Rossi CM, Marinosci A, Lolmede K, Baldissera E, Aiello P, Mueller RB, Herrmann M, Voll RE, Rovere-Querini P, Sabbadini MG, Corti A, Manfredi AA (2009a) Circulating chromogranin A reveals extra-articular involvement in patients with rheumatoid arthritis and curbs TNF-alpha-elicited endothelial activation. J Leukoc Biol 85(1):81–87

Di Comite G, Previtali P, Rossi CM, Dell'Antonio G, Rovere-Querini P, Praderio L, Dagna L, Corti A, Doglioni C, Maseri A, Sabbadini MG, Manfredi AA (2009b) High blood levels of chromogranin A in giant cell arteritis identify patients refractory to corticosteroid treatment. Ann Rheum Dis 68(2):293–295

Doblinger A, Becker A, Seidah NG, Laslop A (2003) Proteolytic processing of chromogranin A by the prohormone convertase PC2. Regul Pept 111(1–3):111–116

Dondossola E, Gasparri A, Bachi A, Longhi R, Metz-Boutigue MH, Tota B, Helle KB, Curnis F, Corti A (2010) Role of vasostatin-1 C-terminal region in fibroblast cell adhesion. Cell Mol Life Sci 67(12):2107–2118. doi:10.1007/s00018-010-0319-5

Dondossola E, Gasparri AM, Colombo B, Sacchi A, Curnis F, Corti A (2011) Chromogranin A restricts drug penetration and limits the ability of NGR-TNF to enhance chemotherapeutic efficacy. Cancer Res 71(17):5881–5890. doi:10.1158/0008-5472.CAN-11-1273

Dondossola E, Crippa L, Colombo B, Ferrero E, Corti A (2012) Chromogranin A regulates tumor self-seeding and dissemination. Cancer Res 72(2):449–459. doi:10.1158/0008-5472.CAN-11-2944

Eskeland NL, Zhou A, Dinh TQ, Wu H, Parmer RJ, Mains RE, O'Connor DT (1996) Chromogranin A processing and secretion: specific role of endogenous and exogenous prohormone convertases in the regulated secretory pathway. J Clin Invest 98(1):148–156

Ferrero E, Scabini S, Magni E, Foglieni C, Belloni D, Colombo B, Curnis F, Villa A, Ferrero ME, Corti A (2004) Chromogranin A protects vessels against tumor necrosis factor alpha-induced vascular leakage. FASEB J 18(3):554–555

Fournier I, Gaucher D, Chich JF, Bach C, Shooshtarizadeh P, Picaud S, Bourcier T, Speeg-Schatz C, Strub JM, Van Dorsselaer A, Corti A, Aunis D, Metz-Boutigue MH (2011) Processing of chromogranins/secretogranin in patients with diabetic retinopathy. Regul Pept 167(1):118–124. doi:10.1016/j.regpep.2010.12.004

Gasparri A, Sidoli A, Sanchez LP, Longhi R, Siccardi AG, Marchisio PC, Corti A (1997) Chromogranin A fragments modulate cell adhesion. Identification and characterization of a pro-adhesive domain. J Biol Chem 272(33):20835–20843

Gayen JR, Saberi M, Schenk S, Biswas N, Vaingankar SM, Cheung WW, Najjar SM, O'Connor DT, Bandyopadhyay G, Mahata SK (2009) A novel pathway of insulin sensitivity in chromogranin A null mice: a crucial role for pancreastatin in glucose homeostasis. J Biol Chem 284(42):28498–28509. doi:10.1074/jbc.M109.020636

Giusti M, Sidoti M, Augeri C, Rabitti C, Minuto F (2004) Effect of short-term treatment with low dosages of the proton-pump inhibitor omeprazole on serum chromogranin A levels in man. Eur J Endocrinol 150(3):299–303

Glattard E, Angelone T, Strub JM, Corti A, Aunis D, Tota B, Metz-Boutigue MH, Goumon Y (2006) Characterization of natural vasostatin-containing peptides in rat heart. FEBS J 273(14):3311–3321

Gregorc V, Spreafico A, Floriani I, Colombo B, Ludovini V, Pistola L, Bellezza G, Vigano MG, Villa E, Corti A (2007) Prognostic value of circulating chromogranin A and soluble tumor necrosis factor receptors in advanced nonsmall cell lung cancer. Cancer 110(4):845–853

Helle KB, Corti A, Metz-Boutigue MH, Tota B (2007) The endocrine role for chromogranin A: a prohormone for peptides with regulatory properties. Cell Mol Life Sci 64(22):2863–2886. doi:10.1007/s00018-007-7254-0

Helpap B, Kollermann J (2001) Immunohistochemical analysis of the proliferative activity of neuroendocrine tumors from various organs. Are there indications for a neuroendocrine tumor-carcinoma sequence? Virchows Archives 438(1):86–91

Hu G, Place AT, Minshall RD (2008) Regulation of endothelial permeability by Src kinase signaling: vascular leakage versus transcellular transport of drugs and macromolecules. Chem Biol Interact 171(2):177–189. doi:10.1016/j.cbi.2007.08.006

Imbrogno S, Angelone T, Corti A, Adamo C, Helle KB, Tota B (2004) Influence of vasostatins, the chromogranin A-derived peptides, on the working heart of the eel (Anguilla anguilla): negative inotropy and mechanism of action. Gen Comp Endocrinol 139(1):20–28

Iyoda A, Hiroshima K, Toyozaki T, Haga Y, Fujisawa T, Ohwada H (2001) Clinical characterization of pulmonary large cell neuroendocrine carcinoma and large cell carcinoma with neuroendocrine morphology. Cancer 91(11):1992–2000

Jakob C, Sterz J, Zavrski I, Heider U, Kleeberg L, Fleissner C, Kaiser M, Sezer O (2006) Angiogenesis in multiple myeloma. Eur J Cancer 42(11):1581–1590. doi:10.1016/j.ejca.2006.02.017

Janson ET, Holmberg L, Stridsberg M, Eriksson B, Theodorsson E, Wilander E, Oberg K (1997) Carcinoid tumors: analysis of prognostic factors and survival in 301 patients from a referral center. Ann Oncol 8(7):685–690

Ji L, Pei ZQ, Ma DF, Zhang J, Su JS, Gao XD, Xue WZ, Chen XP, Wang WS (2012) Prognostic value of circulating catestatin levels for in-hospital heart failure in patients with acute myocardial infarction. Zhonghua Xin Xue Guan Bing Za Zhi 40(11):914–919

Kim T, Tao-Cheng JH, Eiden LE, Loh YP (2001) Chromogranin A, an "on/off"; switch controlling dense-core secretory granule biogenesis. Cell 106(4):499–509

Komarova Y, Malik AB (2010) Regulation of endothelial permeability via paracellular and transcellular transport pathways. Annu Rev Physiol 72:463–493. doi:10.1146/annurev-physiol-021909-135833

Koshimizu H, Kim T, Cawley NX, Loh YP (2010) Chromogranin A: a new proposal for trafficking, processing and induction of granule biogenesis. Regul Pept 160(1–3):153–159. doi:10.1016/j.regpep.2009.12.007

Ligumsky M, Lysy J, Siguencia G, Friedlander Y (2001) Effect of long-term, continuous versus alternate-day omeprazole therapy on serum gastrin in patients treated for reflux esophagitis. J Clin Gastroenterol 33(1):32–35

Liu L, Ding W, Zhao F, Shi L, Pang Y, Tang C (2013) Plasma levels and potential roles of catestatin in patients with coronary heart disease. Scand Cardiovasc J: SCJ 47(4):217–224. doi:10.3109/14017431.2013.794951

Loh YP, Cheng Y, Mahata SK, Corti A, Tota B (2012) Chromogranin A and derived peptides in health and disease. Journal of Molecular Neuroscience: MN 48(2):347–356. doi:10.1007/s12031-012-9728-2

Lugardon K, Raffner R, Goumon Y, Corti A, Delmas A, Bulet P, Aunis D, Metz-Boutigue MH (2000) Antibacterial and antifungal activities of vasostatin-1, the N-terminal fragment of chromogranin A. J Biol Chem 275:10745–10753

Maestroni S, Maestroni A, Ceglia S, Tremolada G, Mancino M, Sacchi A, Lattanzio R, Zucchiatti I, Corti A, Bandello F, Zerbini G (2015) Effect of chromogranin A-derived vasostatin-1 on laser-induced choroidal neovascularization in the mouse. Acta Ophthalmol 93(3):e218–e222. doi:10.1111/aos.12557

Mahata SK, Mahata M, Fung MM, O'Connor DT (2010) Catestatin: a multifunctional peptide from chromogranin A. Regul Pept 162(1–3):33–43

Mandalà M, Stridsberg M, Helle KB, Serck-Hanssen G (2000) Endothelial handling of chromogranin A. Adv Exp Med Biol 482:167–178

Marcucci F, Corti A (2011) How to improve exposure of tumor cells to drugs – promoter drugs increase tumor uptake and penetration of effector drugs. Adv Drug Deliv Rev. doi:10.1016/j.addr.2011.09.007

Marcucci F, Corti A (2012) Improving drug penetration to curb tumor drug resistance. Drug Discov Today 17(19–20):1139–1146. doi:10.1016/j.drudis.2012.06.004

Mehta D, Malik AB (2006) Signaling mechanisms regulating endothelial permeability. Physiol Rev 86(1):279–367. doi:10.1152/physrev.00012.2005

Meng L, Wang J, Ding WH, Han P, Yang Y, Qi LT, Zhang BW (2013) Plasma catestatin level in patients with acute myocardial infarction and its correlation with ventricular remodelling. Postgrad Med J 89(1050):193–196. doi:10.1136/postgradmedj-2012-131060

Metz-Boutigue MH, Garcia-Sablone P, Hogue-Angeletti R, Aunis D (1993) Intracellular and extracellular processing of chromogranin A. Determination of cleavage sites. Eur J Biochem 217(1):247–257

Montesinos MS, Machado JD, Camacho M, Diaz J, Morales YG, Alvarez de la Rosa D, Carmona E, Castaneyra A, Viveros OH, O'Connor DT, Mahata SK, Borges R (2008) The crucial role of chromogranins in storage and exocytosis revealed using chromaffin cells from chromogranin A null mouse. The Journal of Neuroscience: the Official Journal of the Society for Neuroscience 28(13):3350–3358. doi:10.1523/JNEUROSCI.5292-07.2008

Mosley CA, Taupenot L, Biswas N, Taulane JP, Olson NH, Vaingankar SM, Wen G, Schork NJ, Ziegler MG, Mahata SK, O'Connor DT (2007) Biogenesis of the secretory granule: chromogranin A coiled-coil structure results in unusual physical properties and suggests a mechanism for granule core condensation. Biochemistry 46(38):10999–11012

O'Connor DT, Bernstein KN (1984) Radioimmunoassay of chromogranin A in plasma as a measure of exocytotic sympathoadrenal activity in normal subjects and patients with pheochromocytoma. N Engl J Med 311(12):764–770

O'Connor DT, Deftos LJ (1986) Secretion of chromogranin A by peptide-producing endocrine neoplasms. N Engl J Med 314(18):1145–1151

O'Connor DT, Mahata SK, Taupenot L, Mahata M, Livsey Taylor CV, Kailasam MT, Ziegler MG, Parmer RJ (2000) Chromogranin A in human disease. Adv Exp Med Biol 482:377–388

O'Connor DT, Kailasam MT, Kennedy BP, Ziegler MG, Yanaihara N, Parmer RJ (2002) Early decline in the catecholamine release-inhibitory peptide catestatin in humans at genetic risk of hypertension. J Hypertens 20(7):1335–1345

O'Connor DT, Cadman PE, Smiley C, Salem RM, Rao F, Smith J, Funk SD, Mahata SK, Mahata M, Wen G, Taupenot L, Gonzalez-Yanes C, Harper KL, Henry RR, Sanchez-Margalet V (2005) Pancreastatin: multiple actions on human intermediary metabolism in vivo, variation in disease, and naturally occurring functional genetic polymorphism. J Clin Endocrinol Metab 90(9):5414–5425. doi:10.1210/jc.2005-0408

Pieroni M, Corti A, Tota B, Curnis F, Angelone T, Colombo B, Cerra MC, Bellocci F, Crea F, Maseri A (2007) Myocardial production of chromogranin A in human heart: a new regulatory peptide of cardiac function. Eur Heart J 28(9):1117–1127. doi:10.1093/eurheartj/ehm022

Portela-Gomes GM, Grimelius L, Wilander E, Stridsberg M (2010) Granins and granin-related peptides in neuroendocrine tumours. Regul Pept 165(1):12–20. doi:10.1016/j.regpep.2010.02.011

Portel-Gomes GM, Grimelius L, Johansson H, Wilander E, Stridsberg M (2001) Chromogranin A in human neuroendocrine tumors: an immunohistochemical study with region-specific antibodies. Am J Surg Pathol 25(10):1261–1267

Radek KA, Lopez-Garcia B, Hupe M, Niesman IR, Elias PM, Taupenot L, Mahata SK, O'Connor DT, Gallo RL (2008) The neuroendocrine peptide catestatin is a cutaneous antimicrobial and induced in the skin after injury. J Invest Dermatol 128(6):1525–1534. doi:10.1038/sj.jid.5701225

Ramella R, Boero O, Alloatti G, Angelone T, Levi R, Gallo MP (2010) Vasostatin 1 activates eNOS in endothelial cells through a proteoglycan-dependent mechanism. J Cell Biochem 110(1):70–79. doi:10.1002/jcb.22510

Ratti S, Curnis F, Longhi R, Colombo B, Gasparri A, Magni F, Manera E, Metz-Boutigue MH, Corti A (2000a) Structure-activity relationships of chromogranin A in cell adhesion. Identification and characterization of an adhesion site for fibroblasts and smooth muscle cells. J Biol Chem 275(38):29257–29263

Ratti S, Curnis F, Longhi R, Colombo B, Gasparri A, Magni F, Manera E, Metz-Boutigue MH, Corti A (2000b) Structure-activity relationships of chromogranin A in cell adhesion. Identification of an adhesion site for fibroblasts and smooth muscle cells. J Biol Chem 275(38):29257–29263

Ribatti D (2009) Endogenous inhibitors of angiogenesis: a historical review. Leuk Res 33(5):638–644. doi:10.1016/j.leukres.2008.11.019

Ribatti D, Nico B, Vacca A (2006) Importance of the bone marrow microenvironment in inducing the angiogenic response in multiple myeloma. Oncogene 25(31):4257–4266. doi:10.1038/sj.onc.1209456

Russell J, Gee P, Liu SM, Angeletti RH (1994) Inhibition of parathyroid hormone secretion by amino-terminal chromogranin peptides. Endocrinology 135(1):337–342

Sanchez-Margalet V, Gonzalez-Yanes C, Najib S, Santos-Alvarez J (2010) Reprint of: metabolic effects and mechanism of action of the chromogranin A-derived peptide pancreastatin. Regul Pept 165(1):71–77. doi:10.1016/j.regpep.2010.10.004

Sanduleanu S, Stridsberg M, Jonkers D, Hameeteman W, Biemond I, Lundqvist G, Lamers C, Stockbrugger RW (1999) Serum gastrin and chromogranin A during medium- and long-term acid suppressive therapy: a case-control study. Aliment Pharmacol Ther 13(2):145–153

Schneider F, Bach C, Chung H, Crippa L, Lavaux T, Bollaert PE, Wolff M, Corti A, Launoy A, Delabranche X, Lavigne T, Meyer N, Garnero P, Metz-Boutigue MH (2012) Vasostatin-I, a chromogranin A-derived peptide, in non-selected critically ill patients: distribution, kinetics, and prognostic significance. Intensive Care Med 38(9):1514–1522. doi:10.1007/s00134-012-2611-3

Simon JP, Aunis D (1989) Biochemistry of the chromogranin A protein family. Biochem J 262(1):1–13

Steiner HJ, Weiler R, Ludescher C, Schmid KW, Winkler H (1990) Chromogranins A and B are co-localized with atrial natriuretic peptides in secretory granules of rat heart. J Histochem Cytochem 38(6):845–850

Stridsberg M, Eriksson B, Oberg K, Janson ET (2004) A panel of 11 region-specific radioimmunoassays for measurements of human chromogranin A. Regul Pept 117(3):219–227

Syversen U, Ramstad H, Gamme K, Qvigstad G, Falkmer S, Waldum HL (2004) Clinical significance of elevated serum chromogranin A levels. Scand J Gastroenterol 39(10):969–973

Taupenot L, Harper KL, O'Connor DT (2003) The chromogranin-secretogranin family. N Engl J Med 348(12):1134–1149

Tota B, Angelone T, Mazza R, Cerra MC (2008) The chromogranin A-derived vasostatins: new players in the endocrine heart. Curr Med Chem 15(14):1444–1451

Turquier V, Vaudry H, Jegou S, Anouar Y (1999) Frog chromogranin A messenger ribonucleic acid encodes three highly conserved peptides. Coordinate regulation of proopiomelanocortin and chromogranin A gene expression in the pars intermedia of the pituitary during background color adaptation. Endocrinology 140:4104–4112

Veschini L, Crippa L, Dondossola E, Doglioni C, Corti A, Ferrero E (2011) The vasostatin-1 fragment of chromogranin A preserves a quiescent phenotype in hypoxia-driven endothelial cells and regulates tumor neovascularization. FASEB J 25(11):3906–3914. doi:10.1096/fj.11-182410

Waldum HL, Brenna E (2000) Personal review: is profound acid inhibition safe? Aliment Pharmacol Ther 14(1):15–22

Wallez Y, Huber P (2008) Endothelial adherens and tight junctions in vascular homeostasis, inflammation and angiogenesis. Biochim Biophys Acta 1778(3):794–809. doi:10.1016/j.bbamem.2007.09.003

Young RH, Srigley JR, Amin MB, Ulbright TM, Cubilla AL (2000) Carcinoma of the prostate gland (excluding unusual variants and secondary carcinomas). In: Rosai J, Sobin LH (eds) Atlas of tumor pathology, vol 28, 3rd series edn. Armed Forces Institute of Pathology, Washington DC, pp 111–216

Zhang K, Rao F, Wen G, Salem RM, Vaingankar S, Mahata M, Mahapatra NR, Lillie EO, Cadman PE, Friese RS, Hamilton BA, Hook VY, Mahata SK, Taupenot L, O'Connor DT (2006) Catecholamine storage vesicles and the metabolic syndrome: the role of the chromogranin A fragment pancreastatin. Diabetes Obes Metab 8(6):621–633

Zhang D, Lavaux T, Sapin R, Lavigne T, Castelain V, Aunis D, Metz-Boutigue MH, Schneider F (2009) Serum concentration of chromogranin A at admission: an early biomarker of severity in critically ill patients. Ann Med 41(1):38–44

Full Lenght CgA: A Multifaceted Protein in Cardiovascular Health and Disease

Bruno Tota and Maria Carmela Cerra

Abstract The multifunctional protein Chromogranin A (CGA) is a major marker of the sympatho-adrenal neuroendocrine (SAN) activity. Stored in neuroendocrine chromaffin secretory granules with several prohormones and their proteolytic enzymes, with noradrenaline and adrenaline, it is released with catecholamines upon stimulation. It is also present in other cell types, including myocardiocytes of various vertebrates, and humans, particularly in the presence of cardiomyopathy and heart failure. Due to the processing into a number of biologically active peptides, it represents a prohormone with an important modulatory role on endocrine, cardiovascular, metabolic, and immune systems. Circulating CGA increases in the presence of stress-induced excessive SAN activation and of pathologies such as neuroendocrine tumors, and cardiovascular diseases including hypertension, coronary syndrome, and heart failure. Thus, the protein is considered a promising biomarker for a number of severe diseases. Recently, it was found that in the heart of normotensive and hypertensive rats (SHRs), CGA is processed under hemodynamic and excitatory stimuli, and the exogenous full length protein directly affects myocardial and coronary performance by Akt/NOS/NO/cGMP/PKG pathway. We here illustrate the emerging role elicited by CGA in the control of circulatory homeostasis with particular focus on its cardiovascular action under physiological and physiopathological conditions. These actions contribute to extend our knowledge on the sympatho-chromaffin control of the cardiovascular system and its integrated "whip-brake" circuits.

1 Introduction

The spectrum of biological functions attributed to CGA includes the full-length 49-kDa protein implication as a major marker of the sympatho-adrenal neuroendocrine activity (SAN) and, at the same time, its prohormone ability to generate several

---

B. Tota • M.C. Cerra (✉)
Department of Biology, Ecology and Earth Sciences, University of Calabria, 87036 Arcavacata di Rende (CS), Italy
e-mail: maria_carmela.cerra@unical.it

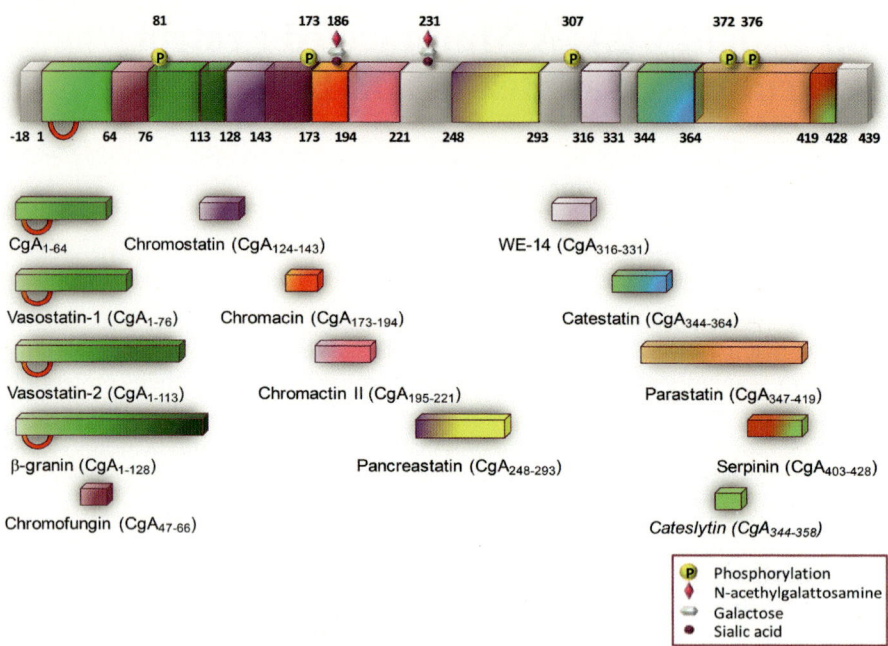

Fig. 1 Human Chromogranin-A (CGA) sequence with post-translational modifications and the derived biologically active peptides. In Italic, the synthetically generated fragment Cateslytin

biologically active peptides through partial processing (Fig. 1). In view of this capability of acting both as sensor of the organism stress- induced perturbations, and homeostatic counter-regulatory effector, CGA appears to posses intriguing cytokine and endocrine properties. Here we aim to illustrate these two facets of full- length CGA with particular reference to the cardio-vascular system under normal and physio-pathological conditions. Since CGA generates at least three peptides with relevant cardiotropic sympatho-adrenergic influence, i.e. Vasostatin 1 (VS-1), Catestatin (CST) and Serpinin (SERP), we will very briefly refer also to their cardio-vascular effects to provide an integrated information on how the CGA system, i.e. the full-length protein and its fragments, may monitor and influence circulatory homeostasis, especially under SAN overactivation. Detailed knowledge on these peptides is reported in other chapters of this Volume.

A physiological hallmark of the CGA system is represented by its involvement in the stress response in which it appears to closely interact with SAN activation. Such interaction is topologically reflected by its subcellular localization. In fact, together with other granins, CGA is stored in the secretory (chromaffin) granules of the diffuse neuroendocrine system and is released with catecholamines (CAs). Within the granules, CGA is also co-stored with neuropeptide Y, cardiac natriuretic peptide hormones, several prohormones and their proteolytic enzymes. The evidence that CGA is also present in the secretory granules of the heart (Tota et al. 2010 and references therein), including the human myocardium (Pieroni et al. 2007), is consistent with an emerging modulatory role of the prohormone and its derived fragments at the cardiac level.

To highlight for the non-expert reader the cardio-vascular interactions between CGA and the sympatho-chromaffin system, we will briefly summarize SAN involvement to maintain circulatory homeostasis under normal and physio-pathological conditions.

2 Physio-Pathological Aspects of SAN Overactivation

The cardiovascular system is intimately linked to the brain through two pathways, the hypothalamus-pituitary-adrenal (HPA) axis and the autonomic nervous system consisting of two limbs, i.e., the sympathetic and parasympathetic pathways. SAN and its end-products, the CAs, play a central role in the stress response ("fight or flight" reaction), characterized by Selye (1936) as the "general adaptation syndrome" (Samuels 2007 and references therein). The peripheral limbs of the stress system, the SAN and the HPA axis, maintain stress-related homeostasis through increased peripheral levels of CAs and glucocorticoids which act synergistically. However, their consequent and prolonged overstimulation can lead to visceral organ dysfunction, experimentally exemplified in the heart by the electrolyte-steroid-cardiopathy with necrosis (Selye and Bajusz 1958; Raab 1969). Therefore, through the SAN-induced overactivation of positive reverberatory networks, e.g., the Renin-Angiotensin System (RAS), in which the activation of one pathway tends to activate another excitatory one, the stress response itself could threaten the homeostasis of target organs and tissues.

The heart and the vasculature work as an integrative interface between the nor-adrenergic nerve terminals, mainly releasing norepinephrine (NE), and the circulating CAs secreted by the adrenal medulla. In addition, similarly to other organs, in the heart, CAs are co- stored and co-released with other neuropeptides and humoral autacoids, in the afferent, efferent, interconnecting short neurons and intracardiac ganglia, as well as in the chromaffin cells and in the cardiomyocytes themselves. The convergence of these SAN excitatory stimuli may contribute to explain why the heart and the vasculature represent a typical paradigm of a stress-threatened organ.

CAs plasma levels induce myocardial excitability, contractility and relaxation (Fig. 2) and its increased concentrations can induce necrotic damage in the heart (Samuels 2007 and references therein). Moreover, the initial heart response to prolonged and excessive stress is represented by cardiac hypertrophy, i.e., a morphological enlargement which tends to compensate or prevent progressive deterioration of cardiac function challenged by the hemodynamic overload (Tota et al. 2008 and references therein). Under extreme or prolonged stress conditions, positive feedback loops can lead the hypertrophied heart to failing processes. A concomitant sustained heightened activation of SNS and RAS tend to control cardiac output and systemic blood pressure.

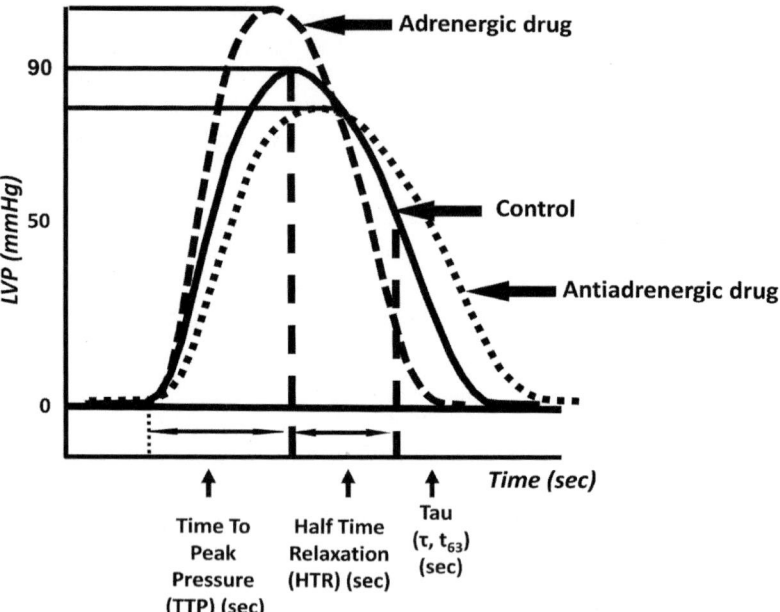

Fig. 2 Representative diagram showing the effects of adrenergic and anti-adrenergic drugs on ventricular contraction and relaxation (expressed as LVP variations)

The spectrum of SAN-elicited cardio-circulatory effects ranges from hyperlipidemia, platelet adhesiveness, blood coagulation, vascular smooth muscle hyperplasia, vasomotor tone, denervated myocardium etc. to immunologic responses induced by cardiovascular changes. Accordingly, SAN overstimulation may involve the actions of numerous different targets, directly impinging the heart and the vasculature (Samuels 2007). Compelling clinical evidence shows that the initial heart response to prolonged SAN over-activity leads to compensatory remodeling, cardiac hypertrophy and, if the stress will overwhelm the system, heart failure (HF) (Triposkiadis et al. 2009 and references therein). Uncontrolled heightened SAN activation may lead to changes more deleterious than those resulting from the actual stress placed on the heart (Samuels 2007). A negative prognostic value in the evolution of human HF has been associated with chronic SAN activation, mainly via CAs, enhancing the pathological processes (Cohn and Yellin 1984). This knowledge has provided the rationale for the anti-adrenergic drug therapy, e.g., beta-adrenergic-blockers.

The emerging cardiovascular importance of the CGA system may widen our knowledge regarding the SAN-orchestrated regulation of the cardio-circulatory system, stimulating, at the same time, potential additional or alternative therapeutic drugs designed for protecting stress-targeted organs.

3 Cardio-Circulatory Implications of Full-Length CGA

It is methodologically relevant to be aware that estimation of plasma levels of CGA and its derived peptides can be obtained by serological determinations that, in addition to processing-independent radioimmunoassay, include region-specific processing-dependent analysis. Noteworthy, only the latter is able to analyse the plasma levels of the various CGA-derived fragments, showing sometime opposite biological effects, functions and prognostic significance (Crippa et al. 2013; Goetze et al. 2014). The CGA in vivo long half-life (~18 min) and its relatively elevated circulating concentrations (including normal conditions), reduce eventual false measurements and facilitate blood collection, pre-analytic handling and final determinations (O'Connor et al. 1989). However, caution is required in CGA detection and quantification, since a variability exists in the methodologies used for determinations and, in many cases, different results may be obtained on the same sample if different methods are used (i.e. RIA vs ELISA). As commented by Angelone et al. (2012), a definitive standardization of the methods for CGA determination is essential to have comparable, uniform, and thus clinically relevant measurements in blood and tissues.

Normal concentrations of circulating CGA range between 0.5 and 2 nM (Helle et al. 2007; Crippa et al. 2013). They increase under stress-induced SAN overstimulation and physio-pathological conditions, e.g., chronic inflammation, neuroendocrine tumors, acute coronary syndromes and chronic HF. Based on this clinical evidence, CGA plasma levels have been employed as prognostic indicators in a number of these diseases (Helle et al. 2007; Angelone et al. 2012; D'Amico et al. 2014).

Plasma CGA concentrations increase up to 10–20 nM (500–1000 ng/ml) in patients with essential hypertension (Takiyyuddin et al. 1995), chronic HF (Ceconi et al. 2002), myocardial infarction (Omland et al. 2003), acute coronary syndromes (Jansson et al. 2009), acute destabilized HF (Dieplinger et al. 2009), and decompensated hypertrophic cardimyopathy (Pieroni et al. 2007). Ceconi et al. (2002) were the first to document that plasma CGA levels significantly parallel the severity of the dysfunction and represent an independent predictor for mortality. This clinical evidence highlights the role of CGA as a potentially new diagnostic and prognostic cardiovascular biomarker independent from conventional markers. Furthermore, various evidences strongly indicate the correlation between CGA and SAN activity. For example, studies by O'Connor and his group have shown in twins that basal plasma CGA level is heritable (Takiyyuddin et al. 1995). In addition, compared with age-matched normotensive counterparts, patients with essential hypertension exhibit augmented plasma CGA and enhanced release of stored CGA in response to adrenal medullary stimulation by insulin-elicited hypoglycemia (Takiyyuddin et al. 1995).

It is of cardiac relevance the observation that targeted ablation of the CGA gene makes CGA-KO mice hypertensive and hyper-adrenergic, with accompanied heart enlargement, increases reactive oxygen species production and consequent nitric oxide (NO) depletion (Gayen et al. 2010). A detailed information on this issue has been provided by Mahata in the present Volume.

Numerous studies, mainly by Corti and his group, revealed that the CGA system works at the interface of endothelial, angiopoietic and blood coagulation processes. It is implicated in the modulation of the endothelial barrier (Ferrero et al. 2002) and tumor-induced vascular remodelling (Veschini et al. 2011). Both CGA and its N-terminal fragment VS-1 are potent inhibitors of the thrombin-induced endothelial permeability and the pro-angiogenic Vascular Endothelial Growth Factor (VEGF) (Ferrero et al. 2002), also inhibiting the Tumor Necrosis Factor (TNF)-induced changes on endothelial cells, i.e. gap formation, disassembly of vascular endothelial-cadherin adherence junctions and vascular leakage (Ferrero et al. 2002; Dondossola et al. 2011). CGA can also affect host/tumor interactions (Colombo et al. 2002). For example, systemic administration of CGA (1 μg) to lymphoma-bearing mice potently reduces the TNF- induced penetration of the patent blue dye in tumor tissues (Dondossola et al. 2011).

In healthy subjects Crippa et al. (2013) detected the presence of biologically relevant plasma levels of full-length CGA, CGA 1-76 (antiangiogenic) and fragments lacking the C-terminal region (proangiogenic). Importantly, they demonstrated that blood coagulation activates a thrombin- dependent almost complete conversion of plasma CGA into fragments lacking the C-terminal region. Thus, the possibility exists that CGA functions as a homeostatic stabilizer of angiogenesis. Under conditions of perturbed angiogenesis (wound healing, cancer, etc.), it can contribute to circulatory and vascular protection via the opposite angiogenic actions of its fragments (possibly VS-1 and Catestatin: CST) produced by tightly spatio-temporally regulated proteolysis.

4 Cardiac CGA: Localization and Processing

Before illustrating the physio-pharmacological aspects of CGA cardio-circulatory activity, we will consider here the intracardiac localization of this granin and its significance.

Imunohistochemical evidence has shown in the myoendocrine granules of the rat heart the co-localization of CGA and Atrial Natriuretic Peptide (ANP) (Steiner et al. 1990). The immunoblotting data were consistent with a more extensive myocardial CGA processing compared to that of the adrenal medulla (Steiner et al. 1990). Possibly, additional source of cardiac CGA and/or CGA- derived peptides may result from the terminal innervations of the heart (Miserez et al. 1992). CGA has been also detected in rat Purkinje conduction fibers, in both rat atrium and ventricle, as well as in H9c2 rat cardiomyocytes (Weiergraber et al. 2000). CGA, Chromogranin B (CGB) and Secretogranin (SG) were found in the secretory granules of the mouse myocardium (Biswas et al. 2010). Of physio-pathological importance, Pieroni et al. (2007) provided histochemical evidence of CGA-positive intracellular staining in the human myocardium. Moreover, using confocal microscopy they showed that in ventricular cardiomyocytes of dilated and hypertrophic human hearts CGA is colocalized with Brain Natriuretic Peptide (BNP). This finding was confirmed by RT-

PCR that documented the myocardial presence of CGA-mRNA; based on ELISA assays with four different monoclonal antibodies, more than 0.5 μg of CGA per gram of left ventricular myocardial tissue was measured. Therefore, it is conceivable that the myocardium under stimulated conditions may constantly release cardiac Natriuretic Peptide Hormones (NPs) together with its co-stored CGA, whose plasma half-life is 18.4 min. These observations emphasize the hypothesis of Corti et al. (1996) and Pieroni et al. (2007) who postulated a significant cardiac contribution to the increased circulating CGA levels reported in their patients. The possible cross-talk between CGA and the NPs system could be implicated in the cardiovascular control counteracting prolonged and reverberating excitatory stimuli, for example, exerting tonic vasodilation, hypotension and cardioprotection against SAN hyperactivation. Dieplinger et al. (2009, and references therein) considered the strong association between plasma CGA/NPs levels and the degree of hemodynamic dysfunction in HF and proposed both hormones as reliable prognostic indicators of the severity of HF. It has been suggested that the significant correlation between BNP levels and left ventricle end diastolic pressure can be indicative of an undefined stretch-elicited release and transcriptional up-regulation mechanisms; this could also be the case for myocardial CGA (Tota et al. 2010 and references therein).

Glattard et al. (2006) provided in the rat heart biochemical characterization of intracardiac CGA and its processing. They submit the RPHPLC purified CGA-immunoreactive fractions from cardiac extracts to Western Blot and MS analysis (TOF/TOF technique) and identified four endogenous N terminal CGA-derived peptides, i.e. CGA4–113, CGA1–124, CGA1–135 and CGA1–199, containing the Vasostatin sequence. Importantly, among these and other C-terminal truncated fragments intact CGA was identified, highlighting the cell-specific proteolytic pattern of CGA, in contrast to the rat adrenal gland in which almost no intact CGA is detected. This and other comparative considerations suggest that in the heart the maturation process can be incomplete and specific. Noteworthy, Pasqua et al. (2013), showed in the rat heart that the cardioactive motif (the VS-1 sequence or a portion of it) is present among the identified low-molecular-mass fragments. These evidences support our view that, under normal or stressful conditions, the heart is able respond to a specific physical (e.g. stretch) or chemical (e.g. CAs, Endothelin-1: ET-1, Angiotensin) stimulus, triggering proteolytic CGA processing with subsequent increase in lower-molecular-mass cardioactive peptides (Pasqua et al. 2013, and references therein). Accordingly, CGA and its derived cardioactive peptides may work as a fine-tuned system, which, at both local and systemic levels, can exert endocrine, autocrine/paracrine cardiovascular modulations.

5 Physio-Pharmacological Profile of CGA Cardiovascular Activity

Evidence regarding the direct myocardial and coronary actions of CGA and its intracardiac stimulus-induced processing has only recently emerged. Pasqua et al. (2013) evaluated for the first time the influence exerted by the full-length human recombinant CGA on the cardiac performance of isolated and Langendorff perfused hearts of normotensive and spontaneously hypertensive rats (SHR), analyzing at the same time its proteolytic processing. The SHR heart is a well-known experimental model which mimics the hypertensive patho-physiological changes of the human heart (Doggrell and Brown 1998), representing an alternative and/or additional tool to the CGA/KO mice mentioned above. It was demonstrated that CGA at concentrations lower, or close to its physiological circulatory levels (1 nM), induces a mild but significant depression of mechanical performance and vasodilates the coronary arteries. These actions are mediated by an endothelium-dependent mechanism that involves PI3K-NO signalling. Moreover, intracardiac CGA appears subjected to proteolytic processing which generates smaller peptides including the cardioactive and vasoactive VS-1, the fragmentation being enhanced by chemical (Isoproterenol: ISO, or ET-1) stimulation. Here below we will detail these findings to highlight their putative physiological and physiopathological implications.

5.1 Myocardial and Coronary Actions

According to the analysis of the cardiac perfusates, exogenous CGA is not cleaved by the heart. Consequently, the myocardial and coronary actions induced by the granin can be attributed to the full-length protein, excluding the involvement of derived fragment, including VS-1.

Exogenous CGA starting from 1 to 4 nM concentrations directly depresses myocardial contractility (negative inotropism) and relaxation (negative lusitropy). The protein (from 100 pM to 4 nM) reduces major inotropic parameters, i.e., LVP and +(LVdP/dt)max, decreasing, at the same time, lusitropy, i.e., −(LVdP/dt)max and T/−t (at 100 pM, 1 and 4 nM). CGA is also able to elicit coronary vasodilation at 1 and 4 nM concentrations without influencing heart rate (HR). Both the inotropic and lusitropic effects disappear at higher concentrations (10–16 nM) (Fig. 3), pointing to a bell-shaped (or U-shaped) concentration/response curve. The underlying mechanism is unknown despite the phenomenon has been previously observed in several experiments with CGA and is common to a number of biological responses, e.g., endostatin-induced anti-tumor activity (Celik et al. 2005), interferon-alpha-mediated inhibition of angiogenesis (Slaton et al. 1999), including the biphasic influence of CGA on blood pressure levels and CAs secretion in mice. It is possible that the bell-shaped curve reflects a counter-regulatory mechanism activated at high concentrations. Another explanation may be related to changes in the quaternary structure of CGA that, at physiological pH, which can exist as a monomer or a dimer (Yoo and Lewis 1996).

CGA-induced negative inotropism and lusitropism are more potent in SHR rats than in the normotensive young counterparts (Fig. 3). Likely, the higher responsiveness of the former to CGA may result from the enhanced sensitivity exhibited by the SHR heart toward inhibitory hormones such as Angiotensin II and NPs (Anand-

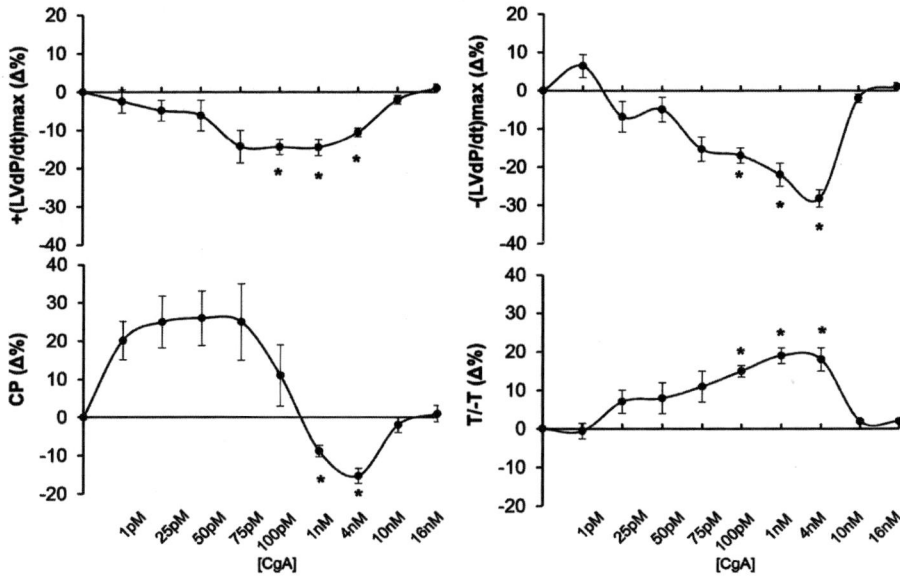

Fig. 3 Dose-dependent response curves of exogenous CGA (1 pM-16 nM) on +(LVdP/dT)max, −(LVdP/dT)max, T/−T, and CP, obtained on isolated and Langendorff perfused young normotensive rat heart preparations (Modified from Pasqua et al. 2013)

Srivastava 1992). The suggested mechanisms include the activation of several inhibitory pathways, e.g., the overexpression of a Gi regulatory protein involved in the depressed vascular tone and impaired myocardial performance occurring in the hypertensive state (Anand-Srivastava 1992). In this context, it is relevant that exogenous CGA causes different coronary effects in young normal rats and SHR: while CGA (1 and 4 nM) elicits vasodilation in the former, the protein appears noneffective in SHR. Of note, CGA/KO mice are hypertensive (Mahapatra et al. 2005), and circulatory CGA protein levels are increased in human hypertension (Takiyyuddin et al. 1995). Therefore, the depressing influence induced by the full length granin might be interpreted as a compensatory response for increased blood pressure.

5.2 Obligatory Endothelium-NO Involvement

As shown by Triton-induced endothelial impairment, the cardiotropic modulatory actions of exogenous full-length CGA require the obligatory role of the endothelium. The fact that the endothelial involvement is also crucial for attaining the similar VS-1- and CST-dependent cardiotropic effects argues for a common signal transduction mechanism of CGA and its two cardio-inhibitory fragments. Results obtained with both whole organ and isolated cell (cultured endothelium, HUVEC) suggest that the intact granin, while perfusing the intracardiac circulatory bed, firstly encounters the endothelial barrier where a sill unidentified binding site is located, thereby triggering a downstream PI3K/Akt-dependent NO signaling that in turn modulates the responses of the myocardiocyte and coronary smooth muscle. Indeed, such endothelium- mediated mechanism appears compatible with the large dimensions of the protein that may prevent it to reach the cell targets subjacent to the endothelium. The involvement of the cardiac endothelium emphasizes the relevance of the interaction between this tissue and CGA mentioned previously and elsewhere in this Volume.

The vascular endothelium is a relevant source of eNOS-produced NO (Balligand et al. 2009). As shown by Pasqua et al. (2013) in ex vivo experiments, the specific chemical inhibition of Akt (an upstream activator of the NO pathway) and the eNOS isoform abolishes CGA cardioactivity. In agreement with this, in the perfused hearts of both normotensive and SHR, as well as in HUVEC, CGA exposure provokes eNOS phosphorylation and its induced actions require an NO-dependent obligatory mechanism. It is well recognized that NO modulates both the beat-to-beat and the long-term contractile performance of the heart (Balligand et al. 2009). Cardiac NOS-produced NO induces fine-tuned tonic depression of myocardial contractility through sGC-PKG signaling, thereby reducing L-type Ca^{2+} current and phosphorylating troponin I (Abi-Gerges et al. 2001). Furthermore, NO-cGMP-PKG activation can also modulate relaxation by inhibiting phospholamban (PLB) phosphorylation, hence attenuating sarcoplasmic reticulum Ca^{2+} reuptake (Stojanovic et al. 2001). In agreement with this knowledge, CGA depresses relaxation in both normotensive and SHR rats. Notably, in the latter, an impaired endothelium-dependent vasodilation has been related to structural changes at the level of myocardial arteries and/or a reduction in both capillary density and eNOS expression (Stojanovic et al. 2001). Intriguingly, Pasqua et al. (2013) have shown different coronary response to full-length CGA between young normotensive and SHR. In the absence of direct experimental explanation, we suggest that the lack of responsiveness observed in the hypertensive rats could result from reduced NO availability/capability in regulating SHR basal coronary flow due to a decreased shear stress-stimulated NO (Crabos et al. 1997; Kojda et al. 1998).

5.3 *Intracardiac CGA Processing*

Pasqua et al. (2013) confirmed in the rat heart the presence of CGA and demonstrated its processing by detecting both in normal and SHR cardiac extracts smaller peptides including the cardioactive and vasoactive VS-1. It is physiologically relevant that the processing is enhanced by chemical (ISO or ET-1) stimulation.

As evidenced by Western Blotting analyses, the major immunoreactive bands of 80–50 kDa detected include the full-length CGA and the truncated fragments lacking the C-terminal region. These data confirm the granin identification and fragmentation reported by Glattard et al. (2006) in the rat heart, as well as the presence of CGA in the human ventricle (Pieroni et al. 2007).

In both normal and diseased SHR hearts, CGA processing appears responsive to physical (perfusion) and chemical (ISO and ET-1) stimuli able to provoke proteolytic fragmentation of cardiac CGA into shorter peptides. Adrenergic (ISO 100 nM) or ET-1 (110 nM) exposure enhances the processing and, compared with the untreated counterparts, the full-length/large fragments appear reduced, while short N-terminal fragments of a size corresponding to that of VS-1 are increased.

The chromaffin granules of the bovine adrenal medulla have provided information regarding the CGA maturation with its starting points at both the N terminus and the C terminus (Metz-Boutigue et al. 1993). The major proteolytic cleavage sites were identified, including the highly conserved 64–65 bond present in the N-terminal moiety of CGA and included in the VS sequence (Metz-Boutigue et al. 1993; Cerra et al. 2008). According to this information, it might be expected that pro-hormone convertase 1/3 (PC1/3), PC2, and carboxypeptidase H/E are implicated in CGA processing in the rat heart. The latter, similarly to the bovine adrenal medulla paradigm, can represent an intriguing experimental model for exploring the major enzymatic events underlining CGA proteolytic processing and how these are regulated to fulfill the requirements of the stimulus (SAN)-CGA proteolysis coupling eventually accomplished by the normal or stressed organ.

6 Conclusions and Perspectives

On the basis of the evidences here examined, it is conceivable that the full-length CGA and its derived cardioactive fragments work as a multilevel integrated system able to sense and, at the same time, counter-regulate overall circulatory homeostasis as well as local organ function, i.e. the beating heart. Therefore, the sensor and effector attitudes of the full-length CGA may indeed represent two faces of the same coin. In particular, the findings that both the full-length CGA and its derived peptides VS-1, CST and SERP exert direct myocardial and coronary effects provide a conceptual link between the granin-induced systemic and intracardiac modulatory influences. They appear to implicate paracrine/autocrine mechanisms, hence expanding the classical concept of the heart as an endocrine organ, especially in relation to elevated SAN outflow and cardiovascular stress (see also Tota et al. 2014). This scenario will be detailed by other chapters of this Volume and is schematically anticipated in Fig. 4.

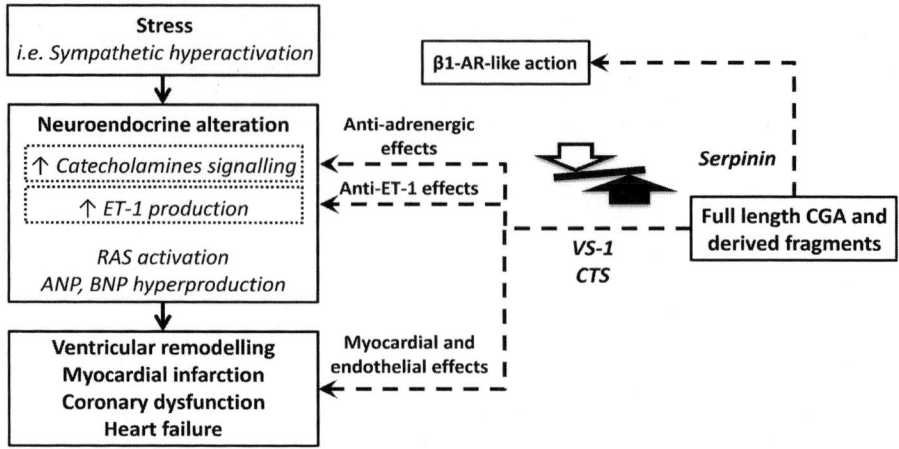

Fig. 4 Scheme of the possible sites of interventions for the CGA system. The protein and its derived fragments may interact at the systemic level with factors such as CAs, ANGII, NPs, etc., thus participating to the stress response, as in the case of the neuroendocrine scenario activated in HF. At the local (heart) level, the direct modulation of the myocardial and the coronary performance may contribute to cardiac homeostasis under basal conditions and in the presence of stress challenges

References

Abi-Gerges N, Fischmeister R, Mery PF (2001) G protein-mediated inhibitory effect of nitric oxide on L-type Ca2current in rat ventricular myocytes. J Physiol 531(Pt 1):117–130

Anand-Srivastava MB (1992) Enhanced expression of inhibitory guanine nucleotide regulatory protein in spontaneously hypertensive rats. Relationship to adenylate cyclase inhibition. Biochem J 288(Pt 1):79–85

Angelone T, Mazza R, Cerra MC (2012) Chromogranin-A: a multifaceted cardiovascular role in health and disease. Curr Med Chem 19(24):4042–4050

Balligand JL, Feron O, Dessy C (2009) eNOS activation by physical forces: from short-term regulation of contraction to chronic remodeling of cardiovascular tissues. Physiol Rev 89(2):481–534

Biswas N, Curello E, O'Connor DT, Mahata SK (2010) Chromogranin/secretogranin proteins in murine heart: myocardial production of Chromogranin A fragment catestatin (Chga(364-384)). Cell Tissue Res 3:353–361

Ceconi C, Ferrari R, Bachetti T, Opasich C, Volterrani M, Colombo B, Parrinello G, Corti A (2002) Chromogranin A in heart failure; a novel neurohumoral factor and a predictor for mortality. Eur Heart J 12:967–974

Celik I, Surucu O, Dietz C, Heymach JV, Force J, Höschele I, Becker CM, Folkman J, Kisker O (2005) Therapeutic efficacy of the endostatin exhibits a biphasic dose-response curve. Cancer Res 65(23):11044–11050

Cerra MC, Gallo MP, Angelone T et al (2008) The homologous rat chromogranin A1–64 (rCGA1–64) modulates myocardial and coronary function in rat heart to counteract adrenergic stimulation indirectly via endothelium-derived nitric oxide. FASEB J 22(11):3992–4004

Cohn JN, Yellin AM (1984) Learned precise cardiovascular control through graded central sympathetic stimulation. J Hypertens Suppl 2:S77–S79

Colombo B, Curnis F, Foglieni C, Monno A, Arrigoni G, Corti A (2002) Chromogranin A expression in neoplastic cells affects tumor growth and morphogenesis in mouse models. Cancer Res 3:941–946

Corti A, Gasparri A, Chen FX, Pelagi M, Brandazza A, Sidoli A, Siccardi AG (1996) Characterisation of circulating Chromogranin A in human cancer patients. Br J Cancer 8:924–932

Crabos M, Coste P, Paccalin M et al (1997) Reduced basal NO-mediated dilation and decreased endothelial NO-synthase expression in coronary vessels of spontaneously hypertensive rats. J Mol Cell Cardiol 29(1):55–65

Crippa L, Bianco M, Colombo B, Gasparri AM, Ferrero E, Loh YP, Curnis F, Corti A (2013) A new Chromogranin A- dependent angiogenic switch activated by thrombin. Blood 2:392–402

D'amico MA, Ghinassi B, Izzicupo P, Manzoli L, Baldassarre A (2014) Biological function and clinical relevance of Chromogranin A and derived peptides. Endocr Connect 2:45–54

Dieplinger B, Gegenhuber A, Haltmayer M, Mueller T (2009) Evaluation of novel biomarkers for the diagnosis of acute destabilized heart failure in patients with shortness of breath. Heart 18:1508–1513

Doggrell SA, Brown L (1998) Rat models of hypertension, cardiac hypertrophy and failure. Cardiovasc Res 39(1):89–105

Dondossola E, Gasparri AM, Colombo B, Sacchi A, Curnis F, Corti A (2011) Chromogranin A restricts drug penetration and limits the ability of NGR-TNF to enhance chemotherapeutic efficacy. Cancer Res 17:5881–5890

Ferrero E, Magni E, Curnis F, Villa A, Ferrero ME, Corti A (2002) Regulation of endothelial cell shape and barrier function by Chromogranin A. Ann N Y Acad Sci 971:355–358

Gayen JR, Zhang K, Ramachandra Rao SP, Mahata M, Chen Y, Kim HS, Naviaux RK, Sharma K, Mahata SK, O'Connor DT (2010) Role of reactive oxygen species in hyperadrenergic hypertension: biochemical, physiological, and pharmacological evidence from targeted ablation of the chromogranin A (Chga) gene. Circ Cardiovasc Genet 3(5):414–425

Glattard E, Angelone T, Strub JM, Corti A, Aunis D, Tota B, Metz-Boutigue MH, Goumon Y (2006) Characterization of natural vasostatin-containing peptides in rat heart. FEBS J 14:3311–3321

Goetze JP, Alehagen U, Flyvbjerg A, Rehfeld JF (2014) Chromogranin A as a biomarker in cardiovascular disease. Biomark Med 1:133–140

Helle KB, Corti A, Metz-Boutigue MH, Tota B (2007) The endocrine role for Chromogranin A: a prohormone for peptides with regulatory properties. Cell Mol Life Sci 22:2863–2886

Jansson AM, Røsjø H, Omland T, Karlsson T, Hartford M, Flyvbjerg A, Caidahl K (2009) Prognostic value of circulating Chromogranin A levels in acute coronary syndromes. Eur Heart J 1:25–32

Kojda G, Kottenberg K, Hacker A, Noack E (1998) Alterations of the vascular and the myocardial guanylate cyclase/cGMP-system induced by long-term hypertension in rats. Pharm Acta Helv 73(1):27–35

Mahapatra NR, O'Connor DT, Vaingankar SM, Hikim AP, Mahata M, Ray S, Staite E, Wu H, Gu Y, Dalton N, Kennedy BP, Ziegler MG, Ross J, Mahata SK (2005) Hypertension from targeted ablation of chromograninAcan be rescued by the human ortholog. J Clin Invest 115(7):1942–1952

Metz-Boutigue MH, Garcia-Sablone P, Hogue-Angeletti R, Aunis D (1993) Intracellular and extracellular processing of chromogranin A. Determination of cleavage sites. Eur J Biochem 217(1):247–257

Miserez B, Annaert W, Dillen L, Aunis D, De Potter W (1992) Chromogranin A processing in sympathetic neurons and release of Chromogranin A fragments from sheep spleen. FEBS Lett 2:122–124

O'Connor DT, Pandlan MR, Carlton E, Cervenka JH, Hslao RJ (1989) Rapid radioimmunoassay of circulating Chromogranin A: in vitro stability, exploration of the neuroendocrine character of neoplasia, and assessment of the effects of organ failure. Clin Chem 35:1631–1637

Omland T, Dickstein K, Syversen U (2003) Association between plasma Chromogranin A concentration and long-term mortality after myocardial infarction. Am J Med 1:25–30

Pasqua T, Corti A, Gentile S, Pochini L, Bianco M, Metz-Boutigue MH, Cerra MC, Tota B, Angelone T (2013) Full- length human chromogranin-A cardioactivity: myocardial, coronary, and stimulus-induced processing evidence in normotensive and hypertensive male rat hearts. Endocrinology 9:3353–3365

Pieroni M, Corti A, Tota B, Curnis F, Angelone T, Colombo B, Cerra MC, Bellocci F, Crea F, Maseri A (2007) Myocardial production of chromogranin A in human heart: a new regulatory peptide of cardiac function. Eur Heart J 28(9):1117–1127

Raab W (1969) Myocardial electrolyte derangement: crucial feature of pluricausal, so-called coronary disease. Ann N Y Acad Sci 147:627–686

Samuels MA (2007) The brain-heart connection. Circulation 1:77–84

Selye H (1936) A syndrome produced by diverse nocuous agents. Nature 138:32

Selye H, Bajusz E (1958) Notes on stress studies in cardiology: cardiac necrosis and its prevention. Schweiz Med Wochenschr 88(46):1147–1155

Slaton JW, Perrotte P, Inoue K, Dinney CP, Fidler IJ (1999) Interferon-mediated down-regulation of angiogenesis- related genes therapy of bladder cancer are dependent on optimization of biological dose and schedule. Clin Cancer Res 5(10):2726–2734

Steiner HJ, Weiler R, Ludescher C, Schmid KW, Winkler H (1990) Chromogranins A and B are colocalized with atrial natriuretic peptides in secretory granules of rat heart. J Histochem Cytochem 6:845–850

Stojanovic MO, Ziolo MT, Wahler GM, Wolska BM (2001) Anti-adrenergic effects of nitric oxide donor SIN-1 in rat cardiac myocytes. Am J Physiol Cell Physiol 281(1):C342–C349

Takiyyuddin MA, Parmer RJ, Kailasam MT, Cervenka JH, Kennedy B, Ziegler MG, Lin MC, Li J, Grim CE, Wright FA, O'Connor DT (1995) Chromogranin A in human hypertension. Influence of heredity. Hypertension 26(1):213–220

Tota B, Angelone T, Mazza R, Cerra MC (2008) The chromogranin A-derived vasostatins: new players in the endocrine heart. Curr Med Chem 15(14):1444–1451

Tota B, Cerra MC, Gattuso A (2010) Catecholamines, cardiac natriuretic peptides and chromogranin A: evolution and physiopathology of a 'whip-brake' system of the endocrine heart. J Exp Biol 213(Pt 18):3081–3103

Tota B, Angelone T, Cerra MC (2014) The surging role of Chromogranin A in cardiovascular homeostasis. Front Chem 2:64

Triposkiadis F, Karayannis G, Giamouzis G, Skoularigis J, Louridas G, Butler J (2009) The sympathetic nervous system in heart failure physiology, pathophysiology, and clinical implications. J Am Coll Cardiol 54(19):1747–1762

Veschini L, Crippa L, Dondossola E, Doglioni C, Corti A, Ferrero E (2011) The vasostatin-1 fragment of Chromogranin A preserves a quiescent phenotype in hypoxia-driven endothelial cells and regulates tumor neovascularization. FASEB J 11:3906–3914

Weiergräber M, Pereverzev A, Vajna R, Henry M, Schramm M, Nastainczyk W, Grabsch H, Schneider T (2000) Immunodetection of alpha1 E voltage-gated Ca (2+) channel in chromogranin-positive muscle cells of rat heart, and in distal tubules of human kidney. J Histochem Cytochem 6:807–819

Yoo SH, Lewis MS (1996) Effects of pH and Ca2 on heterodimer and heterotetramer formation by chromogranin A and chromogranin B. J Biol Chem 271(29):17041–17046

Cardiac Physio-Pharmacological Aspects of Three Chromogranin A-Derived Peptides: Vasostatin, Catestatin, and Serpinin

Tommaso Angelone, Bruno Tota, and Maria Carmela Cerra

Abstract The discovery of a cardiac production of Chromogranin A (CgA) opened a rich field of research whose result was a cardiovascular dimension of this multifunctional protein and the fragments derived by its proteolytic cleavage. In line with its precursor function, at the cardiac level, CgA undergoes proteolytic processing. Moreover, as shown by ex vivo functional studies, it acts on the heart itself. Through the amino terminal (vasostatin: VS) and catestatin (CST) domains, CgA elicits a direct cardiodepressive, antiadrenergic and cardioprotective influence, acting as a cardiac stabilizer under normal conditions and in the presence of stress (i.e. catecholaminergic, and endothelinergic). At the same time, through the C-terminal Serpinin (Serp), CgA elicits a cardiostimulatory beta-adrenergic-like action. In this chapter, the results of more than 10 years of research will be described in order to illustrate the cardiovascular profile of these peptides, as well as their mechanisms of action. The purpose is to highlight the expanding role this protein and its fragments in the neuroendocrine circuits that finely control heart performance in health and disease.

Abbreviations

AD	Adenylate Cyclase
Akt	Protein Kinase B
Au	Aurum
CA	Catecholamine
CaM	Calmodulin
cAMP	Cyclic Adenosine Monophosphate
CgA	Chromogranin A
cGMP	Cyclic Guanosine Monophosphate
Chr	Chromofungin

T. Angelone, PhD (✉) • B. Tota • M.C. Cerra
Department of Biology, Ecology and Earth Sciences, University of Calabria,
87036 Arcavacata di Rende (CS), Italy
e-mail: tommaso.angelone@unical.it; bruno.tota@unical.it; maria_carmela.cerra@unical.it

CO	Cardiac Output
CP	Coronary Pressure
CST	Catestatin
ECM	Myocardial Extracellular Matrix
EE	Endocardial endothelium
ELISA	Enzyme-Linked Immuno Assay
eNOS	Endothelial Nitric Oxide Synthase
ERK1/2	Extracellular Signal–regulated Kinase 1/2
ET-1	Endothelin 1
GC	Guanylate Cyclase
GPCR	G protein-coupled receptor
GSK3β	Glycogen Synthase Kinase 3 beta
HF	Heart Failure
HPLC	High-performance liquid chromatography
HR	Heart rate
hr	human recombinant
Hsp90	Heat Shock Protein 90
HTR	Half Time Relaxation
I/R	Ischemia/Reperfusion
IGF-1	insulin-like growth factor-1
iPLA2	Calcium Independent Phospholipase A_2
IS	Infarct Size
ISO	Isoproterenol
LDH	Lactate Dehydrogenase
L-NAME	Nω-Nitro-L-arginine methyl ester hydrochloride
LVP	Left Ventricular Pressure
miRNA	microRNA
mitoKATP Channel	ATP-Sensitive Potassium Channel
nAChR	Nicotinic Receptor
NO	Nitric Oxide
ODQ	1H-[1,2,4]oxadiazolo[4,3-a]quinoxalin-1-one
PDE2	Phosphodiesterases type 2
pGlu-Serp	pyroglutaminated-Serpinin
PI3K	Phosphatidylinositide 3-Kinase
PKA	Protein Kinase A
PKG	Protein Kinase G
PLN	Phospholamban
PTIO	2-Phenyl-4,4,5,5-tetramethylimidazoline-1-oxyl 3-oxide
PTX	Pertussis Toxin
RISK	Reperfusion Injury Signalling Kinase
ROS	Radical Oxygen Species
RPP	Rate Pressure Product
RyR	Ryanodine receptor
Serp	Serpinin
SHR	Spontaneously Hypertensive Rats

SR   Sarcoplasmic Reticulum
VE   Vascular Endothelium
VS   Vasostatin
WKY   Wistar Kyoto
WT   Wild-Type
β-ARs   β-adrenergic receptors

1 Introduction

The finding that Chromogranin A (CgA) is a substrate for proteolytic cleavage (Helle et al. 2007), and is processed to generate biologically active fragments, provided the basis for a better functional characterization of this ubiquitous endocrine protein (Helle et al. 2007). Due to the presence of multiple pairs of consecutive dibasic and monobasic residues along the amino acid sequence (Helle et al. 2007), CgA is enzymatically fragmented in several short peptides which contribute to the endocrine/paracrine networks that regulate cell, tissue and organ function. These fragments, which derive from either the N terminus, or the central region, or the C terminus, are produced in a tissue-specific manner following appropriate stimuli (Tota et al. 2014) (Fig. 1). They play a significant role in processes such as inflammatory and cardiovascular reactions, innate immunity, energy metabolism, and calcium homeostasis, and are under examination as potential novel clinical biomarkers (Schneider et al. 2012).

Within the specific cardiocirculatory context, several CgA-derived peptides showed intriguing properties, being able to modulate cardiac and vascular function under normal conditions and in the presence of pathological alterations. This is the case of Vasostatin (VS)-1, a potent vasodilator and cardioinhibitory agent

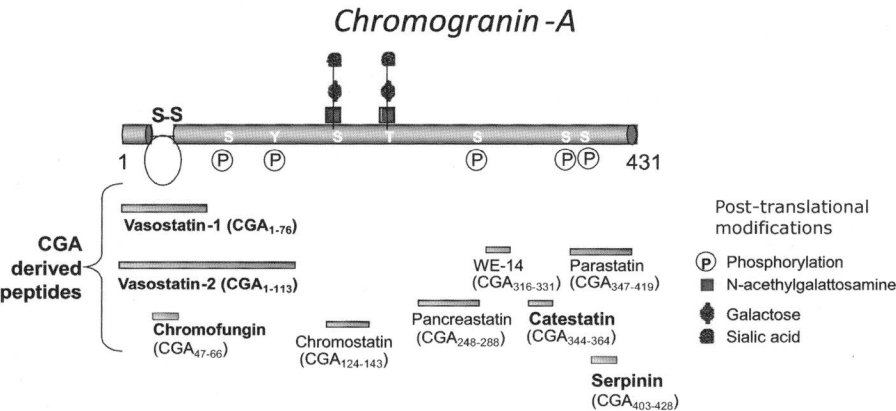

Fig. 1 Schematic illustration of CGA processing (Modified by Tota et al. 2008)

(Cerra et al. 2008), Catestatin (CST), a catecholamine (CA) release inhibitor with antihypertensive and cardioprotective properties (Angelone et al. 2008; Penna et al. 2010), and Serp, a sympathomimetic cardiostimulating peptide (Tota et al. 2012). With reference to the focus of this chapter, we will illustrate the physiopharmacologic cardiovascular properties of these CgA-derived peptides, as well as the molecular cascades that they recruit to modulate the mammalian heart performance under basal conditions and in relation to other endocrines, such as CA and Endothelin-1 (ET-1).

2 Vasostatin-1 as Cardiac Stabilizer

CgA cleavage of at the N- terminus generates fragments of different length, collectively known as Vasostatins (VSs) because of their ability to relax vascular smooth muscle pre-contracted by Endothelin-1 (Aardal et al. 1993). Studies from our research group showed on several animal models (rat, frog and eel) that VSs [VS-1: human recombinant (hr) CgA1-78 (hrVS-1), rat CgA1-64; Vasostatin-2 (hrCgA1-115: hrVS-2)] influence heart performance. The isolated and Langendorff perfused rat heart was also used to better describe the influence of VSs on myocardial contractility and relaxation. It was found that hrVS-1 (containing the rat CgA1-76 sequence) acts as a negative inotrope which reduces contractility and cardiac work in a dose (11–165 nM)-dependent manner (Cerra et al. 2006). These effects are shown by the decreased left ventricular pressure (LVP) and rate pressure product (RPP: HR × LVP), indexes of contractility and work, respectively. In addition, the peptide depresses myocardial relaxation (negative lusitropic effects), by dose-dependently decreasing the maximal rate of the left ventricular pressure decline [(LVdP/dt)min], the half time relaxation (HTR), and T/-t ratio obtained by (LVdP/dt)max/(LVdP/dt)min (Pieroni et al. 2007). Of note, cardiodepression is independent from coronary vasomotility, since the administration of the CgA fragment unaffected Coronary Pressure (CP). Comparable, although less potent, myocardial effects are induced by hrVS-2 that, however, increased CP at 110 and 165 nM (Cerra et al. 2006).

In parallel with the basal negative inotropic and lusitropic effects, VSs counteract the positive inotropism induced by activation of β-adrenergic receptors (β-ARs) by Isoproterenol (ISO) without modifying the β-AR-dependent coronary dilation. This counterbalancing action occurs via a non-competitive type of antagonism, as shown by the percentage of RPP variations evaluated in terms of EC50 values of ISO alone and in presence of hrVS-1 (Cerra et al. 2006). The role of VSs as cardiodepressive peptides in the presence also of adrenergic stimulation is corroborated by evidence obtained by exposing the ex vivo perfused rat heart and papillary muscles to the highly conserved (Metz-Boutigue et al. 1993; Helle et al. 2007) N-terminal native (rat) CgA1-64 (rCgA1-64) fragment (Cerra et al. 2008). The peptide, at concentrations from 33 to 165 nM, similar to those of the precursor CgA in human serum (normal levels: 0.5–4 nM, neuroendocrine tumors and last stages of chronic HF: >10 nM; Helle et al. 2007; see for details Tota and Cerra, Chap. 7, present volume), elicits significant negative inotropic and lusitropic effects without changing HR. Notably, rCgA1-64 elicits coronary activity, since it significantly reduces CP. This effect, obtained on the intact coronary bed, supports the "vasostatin" (*ad litteram*) behavior of this CgA domain, already proposed according to the vasodilation observed on segments of bovine coronary resistance arteries, intrathoracic artery and saphenous vein exposed to the hrCgA 1-78 active domain 1–40 (Brekke et al. 2002).

An interesting characteristic of rCgA1-64 is its counteraction against the stimulatory effect on contractility induced by ISO and ET-1, as well as against the potent ET-1-induced coronary constriction. Experiments on rat isolated papillary muscles, an experimental model in which contractility is analyzed without the influence of HR and coronary flow, confirmed the depressive activity of this CgA domain on basal and ISO-elicited contractility (Cerra et al. 2008). Similar to hrCgA1-78 (Cerra et al. 2006), the rat CgA fragment antagonized β-adrenergic stimulation in a non-competitive manner (Fig. 2). This is shown by the analysis of the percentage of variations of LVP, which provides the EC50 values in the presence of either increasing concentrations of ISO alone or of ISO plus rCgA1-64 (11, 33 and 65 nM) (Fig. 2). Interestingly, as proposed on the bases of results obtained by measuring chronotropism on in situ atria from anesthetized adult dog, the negative lusitropic effect induced by the VSs peptides, together with their antiadrenergic effects, can be associated to alterations of the sinus rate. This may also have clinical relevance since it has been associated to a prolongation of refractoriness and a beneficial stabilizing effect in the presence of arrhythmias (Stavrakis et al. 2012).

The similarity between hrCgA1-78 and rCgA1-76 is extended to the signal transduction mechanism involved in the cardiac effects of the two peptides. In fact, as in the case of hrCgA1-78 (Cappello et al. 2007), the rat fragment signals via a $G_{i/o}$ protein-PI3K-NO-cGMP-PKG-dependent pathway (Cerra et al. 2008). As observed on isolated rat papillary muscles, the mechanism of action implicates a calcium-independent/PI3K-dependent NO release by endothelial cells. In addition, on rat isolated ventricular cells, rCgA1-64 unaffects intracellular calcium concentrations. At the same time, it induces NO release from cultured bovine aortic endothelial cells (BAE-1) via a calcium-independent mechanism (Cerra et al. 2008). Since the coronary dilation induced by rCgA1-64 is abolished by inhibiting the NO-cGMP-PKG pathway, it is possible that the endothelium contributes with the release of vasodilator autacoids, such as NO. It remains unexplored whether also the endocardial endothelium contributes to these NO-mediated mechanisms.

It is extensively documented that myocardial contractility is under the tonic control of the NO/NOS system. On the rat myocardium, NO induced negative inotropism via a sGC-PKG mechanism which decreases L-type Ca^{2+} current (Abi-Gerges et al. 2001) and troponin I phosphorylation (Hove-Madsen et al. 1996). Activation of the endothelial NOS isoform (eNOS) without calcium involvement takes place after stimulation of the endothelium with several humoral mediators such as estrogens, insulin, and insulin-like growth factor-1 (IGF-1) (Hartell et al. 2005), possibly

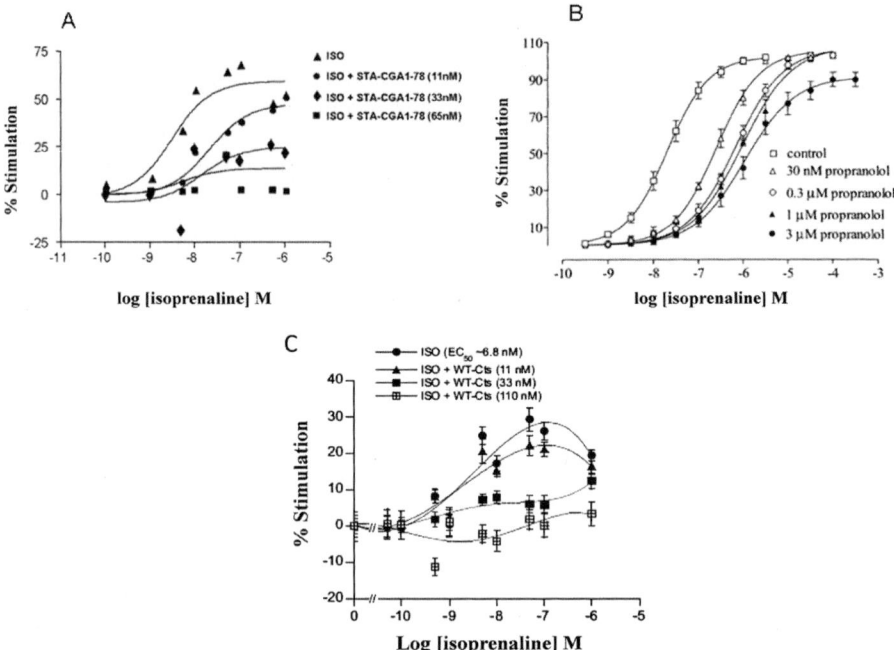

Fig. 2 The sigmoid concentration-response curves of Isoprenaline (ISO)-mediated stimulation (10-10-10-6 M) alone (**a–c**) and of ISO plus a single concentration of human VS1 (STA-CGA1-78) at 11, 33, and 65 nM (**a**) and of ISO plus a single concentration of propranolol at 30 nM, 0.3 μM, 1 μM and 3 μM (**b**) and ISO plus a single concentration of Catestatin (WT-Cst) at 11, 33, and 110 nM (**c**). Concentration is expressed as a percentage [baseline = 0%, peak constriction by ISO and ISO plus VS1 or propranolol or CST = 100%] (Modified by Tota et al. 2008 and Angelone et al. 2008)

through the involvement of Akt dependent eNOS phosphorylation (Shaul et al. 2002). In line with this scenario, experiments on papillary muscles and BAE-1 cells exposed to rCgA1-64, and treated with the PI3K inhibitor Wortmannin, indicate that the NO release induced by the rat peptide depends on PI3K activation (Cerra et al. 2008). Interestingly, Maniatis et al. (2006), proposed that eNOS may be activated in a calcium-independent manner by involving a caveolae-mediated endocytosis elicited by the albumin-binding protein gp60 and activation of downstream Src, Akt and PI3K pathways. It has been hypothesized that VSs interacts with caveolar domain (see for references, Tota et al. 2007), and that endothelial cells internalize CgA1-78 (Ferrero et al. 2004). Therefore, a similar mechanism may explain the VS-1-dependent NOS activation in BAE-1 cells. However, this remains an open field for investigations (Fig. 3).

In the search of possible mechanisms of interaction between VSs peptides and the cell membrane, and in the absence of either a conventional receptor or an action

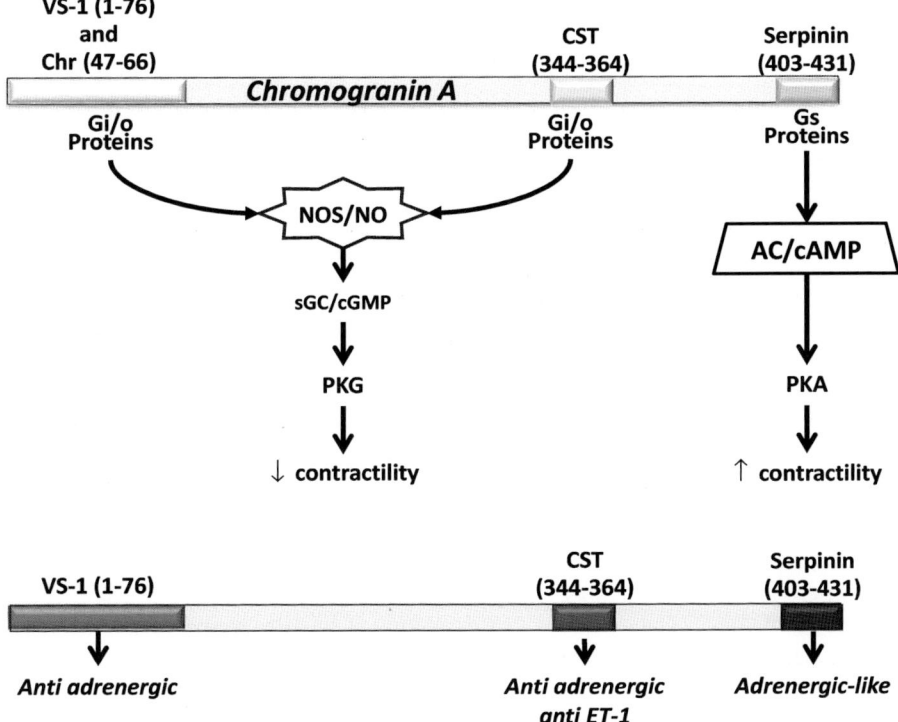

Fig. 3 Representative scheme showing the physiological pathways activated by CgA-derived peptides. Cardioinhibitory effects induced by Human VS1 (1–76), Chromofungin (47–66) and CST involve the NOS/NO/cGMP/PKG pathway. Serpinin-dependent positive inotropism involves the AD/cAMP/PKA pathway (Modified by Tota et al. 2014)

site of VS-1 on the membrane, Di Felice and collaborators (2006) used cardiomyocytes three-dimensionally cultured in Matrigel (myocardial extracellular matrix: ECM) exposed to Aurum conjugated hrCgA1−78 (Au-hrCgA1−78). They observed by immunoglod technique that Au-hrCgA1−78 exclusively localizes outside the plasma membrane of cardiomyocytes, at a distance between 16 and 25 nm. This distance is characteristic of the interactions between ECM proteins and the cell. This suggests that in vivo natural VS-1 acts via cell–ECM interactions. Two putative domains of the human recombinant fragment could be considered for its binding to either the cell membrane or ECM components: an RGD sequence at residues 43–45 (Gasparri et al. 1997) and a net positively charged domain at residues 47–70 (Mandalà et al. 2005). However, the RGD site of CgA is not conserved among different species and its involvement in the regulation of cell–ECM interactions remains to be proved.

2.1 Cardiac Properties of the Vasostatin-Derived Chromofungin

Within the VS-1 domain, a cardiovascular function was recently described for the fragment 47–66. It was identified by Lugardon and collegues in 2001, and named chromofungin (Chr) for its antifungal activity, was identified by Lugardon and collegues (Lugardon et al. 2001). This fragment generates during infections by cleavage by *Staphylococcus aureus* protease Glu-C (Metz-Boutigue et al. 1993). It acts as an immediate protective shield against pathogens (Metz-Boutigue et al. 1998), being able to inhibit microbial cell metabolism (Bartizal et al. 1992), and to penetrate the cell membrane, thus inducing extracellular Ca^{2+} entry by a CaM-regulated iPLA2 pathway (Zhang et al. 2009). According to data found on the in vitro isolated working heart of the frog *Rana esculenta*, Chr also possesses cardioactive properties (Tota et al. 2003; for details, see Gattuso, Imbrogno, Mazza, Chap. 9, present volume). In fact, the Chr sequence was found to depress frog myocardial contractility by eliciting a direct negative inotropic effect (Tota et al. 2003). Very recently, Chr cardiovascular actions were extended also to mammals. By using the isolated and Langendorff perfused rat heart, Filice and co-workers (2015) investigated the influence elicited by Chr on the basal cardiac performance and in the presence of myocardial I/R damage. Data showed that exogenous Chr directly affects the heart by dose-dependently reducing contractility under unstimulated conditions. It reduces LVP, and (LVdP/dt)max (indexes of inotropism), without affecting HR and CP. These effects are obtained by Chr concentrations close to the physiological range of the precursor, CgA, in human serum (Helle et al. 2007). With respect to the data reported on the frog heart (Tota et al. 2003), Filice et al. (2015) demonstrated in the rat that Chr depresses not only contractility (~40% rat vs ~18% in frog), but also relaxation without changing heart rate. It in fact reduces two lusitropic indexes, (LVdP/dt)min and T/-t. Negative inotropic and lusitropic effects induced by Chr (~40%) resemble the cardiodepression elicited on the rat heart by human recombinant (hrCgA1-78) (~20%) which includes the Chr sequence (Cerra et al. 2006). This further stresses the depressive myocardial properties of the CgA N-terminal domain. However, Chr did not change rat coronary reactivity. Although VSs and Chr elicit similar effects on myocardial contractility and relaxation, sequence-specific vascular activities may account for the observed differences in coronary responses. Of note, a strong structure-function relationship characterizes CgA-derived peptides, suggesting that different fragments may display different cardioactivity.

As in the case of the other N-terminal CgA fragments, the presence of receptor-ligand interactions which mediate the cardiac effects of Chr remains elusive. Of note, Metz-Boutigue et al. (2003) reported that Chr exerts antibacterial and antifungal effects by penetrating into, or interacting with, the cell membrane through hydrophobic interactions between specific domains of the peptides and spatially localized regions of the lipid bilayer with consequent modulation of cellular effectors. In particular, an increase of peptide penetration is observed for the presence of ergosterol, the main sterol in yeast and fungus plasma membrane (Metz-Boutigue et al. 2003).

The study by Filice et al. (2015) revealed that, as full length CgA (Pasqua et al. 2013, see for details Tota, Cerra, Chap. 7, present volume), VSs and CST (Cappello et al. 2007; Angelone et al. 2008), the cardiac effects induced by Chr involve the AkT/NOS-NO/cGMP signal transduction pathway. On the rat heart, exposure to Chr increased AkT and eNOS phosphorylation, thus activating the NO-generating cascade. This promotes cGMP production by soluble Guanylate Cyclase (sGC), with final depressant effects on contractility and relaxation (Fig. 3).

Of relevance, Chr also protects the myocardium against I/R injury, acting as a Post conditioning agent (Filice et al. 2015). Given in the early reperfusion, the peptide limits the I/R-dependent myocardial damage. This protection, similar to that obtained by ischemic Post conditioning maneuvers (Vinten-Johansen et al. 2005), is indicated by a significant reduction of infarct size and LDH release. Ischemic hearts exposed to Chr (75 nM) also show a marked improvement of the post-ischemic contractile function expressed as a decrease of contracture development, which is the goal of cardioprotective protocols, due to its inverse relation with the I/R-dependent myocardial damage (Penna et al. 2008). In line with the observed cardioprotection, the inhibition of the PI3K and ERK1/2, upstream kinases of the protective RISK cascade, abolished the systolic recovery induced by Chr. During reperfusion, both PI3K-Akt and ERK1/2 are activated and converge on GSK3β, inducing its phosphorylation/inactivation with a final control on mitoKATP channels, one of the terminal elements of Post conditioning protection (Gomez et al. 2008). Of note, inhibition of mitoK_{ATP} channels abolished myocardial cardioprotective induced by Chr (Filice et al. 2015). Chr-dependent cardioprotection is accompanied by increase of intracellular cGMP, an effect which disappears in hearts co-treated with ODQ, specific GC inhibitor. The importance of cGMP in cardioprotection is well known (Penna et al. 2006). In fact, Post conditioning depends on GC activation via either NOS-dependent or NOS-independent pathways (Penna et al. 2006, 2008).

Lastly, Chr-dependent cardioprotection is accompanied by an increased miRNA-21 expression. This is of relevance since miRNA are small noncoding RNAs that mediate post-transcriptional gene silencing, involved in cardiac physiopathology. Their deregulated expression is linked to the developement of cardiovascular disorders, including infarct size (Zhang 2008; Dong et al. 2009).

3 Cardiac Actions of Catestatin

The CgA352-372 fragment CST, known as a potent endogenous inhibitor of nicotinic receptor (nAChR)-evoked CA secretion (Mahata et al. 1997), is a multifunctional peptide with multiple targets and a relevant vasoactive and anti-hypertensive properties (for ref. Mahata et al. 2010). In mammals, it is present as both wild type peptide (WT-CST, human CgA352-372) and naturally occurring variants (G364S-CST and P370L-CST). As shown on the ex vivo isolated and Langendorff perfused rat heart, wild-type CST, G364S-CST, and P370L-CST affect heart function, also in the absence of the control exerted by SAN over cardiac output (CO) and vascular tone. CST peptides dose-dependently (from 11 to 200 nM) reduce contractility (i.e. LVP, +LVdP/dt), relaxation (−LVdP/dt), and cardiac work (i.e. RPP), while stimulate coronary activity as shown by the increased CP (Angelone et al. 2008). Differently from wild-type CST, P370L-CST only reduces contractility, while G364S-CST is ineffective on the basal mechanical performance. All CST variants act against β-adrenergic (ISO)-induced positive inotropic and lusitropic effects, showing a rank order of potency (for ISO-induced positive inotropism: WT-CST > G364S-CST > P370L-CST; for ISO-induced positive lusitropism: G364S-CST > WT-CST > P370L-CST (Angelone et al. 2008) (Fig. 2). Notably, on both the isolated and Langendorff perfused rat heart (Angelone et al. 2008) and rat papillary muscle (Bassino et al. 2011), CST-evoked major depressive actions are preceded by an early transient positive inotropic effect that disappears 5 min after peptide administration. This positive inotropism disappears if H1 histamine receptors are inhibited, thus suggesting a role for these receptors. This is consistent with previous evidence in rat that activation of myocardial H1 receptors mediates histamine-dependent positive inotropism (Matsuda et al. 2004).

Physio-pharmacologic investigations show that CST-induced negative inotropism and lusitropism involve β2-AR and, with lower affinity, β3-AR, but not β1-AR and cholinergic receptors (Angelone et al. 2008). Interestingly, CST-evoked cardiodepression recruits pertussis toxin (PTX)-sensitive mechanisms, thus calling for a role of Gi/o proteins. It is known that, contrary to the Gs-mediated β1-AR stimulatory effect on contractility and relaxation, β2-AR elicit negative inotropism and lusitropism via Gi/o proteins (Xiao et al. 1999). Thus, the effects induced by CST through a β2-AR/$G_{i/o}$-dependent inhibitory pathway may limit an early Gs-mediated stimulation of the mechanical contractile performance. In the heart, activation of the β2-AR/$G_{i/o}$ cascade recruits PI3K to elicit negative inotropism (Yano et al. 2007). PI3K is an important component of the CST signal transduction mechanism. In fact, exposure of the rat heart to inhibitors of this kinase abolishes the effects induced by the peptide (Angelone et al. 2008, 2012). In the heart, downstream PI3K is the NO/sGC/cGMP cascade (Abi-Gerges et al. 2001). This cascade is involved in the CST-dependent signaling (Angelone et al. 2008, 2012). In fact, in hearts treated with the peptide, cGMP significantly increases and this in turn affects two of its major targets, PKG and phosphodiesterases type 2 (PDE2). Thus, these enzymes may contribute to the CST-dependent depression of myocardial contractility. Via PKG, the peptide may reduce both L-type Ca^{2+} current and troponin C affinity for calcium (Angelone et al. 2008, 2012). Other mediators of the CST-dependent effects are beta-arrestin and phospholamban (PLN) which undergo S-Nitrosylation in the presence of the peptide (Angelone et al. 2012). Beta-arrestin is implicated in G-protein coupled receptors desensitization and internalization by clathrin coated vescicles. When S-nitrosylated, it promotes and accelerates β1-AR desensitization, thus blunting the effects induced by activation of this adrenoceptor (Ozawa et al. 2008). In line with the counter-adrenergic behaviour of CST, the stimulation of beta-arrestin desensitization, via NO-dependent protein S-Nitrosylation, may be of relevance to elicit a more prolonged action of the peptide. This action, in addition to the short-term effects obtained by recruiting kinases, such as PDE2, highlights the presence of a sophisticated double switched counter regulation induced by CST against cardiac beta-adrenergic effects.

In the heart, PLN controls sarcoplasmic reticulum Ca^{2+}-ATPase (SERCA2a) by a phosphorylation/dephosphorylation mechanism. Dephosphorylated PLN inhibits SR Ca^{2+} sequestration via SERCA2a (Reddy et al. 1999), while PLN, phosphorylated at Ser16 by PKA, relieves its inhibition on SERCA2a (Schmidt et al. 2001). In addition to phosphorylation, PLN may be S-nitrosylated. This influences stretch-induced contractile effects on the myocardium (see Garofalo et al. 2009). Accordingly, S-nitrosylated PLN represents a selective downstream target of the NO release induced by CST, which contributes to the modulation of inotropy and lusitropy by regulating SR Ca^{2+} fluxes and Ca^{2+} availability for the contractile apparatus.

Another trait of the signal-transduction mechanism activated by CST and its variants was described by Bassino et al. (2011) by measuring Ca^{2+} transients and myocardial contractility on isolated cardiomyocytes and papillary muscles, respectively. On BAE-1 cells they also evaluated, the effects of CST NO production and eNOS phosphorylation at Ser 1179 ($P^{Ser1179}$eNOS) showing that CST (5–50 nM) reduces in a dose-dependent manner the effects of beta-adrenergic stimulation. Since CST induces a Wortmannin-sensitive, Ca^{2+}-independent increase of NO production and $P^{Ser1179}$eNOS, it is presumable that the effects of the peptide are due to a NO release from endocardial endothelial cells, rather than to a direct myocardial action.

Similar to the wild-type peptide, P370L-CST, but not G364S-CST, induces an anti-adrenergic action and stimulates NO release (Angelone et al. 2008). Interestingly, the Ca^{2+}-independent, caveolae-dependent activation of the Akt-PI3K-eNOS pathway elicited by CST is similar to that suggested for the negative inotrope, anti-adrenergic N-terminal VS-1 (Ramella et al. 2010 and references therein), highlighting common features in the cellular mechanisms regulated by these CgA fragments.

As shown on the rat heart, CST induces a coronary modulation. This coronary activity is non-univocal. Under basal conditions, the peptide dose-dependently increases CP (maximum response at 200 nM) and abolishes the ISO-dependent vasodilation. Contrarily, but in line with the vasodilation observed in humans after CST administration (Fung et al. 2010), it potently dilates coronary vessels pre-constricted by ET-1 (Angelone et al. 2008). The cross-talk ET-1/CST, corroborated also by studies carried out on the frog heart (Mazza et al. 2012), together with the interaction with β adrenoceptors, is consistent with a CST signalling mechanism which starts at the plasma membrane before to converge on the sympatho-inhibitory NO pathway (Fig. 3).

The data here illustrated, together with those reported on other chapters of this volume, propose the negative effects induced by CST on myocardial contractility and relaxation, as well as its coronary vasomotion, as relevant components of the mechanisms that sustain the homeostatic counteraction against sympathetic over-activation. This is a peculiar trait of conditions (e.g., prolonged stress, HF, hypertensive cardiomyopathy, etc.) which are characterized by a potentially harmful spill-over of CAs, ET-1 and other humoral effectors. Accordingly, the cardiac, vascular and antihypertensive properties of CST, suggest the peptide as an autocrine-paracrine modulator that cooperates with full-length CgA and its derived VS-1 in the multilevel processes required for cardio-circulatory homeostasis under healthy and diseased conditions. This is opening to researches of applicative interest, as those aimed to generate CST analogues able to mimic the cardiovascular actions of the peptide, as in the case of the pharmacophore synthesized by Tsigelny et al. (2013). This may be of relevance to enhance the possibilities for clinical treatment of diseases in which an altered Sympatho-Adrenal Neuroendocrine equilibrium is accompanied by a deterioration of the cardiovascular function.

3.1 Myocardial Stretch

Recently, the cardiac actions of CST were extended to the regulation of the myocardial response to stretch (i.e. the Frank-Starling mechanism) (Angelone et al. 2015). It was observed that, on the isolated and Langendorff perfused heart of the normotensive Wistar Kyoto (WKY) rat, CST significantly enhances the preload-induced increases of LVP and (LVdP/dt)max. It potentiates the diastolic response, as shown by the incremented (LVdP/dt) min. Notably, CST improves the myocardial response to stretch (i.e. increased end-diastolic pressures). This occurs not only in normotensive heart, but also in 18-months old Spontaneously Hypertensive Rats (SHR). This is relevant since at 18 months, the SHR heart shows traits that are characteristic of the developing HF, including depressed contractility, ventricular myocardial fibrosis, and reduced Frank-Starling response (Bing et al. 1995). In this context, the potentiation induced by CST on the myocardial response to stretch may be advantageous for the failing heart since it alleviates the functional damage, being also of benefit against HF progression.

The mechanism of action recruited by CST in both normotensive and hypertensive rat to potentiate the Frank-Starling relationship involves the Vascular Endothelium (VE), the AkT/NOSs/NO/cGMP/PKG cascade and is accompanied by an increment in protein S-Nitrosylation. In both WKY and SHR rats the positive effect elicited by CST on the Frank-Starling response was abolished by the functional impairment of the VE induced by perfusion with Triton X-100. Also the exposure to the eNOS inhibitor, L-NAME, and the NO scavenger, PTIO, abolished CST effects on the Frank-Starling response of normotensive WKY and hypertensive SHR. In addition, the effect induced by CST was abolished by specific inhibition of sGC by ODQ and was accompanied by an increase of intracellular cGMP levels.

Together with the Endocardial Endothelium, the VE is an important source of paracrine NO produced by eNOS, while myocytes generates autocrine NO through eNOS and nNOS located within membrane caveole, and in proximity of the SR, respectively (Petroff et al. 2010; Perrelli et al. 2013; Hammond and Balligand 2012). The result is the optimization of LV systolic and diastolic functions, not only under basal conditions, but also in response to stretch (Prendergast et al. 1997).

This is of paramount importance in disease states characterized by LV diastolic dysfunction, including LV hypertrophy and chronic HF. In the failing heart of aged SHR, characterized by contractile and relaxing alteration (Han et al. 2014) the mechanisms for NO generation are impaired because of a dysfunctional endothelium, and a decreased expression of eNOS (Bayraktutan et al. 1998; Zhang et al. 2008). Under these circumstances, a reduced NO availability, and the consequent limitation of myocardial coronary perfusion, elicits detrimental effects on the hemodynamic performance (Bayraktutan et al. 1998).

It is known that NO mediates the inotropic response to sustained stretch through a mechanism that involves S-Nitrosylation of target proteins that include G protein-coupled receptors, Hsp90 (a chaperone involved in eNOS activation), mitochondrial pro-apoptotic and anti-apoptotic proteins (see for references Angelone et al. 2015). Notably, S-Nitrosylation also occurs on SR RyR channels, potassium channels, and L-type calcium channels, with important influences on calcium cycling and contractility (see for references Angelone et al. 2015). In the heart of both WKY and SHR, CST increases NO-dependent S-Nitrosylation mainly of high and low molecular weight proteins. In particular, an augmented S-Nitrosylation is observed at 100 kDa. This corresponds to the apparent molecular weight of dynamin, a GTPase which can be S-Nitrosylated by NO and is emerging in the heart as a regulator of trafficking and function of ion channels, including L-type calcium channels (Angelone et al. 2015).

Of note, CST-induced effect on the Frank-Starling behavior is of the same magnitude as that induced by NO. This supports the notion that this CgA fragment, also in the case of the humoral regulation of the ventricular heterometric response, functions as endogenous NOS activator. In the presence of a deteriorated function, as in the failing heart of aged SHR, this CST-dependent NO production may represent an advantage to the diseased heart since it counteracts the cardiac damage induced by hypertension.

4 Cardiac Actions of Serpinin

CgA cleavage at the penultimate and the last pair of basic residues at C-terminus (CgA403-428) generates a short fragment called Serp (Koshimizu et al. 2010, 2011a). It was described for the first time in mouse pituitary cell line (AtT-20) in which stimulates granules biogenesis (Kim and Loh, 2006). More recently, HPLC and ELISA techniques revealed the presence of Serp peptides in the mammalian (rat) heart (Tota et al. 2012). Three naturally occurring Serp peptides are expressed by the rat cardiac tissue, namely Serp (Ala26Leu), and the two predominant fragments: pyroglutaminated Serp (pGlu-Serp) and a C-terminal extended form, Serp-Ala29Gly (Tota et al. 2012). This suggests that the rodent heart is able to process the C terminal domain of CgA, as it does with the N terminal domain. Exposure of the Langendorff perfused rat heart showed that Serp peptides modulate cardiac performance but, different from the cardiodepressive VS-1 and CST, they stimulate the hemodynamic function (Tota et al. 2012). Within the first 5 min of administration, both Serp and pGlu-Serp dose-dependently (11–165 nM) enhance myocardial contractility and relaxation. pGlu-Serp action starts at 1 nM, showing that this peptide is more potent than Serp. These actions are accompanied by a slight, although non-significant, coronary dilatation (Tota et al. 2012). A similar effect is evident also on isolated rat papillary muscles. By measuring contractility as tension development and muscle length, we found that, while Serp-Ala29Gly is unaffective, pGlu-Serp and Serp induce positive inotropism by increasing the rate and extent of tension development during systole (positive inotropy). This effect, in the whole heart augments stroke volume. In parallel, the peptides accelerate myocardial relaxation (positive lusitropy), hence shortening the overall diastole duration. The analyses carried out on isolated papillary muscle also showed that Serp actions are independent from alterations in HR and coronary flow rate, as well as from norepinephrine release from sympathetic nerve terminals (as demonstrated by tyramine treatment) (Tota et al. 2012).

Interestingly, pGlu-Serp and Serp act as β1-AR-like agonists, mimicking, at nanomolar range, the effects of intracardiac sympathetic neurotransmitters and/or circulating Catecholamines. Consistent with this effect, and opposite to the depressive and anti-adrenergic cascade activated by VS-1 and CST, pGlu-Serp and Serp act via β1-AR/Adenylate Cyclase/cAMP/PKA (Tota et al. 2012) (Fig. 3). This is shown by the results of physio-pharmacologic analyses carried out by using specific inhibitors of the above cascade, by Western Blotting evaluations, and intracellular cAMP evaluation. As in the case of other CgA fragments, no receptor or direct binding partner have been identified for Serp and pGlu-Serp. Thus, the earliest events underpinning the transduction of the Serp signal are unknown. On the basis of physiological and biomolecular evidences, it was presumed that, to activate the β1-adrenergic-induced cascade Serp and pGlu-Serp function as allosteric modulators of β1-AR independent from the ligand binding site (Tota et al. 2012). As other cAMP elevating agonists (Koshimizu et al. 2010), both Serp peptides might bind a G protein-coupled receptor (GPCR) to increase cardiac cAMP. This is indicated by the lack of Serp-induced cardiostimulation after selective inhibition of adenylate cyclase (AD) and PKA, major targets of the AD-cAMP signalling (Tota et al. 2012). Once activated, PKA phosphorylates a number of proteins, including SERCA and PLN, with consequent effects on inotropy and lusitropy. In fact, SERCA activation promotes SR Ca^{2+} uptake, and thus cation removal during diastole. This affects relaxation and the subsequent contraction (Satoh et al. 2011). Ca^{2+} sensitivity for SERCA is enhanced also by PKA-induced phosphorylation of PLN, and this represents a crucial step of the β-adrenergic-PKA cascade (Mattiazzi et al. 2007). Of note, positive inotropism and lusitropism induced by pGlu-Serp are accompanied by an increased PNL phosphorylation at Ser16, and are abolished by SERCA inhibition by thapsigargin (Tota et al. 2012). This suggests, in agreement with the sympathomimetic function proposed for Serp peptides, that their action occurs by modulating myocardial Ca^{2+} fluxes with the resulting stimulation of inotropy and relaxation.

The cardiostimulation induced by Serp peptides is accompanied by an enhancement of ERK1/2 and GSK3β phosphorylation (Tota et al. 2012). In the mammalian heart, ERK1/2 and GSK3β, two targets of PKA phosphorylation, are components of the RISK (*reperfusion injury signalling kinase*) cascade involved in myocardial protection against ischemia-reperfusion injury (Hausenloy et al. 2011). They presumably contribute also to the anti-apoptotic effect induced by pGlu-Serp against radical oxygen species (ROS) in cultured cerebral neurons (Koshimizu et al. 2011b). Very recently, it was observed that pGlu-Serp reduces infarct size and preserves the hemodynamic function of normotensive and SHR hearts, given in pre- and post-conditioning (Pasqua et al. 2015; for details see Pagliaro & Penna, Chap. 11, present volume). In pre-conditioning, the cardioprotection elicited by the peptide is mild, but it is streaking in post-conditioning. Administered at the reperfusion, pGlu-Serp induced a more potent cardioprotection in SHR than in normotensive rats, as shown by the better post-ischemic hemodynamic recovery (evaluated as LVDevP) observed in WKY. In both SHR and WKY, the peptide also reduces contracture and the infarct size (IS). However, the effect of pGlu-Serp on IS is less pronounced in SHR, which show a larger area of damaged tissue. Analyses of the mechanism of action recruited by pGlu-Serp to elicit protection showed the involvement of the RISK cascade. In fact, co-infusion of the ischemic heart to pGlu-Serp and inhibitors of PI3K/Akt, MitoKATP channels and PKC abolished cardioprotection (Pasqua et al. 2015).

5 Conclusion

On the whole, results obtained in the last 10 years of research convincingly revealed the cardiovascular activities of the CgA-derived VSs, CST and serpinin, along with their striking adreno-sympathetic regulatory influences. The cardiovascular role of the CgA-derived peptides represents an important component of the finely integrated local and systemic neuroendocrine networks, particularly in the presence of a perturbed cardio circulatory homeostasis. This knowledge, which has widened the prohormone/cytokine profile of CgA, has been paralleled and integrated by a growing number of clinical studies on the biomedical implications of both full length protein and its derived peptides - particularly the antihypertensive CST - in various cardiovascular diseases, in relation to their diagnostic and prognostic value (see other chapters of the present book).

References

Aardal S, Helle KB, Elsayed S, Reed RK, Serck-Hanssen G (1993) Vasostatins, comprising the N-terminal domains of chromogranin A, suppress tension in isolated human blood vessel segments. J Neuroendocrinol 5:105–112

Abi-Gerges N, Fischmeister R, Meiry PF (2001) G protein-mediated inhibitoryeffect of a nitric oxide donor on the L-type Ca2+ current in rat ventricularmyocytes. J Physiol 531:117–130

Angelone T, Quintieri AM, Brar BK, Limchaiyawat PT, Tota B, Mahata SK, Cerra MC (2008) The antihypertensive chromogranin a peptide catestatin acts as a novel endocrine/paracrine modulator of cardiac inotropism and lusitropism. Endocrinology 149(10):4780–4793

Angelone T, Quintieri AM, Pasqua T, Gentile S, Tota B, Mahata SK, Cerra MC (2012) Phosphodiesterase type-2 and NO-dependent S-nitrosylation mediate the cardioinhibition of the antihypertensive catestatin. Am J Physiol Heart Circ Physiol 302(2):H431–H442

Angelone T, Quintieri AM, Pasqua T, Filice E, Cantafio P, Scavello F, Rocca C, Mahata SK, Gattuso A, Cerra MC (2015) The NO stimulator, Catestatin, improves the Frank-Starling response in normotensive and hypertensive rat hearts. Nitric Oxide 50:10–19

Bartizal K, Abruzzo G, Trainor C, Krupa D, Nollstadt K, Schmatz D, Schwartz R, Hammond M, Balkovec J, Vanmiddlesworth F (1992) In vitro antifungal activities and in vivo efficacies of 1,3-beta-D-glucan synthesis inhibitors L-671,329, L-646,991, tetrahydroechinocandin B, and L-687,781, a papulacandin. Antimicrob Agents Chemother 36(8):1648–1657

Bassino E, Fornero S, Gallo MP, Ramella R, Mahata SK, Tota B, Levi R, Alloatti G (2011) A novel catestatin-induced antiadrenergic mechanism triggered by the endothelial PI3K-eNOS pathway in the myocardium. Cardiovasc Res 91(4):617–624

Bayraktutan U, Yang ZK, Shah AM (1998) Selective dysregulation of nitric oxide synthase type 3 in cardiac myocytes but not coronary microvascular endothelial cells of spontaneously hypertensive rat. Cardiovasc Res 3:719–726

Bing OHL, Brooks WW, Robinson KG, Slawsky MT, Hayes JA, Litwin SE, Sen S, Conrad CH (1995) The spontaneously hypertensive rat as a model of the transition from compensated left ventricular hypertrophy to failure. J Mol Cell Cardiol 27:383–396

Brekke JF, Osol GJ, Helle KB (2002) N-terminal chromogranin-derived peptides as dilators of bovine coronary resi stance arteries. Regul Pept 105:93–100

Cappello S, Angelone T, Tota B, Pagliaro P, Penna C, Rastaldo R, Corti A, Losano G, Cerra MC (2007) Human recombinant chromogranin A-derived vasostatin-1 mimics preconditioning via an adenosine/nitric oxide signaling mechanism. Am J Physiol Heart Circ Physiol 293(1):H719–H727

Cerra MC, De Iuri L, Angelone T, Corti A, Tota B (2006) Recombinant N-terminal fragments of chromogranin-A modulate cardiac function of the Langendorff-perfused rat heart. Basic Res Cardiol 101(1):43–52

Cerra MC, Gallo MP, Angelone T, Quintieri AM, Pulerà E, Filice E, Guérold B, Shooshtarizadeh P, Levi R, Ramella R, Brero A, Boero O, Metz-Boutigue MH, Tota B, Alloatti G (2008) The homologous rat chromogranin A1-64 (rCGA1-64) modulates myocardial and coronary function in rat heart to counteract adrenergic stimulation indirectly via endothelium-derived nitric oxide. FASEB J 22(11):3992–4004

Di Felice V, Cappello F, Montalbano A, Ardizzone N, Campanella C, De Luca A, Amelio D, Tota B, Corti A, Zummo G (2006) Human recombinant vasostatin-1 may interfere with cell-extracellular matrix interactions. Ann N Y Acad Sci 1090:305–310

Dong S, Cheng Y, Yang J, Li J, Liu X, Wang X, Wang D, Krall TJ, Delphin ES, Zhang C (2009) MicroRNA expression signature and the role of microRNA-21 in the early phase of acute myocardial infarction. J Biol Chem 284:29514–29525

Ferrero E, Scabini S, Magni E, Foglieni C, Belloni D, Colombo B, Curnis F, Villa A, Ferrero ME, Corti A (2004) Chromogranin A protects vessels against tumor necrosis factor alpha-induced vascular leakage. FASEB J 18:554–556

Filice E, Pasqua T, Quintieri AM, Cantafio P, Scavello F, Amodio N, Cerra MC, Marban C, Schneider F, Metz-Boutigue MH, Angelone T (2015) Chromofungin, CgA47-66-derived peptide, produces basal cardiac effects and postconditioning cardioprotective action during ischemia/reperfusion injury. Peptides 71:40–48

Fung MM, Salem RM, Mehtani P, Thomas B, Lu CF, Perez B, Rao F, Stridsberg M, Ziegler MG, Mahata SK, O'Connor DT (2010) Direct vasoactive effects of the Chromogranin A (CHGA) peptide catestatin in humans in vivo. Clin Exp Hypertens 5:278–287

Garofalo F, Parisella ML, Amelio D, Tota B, Imbrogno S (2009) Phospholamban S-nitrosylation modulates Starling response in fish heart. Proc Biol Sci 276(1675):4043–4052

Gasparri A, Sidoli A, Sanchez LP, Longhi R, Siccardi AG, Marchisio PC, Corti A (1997) Chromogranin A fragments modulate cell adhesion. Identification and characterization of a pro-adhesive domain. J Biol Chem 272(33):20835–20843

Gomez L, Paillard M, Thibault H, Derumeaux G, Ovize M (2008) Inhibition of GSK3beta by postconditioning is required to prevent opening of the mitochondrial permeability transition pore during reperfusion. Circulation 117(21):2761–2768

Hammond J, Balligand JL (2012) Nitric oxide synthase and cyclic GMP signaling in cardiac myocytes: from contractility to remodeling. J Mol Cell Cardiol 52:330–340

Han JC, Tran K, Johnston CM, Nielsen PMF, Barrett CJ, Taberner AJ, Loiselle DS (2014) Reduced mechanical efficiency in left ventricular trabeculae of the spontaneously hypertensive rat. Physiol Rep 2:e12211

Hartell NA, Archer HE, Bailey CJ (2005) Insulin-stimulated endothelial nitric oxide release is calcium independent and mediated via protein kinase B. Biochem Pharmacol 69(5):781–790

Hausenloy DJ, Lecour S, Yellon DM (2011) Reperfusion injury salvage kinase and survivor activating factor enhancement prosurvival signaling pathways in ischemic postconditioning: two sides of the same coin. Antioxid Redox Signal 14(5):893–907

Helle KB, Corti A, Metz-Boutigue MH, Tota B (2007) The endocrine role for chromogranin A: a prohormone for peptides with regulatory properties. Cell Mol Life Sci 64(22):2863–2886

Hove-Madsen L, Méry PF, Jurevicius J, Skeberdis AV, Fischmeister R (1996) Regulation of myocardial calcium channels by cyclic AMP metabolism. Basic Res Cardiol 91(Suppl 2):1–8. Review

Kim T, Loh YP (2006) Protease nexin-1 promotes secretory granule biogenesis by preventing granule protein degradation. Mol Biol Cell 17(2):789–798

Koshimizu H, Kim T, Cawley NX, Loh YP (2010) Chromogranin A: a new proposal for trafficking, processing and induction of granule biogenesis. Regul Pept 160(1-3):153–159

Koshimizu H, Cawley NX, Kim T, Yergey AL, Loh YP (2011a) Serpinin: a novel chromogranin A-derived, secreted peptide up-regulates protease nexin-1 expression and granule biogenesis in endocrine cells. Mol Endocrinol 25(5):732–744

Koshimizu H, Cawley NX, Yergy AL, Loh YP (2011b) Role of pGlu-serpinin, a novel chromogranin A-derived peptide in inhibition of cell death. J Mol Neurosci 45(2):294–303

Lugardon K, Chasserot-Golaz S, Kieffer AE, Maget-Dana R, Nullans G, Kieffer B, Aunis D, Metz-Boutigue MH (2001) Structural and biological characterization of chromofungin, the antifungal chromogranin A-(47–66)-derived peptide. J Biol Chem 276(38):35875–35882

Mahata SK, O'Connor DT, Mahata M, Yoo SH, Taupenot L, Wu H, Gill BM, Parmer RJ (1997) Novel autoendocrine feedback control of catecholamine release. A discrete chromogranin A fragment is a non competitive nicotinic cholinergic antagonist. J Clin Invest 100:1623–1633

Mahata SK, Mahata M, Fung MM, O'Connor DT (2010) Catestatin: a multifunctional peptide from chromogranin A. Regul Pept 162(1-3):33–43

Mandalà M, Brekke JF, Serck-Hanssen G, Metz-Boutigue MH, Helle KB (2005) Chromogranin A-derived peptides: interaction with the rat posterior cerebral artery. Regul Pept 124(1-3):73–80

Maniatis NA, Brovkovych V, Allen SE, John TA, Shajahan AN, Tiruppathi C, Vogel SM, Skidgel RA, Malik AB, Minshall RD (2006) Novel mechanism of endothelial nitric oxide synthase activation mediated by caveolae internalization in endothelial cells. Circ Res 99(8):870–877

Matsuda N, Jesmin S, Takahashi Y, Hatta E, Kobayashi M, Matsuyama K, Kawakami N, Sakuma I, Gando S, Fukui H, Hattori Y, Levi R (2004) Histamine H1 and H2 receptor gene and protein levels are differentially expressed in the hearts of rodents and humans. J Pharmacol Exp Ther 309(2):786–795

Mattiazzi A, Vittone L, Mundiña-Weilenmann C (2007) Ca2+/calmodulin-dependent protein kinase: a key component in the contractile recovery from acidosis. Cardiovasc Res 73(4):648–656

Mazza R, Pasqua T, Gattuso A (2012) Cardiac heterometric response: the interplay between Catestatin and nitric oxide deciphered by the frog heart. Nitric Oxide 27(1):40–49

Metz-Boutigue MH, Garcia-Sablone P, Hogue-Angeletti R, Aunis D (1993) Intracellular and extracellular processing of chromogranin A. Determination of cleavage sites. Eur J Biochem 217(1):247–257

Metz-Boutigue MH, Goumon Y, Lugardon K, Strub JM, Aunis D (1998) Antibacterial peptides are present in chromaffin cell secretory granules. Cell Mol Neurobiol 18(2):249–266

Metz-Boutigue MH, Kieffer AE, Goumon Y, Aunis D (2003) Innate immunity: involvement of new neuropeptides. Trends Microbiol 11(12):585–592

Ozawa K, Whalen EJ, Nelson CD, Mu Y, Hess DT, Lefkowitz RJ, Stamler JS (2008) S-nitrosylation of beta-arrestin regulates beta-adrenergic receptor trafficking. Mol Cell 3:395–405

Pasqua T, Corti A, Gentile S, Pochini L, Bianco M, Metz-Boutigue MH, Cerra MC, Tota B, Angelone T (2013) Full-length human chromogranin-A cardioactivity: myocardial, coronary, and stimulus-induced processing evidence in normotensive and hypertensive male rat hearts. Endocrinology 154(9):3353–3365

Pasqua T, Tota B, Penna C, Corti A, Cerra MC, Loh YP, Angelone T (2015) pGlu-serpinin protects the normotensive and hypertensive heart from ischemic injury. J Endocrinol 227(3):167–178

Penna C, Cappello S, Mancardi D, Raimondo S, Rastaldo R, Gattullo D, Losano G, Pagliaro P (2006) Post-conditioning reduces infarct size in the isolated rat heart: role of coronary flow and pressure and the nitric oxide/cGMP pathway. Basic Res Cardiol 101(2):168–179

Penna C, Mognetti B, Tullio F, Gattullo D, Mancardi D, Pagliaro P, Alloatti G (2008) The platelet activating factor triggers preconditioning-like cardioprotective effect via mitochondrial K-ATP channels and redox-sensible signaling. J Physiol Pharmacol 59:47–54

Penna C, Alloatti G, Gallo MP, Cerra MC, Levi R, Tullio F, Bassino E, Dolgetta S, Mahata SK, Tota B, Pagliaro P (2010) Catestatin improves post-ischemic left ventricular function and decreases ischemia/reperfusion injury in heart. Cell Mol Neurobiol 30(8):1171–1179

Perrelli MG, Tullio F, Angotti C, Cerra MC, Angelone T, Tota B, Alloatti G, Penna C, Pagliaro P (2013) Catestatin reduces myocardial ischaemia/reperfusion injury: involvement of PI3K/Akt, PKCs, mitochondrial KATP channels and ROS signaling. Pflugers Arch 465(7):1031–1040

Petroff MG, Kim SH, Pepe S, Dessy C, Marbán E, Balligand JL, Sollott SJ (2010) Endogenous nitric oxide mechanisms mediate the stretch dependence of Ca2+ release in cardiomyocytes. Nat Cell Biol 3(10):867–873

Pieroni M, Corti A, Tota B, Curnis F, Angelone T, Colombo B, Cerra MC, Bellocci F, Crea F, Maseri A (2007) Myocardial production of chromogranin A in human heart: a new regulatory peptide of cardiac function. Eur Heart J 28(9):1117–1127

Prendergast BD, Sagach VF, Shah AM (1997) Basal release of nitric oxide augments the Frank-Starling response in the isolated heart. Circulation 96(4):1320–1329

Ramella R, Boero O, Alloatti G, Angelone T, Levi R, Gallo MP (2010) Vasostatin 1 activates eNOS in endothelial cells through a proteoglycan-dependent mechanism. J Cell Biochem 110(1):70–79

Reddy LG, Shi Y, Kutchai H, Filoteo AG, Penniston JT, Thomas DD (1999) An autoinhibitory peptide from the erythrocyte Ca-ATPase aggregates and inhibits both muscle Ca-ATPase isoforms. Biophys J 76(6):3058–3065

Satoh K, Matsu-Ura T, Enomoto M, Nakamura H, Michikawa T, Mikoshiba K (2011) Highly cooperative dependence of sarco/endoplasmic reticulum calcium ATPase(SERCA) 2a pump activity on cytosolic calcium in living cells. J Biol Chem 286:20591–20599

Schmidt AG, Edes I, Kranias EG (2001) Phospholamban: a promising therapeutic target in heart failure? Cardiovasc Drugs Ther 15(5):387–396

Schneider F, Bach C, Chung H, Crippa L, Lavaux T, Bollaert PE, Wolff M, Corti A, Launoy A, Delabranche X, Lavigne T, Meyer N, Garnero P, Metz-Boutigue MH (2012) Vasostatin-I, a chromogranin A-derived peptide, in non-selected critically ill patients: distribution, kinetics, and prognostic significance. Intensive Care Med 38(9):1514–1522

Shaul PW, Afshar S, Gibson LL, Sherman TS, Kerecman JD, Grubb PH, Yoder BA, McCurnin DC (2002) Developmental changes in nitric oxide synthase isoform expression and nitric oxide production in fetal baboon lung. Am J Physiol Lung Cell Mol Physiol 283(6):L1192–L1199

Stavrakis S, Scherlag BJ, Fan Y, Liu Y, Liu Q, Mao J, Cai H, Lazzara R, Po SS (2012) Antiarrhythmic effects of vasostatin-1 in a canine model of atrial fibrillation. J Cardiovasc Electrophysiol 23(7):771–777

Tota B, Mazza R, Angelone T, Nullans G, Metz-Boutigue MH, Aunis D, Helle KB (2003) Peptides from the N-terminal domain of chromogranin A (vasostatins) exert negative inotropic effects in the isolated frog heart. Regul Pept 114(2–3):123–130

Tota B, Quintieri AM, Di Felice V, Cerra MC (2007) New biological aspects of chromogranin A-derived peptides: focus on vasostatins. Comp Biochem Physiol A Mol Integr Physiol 147(1):11–18

Tota B, Angelone T, Mazza R, Cerra MC (2008) The chromogranin A-derived vasostatins: new players in the endocrine heart. Curr Med Chem 15(14):1444–1451

Tota B, Gentile S, Pasqua T, Bassino E, Koshimizu H, Cawley NX, Cerra MC, Loh YP, Angelone T (2012) The novel chromogranin A-derived serpinin and pyroglutaminated serpinin peptides are positive cardiac β-adrenergic-like inotropes. FASEB J 26(7):2888–2898

Tota B, Angelone T, Cerra MC (2014) The surging role of Chromogranin A in cardiovascular homeostasis. Front Chem 2:64

Tsigelny IF, Kouznetsova VL, Biswas N, Mahata SK, O'Connor DT (2013) Development of a pharmacophore model for the catecholamine release-inhibitory peptide catestatin: virtual screening and functional testing identify novel small molecule therapeutics of hypertension. Bioorg Med Chem 21(18):5855–5869

Vinten-Johansen J, Zhao ZQ, Jiang R, Zatta AJ (2005) Myocardial protection in reperfusion with postconditioning. Expert Rev Cardiovasc Ther 3(6):1035–1045

Xiao RP, Cheng H, Zhou YY, Kuschel M, Lakatta EG (1999) Recent advances in cardiac beta(2)-adrenergic signal transduction. Circ Res 11:1092–1100

Yano N, Ianus V, Zhao TC, Tseng A, Padbury JF, Tseng YT (2007) A novel signaling pathway for beta-adrenergic receptor-mediated activation of phosphoinositide 3-kinase in H9c2 cardiomyocytes. Am J Physiol Heart Circ Physiol 1:H385–H393

Zhang C (2008) MicroRNAs: role in cardiovascular biology and disease. Clin Sci 114(12):699–706

Zhang YH, Zhang MH, Sears CE, Emanuel K, Redwood C, ElArmouche A, Kranias EG, Casadei B (2008) Reduced phospholamban phosphorylation is associated with impaired relaxation in left ventricular myocytes from neuronal NO synthase decient mice. Circ Res 102:242–249

Zhang D, Shooshtarizadeh P, Laventie BJ, Colin DA, Chich JF, Vidic J, de Barry J, Chasserot-Golaz S, Delalande F, Van Dorsselaer A, Schneider F, Helle K, Aunis D, Prévost G, Metz-Boutigue MH (2009) Two chromogranin a-derived peptides induce calcium entry in human neutrophils by calmodulin-regulated calcium independent phospholipase A2. PLoS One 4(2):e4501

Comparative Aspects of CgA-Derived Peptides in Cardiac Homeostasis

Alfonsina Gattuso, Sandra Imbrogno, and Rosa Mazza

Abstract This chapter is an overview of the cardiotropic actions of the Chromogranin A-derived peptides, vasostatins and catestatin on the isolated and perfused eel (*Anguilla anguilla*) and frog (*Rana esculenta*) hearts, used as paradigms of fish and amphibian hearts. Our studies highlight important cardiotropic features of the two peptides both at basal (negative inotropism) and stimulated (anti-adrenergic effect: eel and frog; anti-endothelin action: frog) conditions. In addition, catestatin positively modulates the Frank-Starling response both in eel and frog hearts. Overall, the comparison of vasostatins and catestatin-mediated role in cardiac homeostasis of fish and amphibians illustrates aspects of uniformity and species specific differences in the mechanism of action of the peptides.

Abbreviations

Akt	Protein kinase B
ANP	Atrial natriuretic peptide
bCgA4–16	bovine CgA4–16
bCgA47–66	bovine CgA47–66
CA	Catecholamines
CgA	Chromogranin A
CgA1–40SH	CgA1–40 without an intact disulfide bridge
CgA1–40SS	CgA1–40 with an intact disulfide bridge
CST	Catestatin
EE	Endocardial endothelium
eNOS	Endothelial nitric oxide synthase
ET-1	Endothelin-1
ETAR	Endothelin-1 A subtype receptor
ETBR	Endothelin-1 B subtype receptor

A. Gattuso, PhD (✉) • S. Imbrogno • R. Mazza
Department of Biology, Ecology, and Earth Science, University of Calabria, 87036 Arcavacata di Rende, Italy
e-mail: alfonsina.gattuso@unical.it; sandra.imbrogno@unical.it; rosa.mazza@unical.it

fCgA4–16	frog CgA4–16
fCgA47–66	frog CgA47–66
ISO	Isoproterenol
NO	Nitric Oxide
NOS	Nitric Oxide Synthase
PI3K	Phosphatidyl 3-kinase
PKG	Protein Kinase G
PLN	Phospholamban
PTx	Pertussis toxin
SERCA2a	Sarcoplasmic Reticulum Ca^{2+}-ATPase
VS-1	Vasostatin 1 (CgA_{1-76})
VS-2	Vasostatin 2 (CgA_{1-113})
VSs	Vasostatins
W7	N-(6-aminohexil)-5-chloro-1-naphthalenesulfonamide

1 Introduction

Chromogranins A (CgA) is a member of the granin family, a group of acidic soluble proteins, found in secretory granules of endocrine, neuroendocrine, and neuronal cells, which are co-stored and co-released with hormones, neurotransmitters, and/or amines in response to specific stimuli. CgA, the first member of the family to be isolated and characterized (Banks and Helle 1965), is a protein of 48 kDa identified almost five decades ago, extensively studied as to its expression, structure and function. Soon after the first sequencing of bovine CgA (Iacangelo et al. 1986), many immunological and sequence studies have shown its ubiquitous distribution throughout the animal world, from invertebrates to mammals, remarking its notable phylogenetic conservation. CgA is present in teleost fish (Deftos et al. 1987), amphibians (Reinecke et al. 1991), reptiles (Trandaburu et al. 1999), birds (Reinecke et al. 1991), humans, pigs and rats (Tota et al. 2007; Helle et al. 2007). Among invertebrates, the occurrence of CgA has been reported in the nematode parasite *Ascaris suum* (Smart et al. 1992), the protozoan *Paramecium tetraurelia* (Peterson et al. 1987), and in coelenterates (Barkatullah et al. 1997).

Studies performed during the last 25 years evidenced the presence of CgA in the heart of several mammalian and non mammalian vertebrates. It was identified in the cardiac conduction system and in the atrial myoendocrine granules of the rat heart (Steiner et al. 1990; Weiergraber et al. 2000) and in the ventricular myocardium of the human heart (Pieroni et al. 2007). Both in humans and rodents, it was found to be co-stored and co-secreted with catecholamines (CAs) and natriuretic peptides (Steiner et al. 1990; Pieroni et al. 2007; Biswas et al. 2010). In non-mammalian vertebrates, CgA expression was detected in the secretory granules of frog atrial myocytes (Krylova 2007).

CgA cleavage, at the level of the multiple pairs of dibasic sites which enrich its sequence, gives rise several bioactive fragments, which exert a broad spectrum of regulatory activities by influencing endocrine, cardiovascular, and immune systems. These include the dysglycemic hormone pancreastatin (Tatemoto et al. 1986), the vasodilators vasostatin 1 (VS-1) and vasostatin 2 (VS-2) (Aardal and Helle 1992; Aardal et al. 1993), the antimicrobial agent chromacin (Strub et al. 1996), the catecholamine release-inhibitory peptide catestatin (CST) (Mahata et al. 1997, 2003, 2004), the antifungal fragment chromofungin (Lugardon et al. 2001) and its well-conserved domains WE-14, parastatin, and GE-25, whose role is currently under searching, and the recently identified serpinin peptides (Serp and pGlu-Serp) with important roles in neuroendocrine cells granule biogenesis and cell death (Koshimizu et al. 2011).

For cardiovascular interest, the N-terminal CgA derived peptides VS-1 and VS-2, corresponding to the CgA amino acids 1–76 (VS-1) and 1–113 (VS-2), respectively, and the middle domain CST (CgA344–364) are under intensive investigations for their role as anti-adrenosympathetic stabilizers in cardiovascular homeostasis and anti-ischemic cardioprotection. In fact, it has been demonstrated that both in mammalian and non mammalian vertebrates they are able to modulate basal cardiac performance and to exert anti-adrenergic cardio-suppressive actions which would protect the heart against excessive systemic and/or intra-cardiac excitatory stimuli (Tota et al. 2003; Corti et al. 2002, 2004; Imbrogno et al. 2004, 2010; Cerra et al. 2006; Angelone et al. 2008; Mazza et al. 2008, 2015). These experimental evidences have lead to propose CgA as precursor of different hormone peptides which function as novel cardiac modulators also able to protect the heart against excessive systemic and/or intra-cardiac excitatory stimuli.

On the basis of these premises, in this review we will summarize present knowledge regarding the influence exerted by the CgA-derived peptides VS-1, VS-2 and CST on fish and amphibian hearts, which, in the absence of other data, is mostly based on our own studies. By examining mechanistically the transduction pathways activated, we will consider aspects of uniformity and species-specific diversity in the exerted cardiotropism. Moreover, the role of these peptides as cardio-protective agents able to counteract the effects of excessive systemic and/or intra-cardiac excitatory stimuli will be discussed.

The purpose is to provide the basis for a comprehensive picture of the potentials of the CgA-derived peptides in the cardiac homeostasis of non mammalian vertebrates.

2 Vasostatins

Vasostatins (VSs), together with pancreastatin, parastatin, and catestatin, are the main biologically active peptides generated by the proteolytic processing of CgA. Named 'vasostatins' for their ability to relax vessels precontracted by high endothelin-1 (ET-1) and potassium concentrations (Aardal and Helle 1992), they have been identified in both poikilotherm (frog) and homeotherm (rat, pig, bovine, human) vertebrates. In all species examined so far, they exhibit a very high percentage identity and share important common traits such as the sequence 50–62 (unchanged) and the disulfide bridge C17-C38 that appears crucial for their biological activity (Helle et al. 2007). The main VSs consist of the highly conserved vertebrate domain vasostatin 2 (VS-2) and the less conserved vasostatin 1 (VS-1), which correspond to the N-terminal peptides CgA1–113 and CgA1–76, respectively. Beside these, the shorter VS peptides CgA1–40, CgA4–57, CgA47–66 (chromofungin), and CgA67–76 are also formed within the matrix of chromaffin granules and are co-released with CAs following chromaffin cell stimulation, for example by ACh (Metz-Boutigue et al. 1993) and also secreted upon stimulation of the isolated retrogradely perfused bovine adrenal gland (Helle et al. 1993). N-terminal fragments containing the VS-1 domain (CgA4–113, CgA1–124, CgA1–135, and CgA1–199) have been detected in rat heart extracts, together with a larger fragment presumably corresponding to the intact CgA, suggesting an intracardiac production of VSs (Glattard et al. 2006). VSs act as multifunctional regulatory peptides which, through autocrine, paracrine and/or endocrine mechanisms depending, among other factors, on cell and tissue targets as well as on the local concentration of the peptides, modulate several physiological processes (Helle et al. 2001, 2007).

2.1 Cardiac Effects Under Basal and ISO-Stimulated Conditions

Works performed in the last decade by our research group have indicated that VS-1, VS-2, and the human CgA7–57 synthetic peptide can act as cardiostatins through negative modulation of myocardial performance. By using the isolated and perfused working heart of eel and frog, a preparation suitable for assessing the direct effects of cardiotropic substances on myocardial performance, we observed that the human recombinant VS-1, VS-2, and the human CgA7–57 synthetic peptide elicit an inhibitory modulation of the basal cardiac performance (Corti et al. 2002, 2004; Imbrogno et al. 2004; Tota et al. 2004). VS-1 and CgA7–57 appeared more potent than VS-2 as inhibitory inotropic agents, allowing to hypothesize that the region 7–57 of VS-1 contains the structural determinants for this activity. Interestingly, both in the eel and frog heart, VSs peptides counteracted the positive inotropism elicited by the β–adrenergic agonist isoproterenol (ISO) (Corti et al. 2004; Imbrogno et al. 2004; Tota et al. 2004). Of note, these VSs–dependent cardiosuppressive and antiadrenergic effects resulted similar to those reported in the Langendorff perfused rat heart (Cerra et al. 2006), consistent with an ubiquitous cardio-inhibitory role of VSs in vertebrates which protect the heart by excessive excitatory stimulations. This allows to consider VSs as components of the "zero steady-state error" counter-regulatory homeostatic system postulated by Koeslag et al. (1999) for other CgA-derived peptides.

By using the isolated working heart preparation of frog as bioassay, specific sequences included in the VS-1 peptide [frog and bovine CgA4–16 and CgA47–66, and bovine CgA1–40 with (CgA1– 40SS) and without an intact disulfide bridge (CgA1–40SH)] have been tested both under basal and β-adrenergic stimulated conditions. All the fragments studied were able to modulate the cardiac performance in frog, emphasizing the high phylogenetic conservation of the CgA1–76 sequence. The most potent fragments in exerting intrinsic negative inotropy and counteraction against ISO-mediated positive inotropism were the intact VS-1, the fragment CgA7–57 and the shorter peptide CgA1–40 with the intact disulfide-bridge loop (Tota et al. 2003, 2004; Corti et al. 2004). CgA4–16 and CgA47–66 sequences of frog and bovine showed lower potencies than those reported for the human recombinant VS-1 and CgA7–57 and for the natural configuration of the bovine CgA1–40, i.e. CgA1–40SS. Frog CgA4–16 (fCgA4–16) resulted the least effective sequence (Tota et al. 2003). Of note, this peptide shows a higher net negative charge than the bovine counterpart. The ISO-evoked positive inotropy was counteracted by the bovine CgA1–40SS but not CgA1–40SH. The N- and C-terminal sequences of both frog and bovine CgA4–16 and CgA47–66 exerted their anti-adrenergic action at concentrations higher than those required by the fragment which include disulfide-bridge (CgA1–40SS) (Tota et al. 2003). These data suggest that the disulfide-bridge region of VS-1 is crucial for the cardio-suppression elicited by VS-1 and CgA7–57, both under basal or adrenergically stimulated conditions (Table 1).

2.2 Mechanisms of Action

On the isolated and perfused heart preparation from both eel and frog, the inhibitory actions of VS-1 and CgA7–57 depend on calcium and potassium channels, being abolished by lanthanum (a non-specific Ca^{2+} channel inhibitor), diltiazem (a specific L-type Ca^{2+} channel blocker), tetraethylammonium chloride (an inhibitor of the calcium activated-channels), and glibenclamide (an inhibitor of both the sarcolemmal and mitochondrial ATP-potassium channels), all used at concentrations that *per se* do not affect myocardial mechanical performance (Corti et al. 2004; Imbrogno et al. 2004). These data are in agreement with Aardal and Helle (1992) which reported a Ca^{2+}-dependent vasoinhibitory action of VS in human thoracic arteries, and Brekke and coworkers (Brekke et al. 2002) which observed that the vasodilatatory action induced by CgA1–40 in the bovine coronary arteries is abolished by K^+ channels blockers.

Moreover, in both eel and frog hearts, the VSs-mediated-negative inotropy was blocked by pre-treatment with inhibitors of cytoskeleton reorganization, such as cytochalasin-D (a blocker of actin polymerization), wortmannin (an inhibitor of PI3-kinase/protein kinase B signal-transduction cascade), butanedione 2-monoxime (an antagonist of myosin ATPase), and N-(6-aminohexil)-5-chloro-1-naphthalenesulfonamide (W7) (a calcium-calmodulin antagonist), pointing the

Table 1 Cardiac effects and mechanism of action of CGA-derived peptides (*ND* = Not Detected)

CgA-derived peptides	Basal condition	Adrenergic stimulation	Signalling principles involved	Frank-Starling response	References
Frog					
VS-1	Negative inotropism	Anti-adrenergic action	Ca^{2+} and K^+ channels	ND	Corti et al. (2002, 2004) Tota et al. (2004)
VS-2	Negative inotropism	Anti-adrenergic action	ND	ND	Corti et al. (2002)
CgA $_{7-57}$	Negative inotropism	Anti-adrenergic action	Ca^{2+} and K^+ channels cytoskeleton	ND	Corti et al. (2002, 2004) Tota et al. (2004) Mazza et al. (2007)
bCgA $_{1-40SS}$	Negative inotropism	Anti-adrenergic action	ND	ND	Tota et al. (2003)
bCgA $_{1-40SH}$	No effect	Anti-adrenergic action	ND	ND	Tota et al. (2003)
bCgA $_{4-16}$	Negative inotropism	Anti-adrenergic action	ND	ND	Tota et al. (2003)
fCGA $_{4-16}$	No effect	Anti-adrenergic action	ND	ND	Tota et al. (2003)
bCgA $_{47-66}$	Negative inotropism	Anti-adrenergic action	ND	ND	Tota et al., (2003)
fCgA $_{47-66}$	Negative inotropism	Anti-adrenergic action	ND	ND	Tota et al. (2003)
CST	Negative inotropism	Anti-adrenergic action	EE-NOS–NO–cGMP cascade, ETB receptors	Positive modulation	Mazza et al. (2008, 2012)
Eel					
VS-1	Negative inotropism	Anti-adrenergic action	Ca^{2+} and K^+ channels EE-NO-cGMP cascade, cytoskeleton	ND	Imbrogno et al. (2004) Tota et al. (2004) Mazza et al. (2007)
VS-2	Negative inotropism	Anti-adrenergic action	ND	ND	Imbrogno et al. (2004)
CgA $_{7-57}$	Negative inotropism	Anti-adrenergic action	Ca^{2+} and K^+ channels	ND	Imbrogno et al, (2004) Tota et al. (2004)
CST	Negative inotropism	Anti-adrenergic action	β3-AR-Gi/o--NO-cGMP cascade	Positive modulation	Imbrogno et al. (2010)

cytoskeleton as an important determinant of the signal-transduction cascade which underlies the inhibitory influence exerted by VSs on the contractile myocardial machinery (Mazza et al. 2007). In both eel and frog heart preparations, the treatment with pertussis toxin (PTx), a toxin which uncouples signal transduction between several families of receptors and Gi or Go proteins (Ai et al. 1998), did not modify basal cardiac performance; however, it abolished the inotropic effect of VSs in the eel heart (Imbrogno et al. 2004), without influencing it in the frog heart (Corti et al. 2004). So far, except for CST, which is known to interact with nicotinic receptors, functioning as a non-competitive antagonist (Mahata et al. 1997; Kraszewski et al. 2015), no conclusive evidence is available concerning the presence of specific receptors for CgA fragments, including VS peptides. Therefore, whether VSs act on the heart *via* classic receptor-ligand interactions still remains to be elucidated. Apart from specific receptor interactions, in the eel heart VSs could activate G proteins through spatially localized cell membrane perturbation caused by the interaction of the lipophylic portion of the peptide with a lipidic bilayer domain. Indeed, such a mechanism was suggested to explain the antimicrobic action of some VS-derived fragments (Lugardon et al. 2002; Maget-Dana et al. 2002). Conceivable, VS peptides might exert their action, including the myocardial inotropic effects, by modulating the activities of signalling principles, regulatory proteins, such as ion channels, and cell membrane elements, implicated in the regulation of intercellular communication. Aspects of diversity have been also reported in relation to the involvement of cholinergic and adrenergic systems in the determinism of VSs induced cardiosuppression. In fact, in the eel heart, pretreatments with either atropine (a non-specific muscarinic antagonist) or phentolamine and propanolol (α and β adrenergic antagonists, respectively), which *per se* do not modify basal cardiac parameters, abolish the VSs-mediated inotropism (Imbrogno et al. 2004); in contrast, in the frog heart the VS-induced negative inotropism is unchanged by these treatments (Corti et al. 2004). This different response pattern between teleost and amphibian hearts is also revealed by the involvement of the endocardial endothelium (EE) and the Nitric Oxide (NO)-cGMP signal-transduction pathway. While in the eel heart both the EE and the NO-cGMP signaling transduce the intracavitary VSs signal to the beating myocardium (Imbrogno et al. 2004), in the frog heart, neither the EE nor the NO-cGMP mechanism appear relevant for eliciting the VSs negative effect (Corti et al. 2004). These differences in the transduction pathways in frog *vs* eel heart may be explained by species-specific differences (from ultrastructural to biochemical and molecular levels), which may affect peptide binding, internalization, trans-endocardial transport, etc. For example, scavenger receptors have been characterized in the EE of the teleost heart (Seternes et al. 2001). Thus, it can be hypothesized that in teleost blood-borne endoluminal VS peptides interact with scavenger receptors, thereby triggering an EE-mediated mechanism, which in turn affects myocardial inotropy (Table 1).

3 Catestatin

CST (human CgA352–372, bovine CgA344–364) is a 21 amino acid cationic and hydrophobic peptide product of CgA proteolysis, co-secreted and co-released with CgA and CAs. It acts as an endogenous non-competitive inhibitor of nicotine-evoked catecholamine secretion (Mahata et al. 1997; Herrero et al. 2002). In addition, it prevents the nicotinic desensitization of catecholamine release from chromaffin cells (Mahata et al. 1999) and, in vivo, it blocks stimulation of both secretion and transcription of its precursor CgA (Mahata et al. 2003). Plasma level of CST are significantly lower in patients with essential hypertension and in normotensive subjects with family history of hypertension (O'Connor et al. 2002). Consistent with human studies, genetic ablation of ChgA gene results in high blood pressure in mice (Mahapatra et al. 2005). Three biologic variants of CST with varying effects on adrenergic inhibition have been identified with the following order of potency: Pro370Leu > wildtype(WT) > Gly364Ser > Arg374Gln (Mahata et al. 2004; Wen et al. 2004). In vivo, the CST Gly364Ser variant causes profound changes in human autonomic activity, both at the level of the parasympathetic and sympathetic branches, and seems to reduce risk of developing hypertension, especially in men (Rao et al. 2007).

CST is highly conserved in mammals, with a great homology between human and mouse (approximately 86%); moreover, human CST shows a moderate homology with non mammalian vertebrates: 38% with jungle fowl, 33% with frog, and approximately 19% with zebrafish (Bartolomucci et al. 2011).

3.1 Cardiac Effects Under Basal and ISO-Stimulated Conditions

On the perfused working frog (Mazza et al. 2008) and eel (Imbrogno et al. 2010) hearts, CST (bovine CgA344–364) acts as a cardio-depressing agent in a dose-dependent manner (11–165 nmol/l) and counteracts the ISO-induced cardio-stimulatory effect through a non-competitive antagonism. In addition, in the frog heart CST is able to abolish the endothelin-1 (ET-1)-induced positive inotropism, but does not affect the ET-1-dependent negative effect, the latter being consistent with a convergent signalling pathway for both ET-1 and CST (Table 1). As in eel and frog hearts, also in the rat heart (Angelone et al. 2008), CST directly suppresses the mechanical performance of both non-stimulated and adrenergically stimulated preparations, supporting the ubiquitous cardio-depressive action of this peptide in vertebrates. Of note, in the Langendorff perfused rat heart, human CST variants counteracted the positive inotropic and lusitropic effects of ISO with different potency and counteracted the positive inotropism and lusitropism and the coronary constriction of ET-1, pointing them as important components of a homeostatic counteraction against excessive SAN activation, e.g., prolonged stress, HF, hypertensive cardiomyopathy (Angelone et al. 2008).

The CST-induced cardiosuppressive and anti-adrenergic effects observed in eel and frog resulted comparable to those elicited by VSs both in mammalian and non mammalian vertebrates (Tota et al. 2004). The biological significance of such apparently redundant molecular strategy remains to be clarified. Conceivably, processing more than one cardioactive peptide from the same prohormone might be advantageous in maintaining homeostasis. It can be hypothesized that, by acting on overlapping or different sites, VS-1 and CST may achieve selective intracardiac actions in a spatiotemporal and tissue-specific manner (e.g. summation and synergism or potentiation of the target cell responses). A similar cardiovascular case is represented by the proteolytic cleavage of proANF precursor that in mammals gives rise to the major form of circulating ANP (ANP1–28) and several biologically active peptides (proANF1–30, long-acting sodium stimulator; proANF31–67, vessel dilator; and proANF79–98, kaliuretic stimulator) (Vesely et al. 1994). At least two of them, i.e. vessel dilator and ANP, show almost overlapping properties, being vasodilatory, diuretic, and natriuretic (Vesely 2006). The examples provided by both CgA and proANF illustrate the striking cardiac potential for multilevel interactions between endocrine precursors and their derived peptides.

3.2 Mechanism of Action

In the frog heart, the CST cardiosuppressive action appears EE-NOS/NO/cGMP-dependent, being abolished by functional EE damage with Triton X-100 and pretreatment with NOS or guanylate cyclase inhibitors. Of note, this contrasts with the VS-1-induced negative inotropism in the frog heart, which, unlike the pattern demonstrated in eel and rat hearts (Imbrogno et al. 2004; Cappello et al. 2007), involved neither the EE nor the G protein nor the NO-cGMP-protein kinase G mechanism (Corti et al. 2004). The question as to how and why the two peptides, representing NH_2-and COOH-terminal domains of CgA, respectively, utilize divergent signaling pathways to converge on similar cardiotropic actions, remains open.

Moreover, in the frog heart, the CST negative inotropism is mediated by ETB receptor (ETBR) subtype, mainly expressed in endothelial cells and linked to eNOS phosphorylation and NO synthesis, but not by ETAR (mostly expressed in cardiomyocytes), thus emphasizing the paracrine role of EE in regulating the contractility of the subjacent myocardium (Mazza et al. 2008). This role appears of particular importance in the frog heart, where the EE lining the very extensive lacunae of the avascular spongy ventricle represents a relevant source of bioactive NO (Sys et al. 1997; Gattuso et al. 1999; Mazza et al. 2010, 2012, 2013). The crucial role of ETBR in the CST signaling is further supported by the evidence that, in the presence of its selective inhibition, CST fails to inhibit both the ISO-and ET-1-elicited positive inotropic effects as well as its inhibitory and stimulatory effects on phospholamban (PLN) and ERK1/2 phosphorilation, respectively (Mazza et al. 2008). The finding that all the CST effects are mediated by ETBR, but not by ETAR, reinforces the hypothesis that the peptide actions involve the ET-1 subtype receptors mainly located on the EE. Of note, also in the rat heart, the CST cardio-suppressive action requires the functional integrity of the EE, sustaining the EE-myocardial interaction in the CST action (Mazza et al. 2008).

In the eel heart, the CST-induced cardio-depressive effects are achieved through a pertussis toxin-sensitive (PTX) activation, independent from receptor, of Gi/o proteins. In the same way, atropine pretreatment does not affect the CST response in the eel heart, pointing to a muscarinic receptor independent effect (Imbrogno et al. 2010). Being CST a cationic and hydrophobic peptide, it is improbable that it directly modulates receptors. A PTX-sensitive mechanism has been also reported by Krüger et al. (2003) in rat mast cells where the active domain of bovine cateslytin (CgA344–358) induced histamine release through aggregation on negatively charged membranes (Jean-François et al. 2008). This alternative aggregation mechanism may be also hypothesized for CST even if further research is needed to deepen this issue. In addition, in the eel heart, the CST negative inotropic action is also mediated by β1/β2/β3-adrenergic receptors and involves a NO-cGMP-dependent mechanism (Table 1). Of note, in the rat, the negative inotropic and lusitropic effects induced by CST involve both β2/β3-ARs, with a higher affinity for the first one, but not β1-AR, while they are reduced by α-AR and unaffected by cholinergic receptors inhibition also recruiting a PI3K/Akt/eNOS/NO/cGMP-dependent mechanism (Angelone et al. 2008).

3.3 *Effects Under Loading Stimulated Conditions*

The Frank–Starling's law (intrinsic regulation) operates in all classes of vertebrates. It varies among vertebrates particularly between mammalian and amphibian/fish species, the latter showing an elevated sensitivity to the heterometric response in part ascribed to a greater extensibility of the thin myocardial trabeculum (Shiels and White 2008). While cardiac output in fish and amphibians is increased, to a large extent, by changing the volume of blood pumped by the heart, in mammals it is modulated by increasing the heart rate rather than volume. As largely shown, the Frank-Starling response is modulated by many endogenous and exogenous substances. Haemodynamic forces, such as stretch, can activate major cardiac homeometric autoregulatory mechanisms that include autocrine and paracrine molecules released from endothelial/endocardial cells and cardiac myocytes that modulate myocytes contractility. The importance of this control has been firstly emphasized by the discovery that the cardiac natriuretic peptides adapt myocardial contractility and relaxation in a stretch-dependent manner (de Bold et al. 1981). This scenario was further expanded by the role that the stretch-induced release of Nitric Oxide from cardiomyocytes, vascular, and endocardial cells, exerts on pumping performance enhancing the Frank–Starling response (Prendergast et al. 1997; Imbrogno et al. 2001; Garofalo et al. 2009; Mazza et al. 2012). More recently, studies both in mammalian and non-mammalian vertebrates have provided evidence that CST, by functioning as an endogenous NOS activator, also contributes to the humoral regulation of the heterometric response (Imbrogno et al. 2010; Mazza et al. 2012; Angelone et al. 2015).

In the frog heart, Mazza and co-workers (2012) demonstrated that CST positively modulates the heterometric response through an EE-dependent NO release which involved a PI3K–NOS–cGMP pathway confirming, also under stretch conditions, CST as a paracrine stimulus for NO release from EE. This effect appears mediated by ETBR since it is blocked by BQ788 but not by BQ123. The evidence that CST-induced effect on the Frank-Starling response was of the same magnitude of that induced by NO, strongly supports the idea that CST contributes to the endocrine/paracrine regulation of the heterometric response of the vertebrate heart by functioning as an endogenous NOS activator. This view is also reinforced by the evidence obtained in the eel heart in which, mimicking the NO modulation, CST improves the Frank–Starling response (Imbrogno et al. 2010; Garofalo et al. 2009), as supported by the increased p-eNOS expression in CST-treated hearts (Table 1). A reduction of the cardiac response to preload was observed after NOS, but not guanylate cyclase, inhibition suggesting an action mechanism dependent on NO but independent on its classical cGMP signaling. Of note, the CST-induced increase of the heterometric response was significantly reduced by inhibition of SERCA2a pumps (Imbrogno et al. 2010), indicating that CST-induced NO release directly regulates SR Ca^{2+} reuptake.

On the basis of these data it appears that exogenous CST, acting as a paracrine stimulus for NO release, is able to modulate cardiac function under both basal and stretch hemodynamic conditions.

4 Cardiac CST and VSs Profiles

Summarizing work from the last decade, here we have documented in a comparative context, the cardiotropic features of CgA-derived peptides (VS and CST) on eel and frog hearts used as bioassays of non mammalian vertebrates, also highlighting aspects of uniformity and species-specific diversity in the cardiotropism of VSs and CST. Apart from the antihypertensive profile of CST, all peptides induce intrinsic negative inotropy and counteract the β-adrenergic-mediated positive inotropism exerted by isoproterenol, which might be important under sympathetic overstimulation conditions (Table 1).

It is of relevance that both in eel and frog heart the minimally effective peptide concentrations match with the nanomolar range corresponding to the circulating level of CgA normally found in man. In an integrated perspective, this points to VSs and CST as components of a wider orchestration of their precursor CgA in regulating circulatory homeostasis. In this context, CgA and its derived peptides, VSs and CST, appear as new players in the scenario of the endocrine heart, functioning as cardiac counter-regulators under intense excitatory stimuli (e.g. CA-induced myocardial stress).

In particular, while in the eel the VS-1-elicited negative inotropism implicates the functional involvement of β-ARs receptors, as well as a PTX sensitive G-protein-NO–cGMP–PKG pathway, in the frog heart its cardio-inhibition is unaffected by both adrenergic receptor and G proteins inhibition and recruits a NO-independent pathway. Furthermore, in both species, the structure-function relationships analysis of several sequences of VS-1 reveals the importance of highly conserved domains and functionally important regions.

In both fish and frog hearts, the heterometric regulation of cardiac function (Frank-Starling response) was positively affected by CST through an EE-induced NO release. Moreover, the evidence that CST-induced effect on the Frank-Starling response is of the same magnitude of that induced by NO, strongly supports the idea that CST contributes to the intrinsic regulation of the vertebrate heart by functioning as an endogenous NOS activator. In this context, the comparative analysis of CgA-derived peptides cardioactivity in fish and frog, characterized by different ventricular myoangioarchitecture (avascular heart: frog; poorly vascularized heart: eel), has been helpful to identify the EE-myocardial interactions underpinning VSs and CST actions.

Furthermore, in addition to its antiadrenergic action, in the frog heart CST acts as a direct cardiac modulator under ET-1 stimulation, suggesting a role in the homeostatic counteraction not only against excessive SAN activation, but also against hypertensive cardiomyopathy, as established in mammalian counterpart (Angelone et al. 2008).

Looking toward mammalian vertebrates, CST dose-dependently increases coronary pressure and abolishes the ISO-dependent vasodilation under basal conditions, while it potently vasodilates the ET-1 preconstricted coronaries (Angelone et al. 2008), thus reinforcing previous evidence of vasodilation promoted by both endogenous and exogenous CST in human subjects (O'Connor et al. 2002; Fung et al. 2010). Of note, also the native (rat) CgA 1–64 (rCgA1–64), corresponding to human VS-1 (Metz-Boutigue et al. 1993) is able to counteract ET-1-elicited positive contractility in rat, as well as the ET-1-induced coronary constriction (Cerra et al. 2006).

Finally, VS-1 and CST are able to mediate cardioprotective effects, primarily through a direct action on the myocardium, rather than endothelium-mediated effects. The comparison of the cardioprotective effects of VS-1 and CST in ischemic conditioning highlights, at the same time, a remarkable similarity and subtle difference, VS-1 appearing as a pre-conditioning inducer while CST emerging as a post-conditioning agent (Penna et al. 2012).

In conclusion, the experimental evidence regarding VS-1 and CST as cardiocirculatory homeostatic stabilizers makes wider the view of the neurovisceral control of the heart, particularly in relation to the concept of counter-regulatory hormones in "zero steady-state error" homeostasis elaborated by Koeslag et al. (1999). At the same time, the use of fish and amphibian paradigms adds a new piece to the expanding puzzle of neuroendocrine control of cardiac function both under basal and stress conditions, allowing to illustrate how these neuro-endocrine agents have evolved as components of the homeostatic control of cardiac function in vertebrates during evolutionary transitions.

References

Aardal S, Helle KB (1992) The vasoinhibitory activity of bovine chromogranin A fragment (vasostatin) and its independence of extracellular calcium in isolated segments of human blood vessels. Regul Pept 41:9–18
Aardal S, Helle KB, Elsayed S, Reed RK, Serck-Hanssen G (1993) Vasostatins, comprising the N-terminal domain of chromogranin A, suppress tension in isolated human blood vessel segments. J Neuroendocrinol 5:405–412
Ai T, Horie M, Obayashi K, Sasayama S (1998) Accentuated antagonism by angiotensin II on guinea-pig cardiac L-type Ca-currents enhanced by b-adrenergic stimulation. Eur J Phys 436:168–174
Angelone T, Quintieri AM, Brar BK, Limchaiyawat PT, Tota B, Mahata SK, Cerra MC (2008) The antihypertensive chromogranin a peptide catestatin acts as a novel endocrine/paracrine modulator of cardiac inotropism and lusitropism. Endocrinol 149:4780–4793
Angelone T, Quintieri AM, Pasqua T, Filice E, Cantafio P, Scavello F, Rocca C, Mahata SK, Gattuso A, Cerra MC (2015) The NO stimulator, Catestatin, improves the Frank-Starling response in normotensive and hypertensive rat hearts. Nitric Oxide 50:10–19
Banks P, Helle K (1965) The release of protein from the stimulated adrenal medulla. Biochem J 97:40C–41C
Barkatullah SC, Curry WJ, Johnston CF, Hutton JC, Buchanan KD (1997) Ontogenetic expression of chromogranin A and its derived peptides, WE-14 and pancreastatin, in the rat neuroendocrine system. Histochem Cell Biol 107:251–257
Bartolomucci A, Possenti R, Mahata SK, Fischer-Colbrie R, Loh YP, Salton SR (2011) The extended granin family: structure, function, and biomedical implications. Endocr Rev 32:755–797
Biswas N, Curello E, O'Connor DT, Mahata SK (2010) Chromogranin/secretogranin proteins in murine heart: myocardial production of chromogranin A fragment catestatin (Chga(364–384)). Cell Tissue Res 342:353–361
de Bold AJ, Borenstein HB, Veress AT, Sonnenberg H (1981) A rapid and potent natriuretic response to intravenous injection of atrial myocardial extract in rats. Life Sci 28:89–94
Brekke JF, Osol GJ, Helle KB (2002) N-terminal chromogranin derived peptides as dilators of bovine coronary resistance arteries. Regul Pept 105:93–100
Cappello S, Angelone T, Tota B, Pagliaro P, Penna C, Rastaldo R, Corti A, Losano G, Cerra MC (2007) Human recombinant chromogranin A-derived vasostatin-1 mimics preconditioning via an adenosine/nitric oxide signalling mechanism. Am J Physiol Heart Circ Physiol 293:H719–H727
Cerra MC, De Iuri L, Angelone T, Corti A, Tota B (2006) Recombinant N-terminal fragments of chromogranin-A modulate cardiac function of the Langendorff-perfused rat heart. Basic Res Cardiol 101:43–52
Corti A, Mannarino C, Mazza R, Colombo B, Longhi R, Tota B (2002) Vasostatin exert negative inotropism in the working heart of the frog. Ann N Y Acad Sci 971:362–365
Corti A, Mannarino C, Mazza R, Angelone T, Longhi R, Tota B (2004) Chromogranin A N-terminal fragments vasostatins-1 and the synthetic CgA 7–57 peptide act as cardiostatins on the isolated working frog heart. Gen Comp Endocrinol 136:217–224
Deftos LJ, Björnsson BT, Burton DW, O'Connor DT, Copp DH (1987) Chromogranin A is present in and released by fish endocrine tissue. Life Sci 40:2133–2136

Fung MM, Salem RM, Mehtani P, Thomas B, Lu CF, Perez B, Rao F, Stridsberg M, Ziegler MG, Mahata SK, O'Connor DT (2010) Direct vasoactive effects of the chromogranin A (CHGA) peptide catestatin in humans in vivo. Clin Exp Hypertens 32:278–287

Garofalo F, Parisella ML, Amelio D, Tota B, Imbrogno S (2009) Phospholamban S-nitrosylation modulates Starling response in fish heart. Proc Biol Sci 276:4043–4052

Gattuso A, Mazza R, Pellegrino D, Tota B (1999) Endocardial endothelium mediates luminal ACh-NO signaling in isolated frog heart. Am J Phys 276:H633–H641

Glattard E, Angelone T, Strub JM, Corti A, Aunis D, Tota B, Metz-Boutigue MH, Goumon Y (2006) Characterization of natural vasostatin containing peptides in rat heart. FEBS J 273:3311–3321

Helle KB, Marley PD, Angeletti RH, Aunis D, Galindo E, Small DH, Livett BG (1993) Chromogranin A: secretion of processed products from the stimulated retrogradely perfused bovine adrenal gland. J Neuroendocrinol 5:413–420

Helle KB, Metz-Boutigue MH, Aunis D (2001) Chromogranin A as a calcium-binding precursor for a multitude of regulatory peptides for immune, endocrine and metabolic system. Curr Med Chem 1:119–140

Helle KB, Corti A, Metz-Boutigue MH, Tota B (2007) The endocrine role for chromogranin A: a prohormone for peptides with regulatory properties. Cell Mol Life Sci 64:2863–2886

Herrero CJ, Alés E, Pintado AJ, López MG, García-Palomero E, Mahata SK, O'Connor DT, García AG, Montiel C (2002) Modulatory mechanism of the endogenous peptide catestatin on neuronal nicotinic acetylcholine receptors and exocytosis. J Neurosci 22:377–388

Iacangelo A, Affolter HU, Eiden LE, Herbert E, Grimes M (1986) Bovine chromogranin A sequence and distribution of its messenger RNA in endocrine tissues. Nature 323:82–86

Imbrogno S, De Iuri L, Mazza R, Tota B (2001) Nitric oxide modulates cardiac performance in the heart of *Anguilla anguilla*. J Exp Biol 204:1719–1727

Imbrogno S, Angelone T, Corti A, Adamo C, Helle KB, Tota B (2004) Influence of vasostatins, the chromogranin A-derived peptides, on the working heart of the eel (*Anguilla anguilla*): negative inotropy and mechanism of action. Gen Comp Endocrinol 139:20–28

Imbrogno S, Garofalo F, Cerra MC, Mahata SK, Tota B (2010) The catecholamine release-inhibitory peptide catestatin (chromogranin A344–363) modulates myocardial function in fish. J Exp Biol 213:3636–3643

Jean-François F, Castano S, Desbat B, Odaert B, Roux M, Metz-Boutigue MH, Dufourc EJ (2008) Aggregation of cateslytin beta-sheets on negatively charged lipids promotes rigid membrane domains. A new mode of action for antimicrobial peptides? Biochemistry 47:6394–6402

Koeslag JH, Saunders PT, Wessels JA (1999) The chromogranins and the counterregulatory hormones: do they make homeostatic sense? J Physiol 517:643–649

Koshimizu H, Cawley NX, Yergy AL, Loh YP (2011) Role of pGlu-Serpinin, a novel chromogranin A-derived peptide in inhibition of cell death. J Mol Neurosci 45:294–303

Kraszewski S, Drabik D, Langner M, Ramseyer C, Kembubpha S, Yasothornsrikul S (2015) A molecular dynamics study of catestatin docked on nicotinic acetylcholine receptors to identify amino acids potentially involved in the binding of chromogranin A fragments. Phys Chem Chem Phys 17:17454–17460

Krüger PG, Mahata SK, Helle KB (2003) Catestatin (CgA344–364) stimulates rat mast cell release of histamine in a manner comparable to mastoparan and other cationic charged neuropeptides. Regul Pept 114:29–35

Krylova MI (2007) Chromogranin A: immunocytochemical localization in secretory granules of frog atrial cardiomyocytes. Tsitologiia 49:538–543

Lugardon K, Chasserot-Golaz S, Kieffer AE, Maget-Dana R, Nullans G, Kieffer B, Aunis D, Metz-Boutigue MH (2001) Structural and biological characterizationof chromofungin, the antifungal chromogranin A-(47 – 66)-derived peptide. J Biol Chem 276:35875–35882

Lugardon K, Chasserot-Golaz S, Kieffer AE, Maget-Dana R, Nullans G, Kieffer B, Aunis D, Metz-Boutigue MH (2002) Structural and biological characterization of chromofungin, the antifungal chromogranin A (47–66)-derived peptide. Ann N Y Acad Sci 971:359–361

Maget-Dana R, Metz-Boutigue MH, Helle KB (2002) The N-terminal domain of chromogranin A (CgA1–40) interacts with monolayers of membrane lipids of fungal and mammalian compositions. Ann N Y Acad Sci 971:352–354

Mahapatra NR, O'Connor DT, Vaingankar SM, Sinha Hikim AP, Mahata M, Ray S, Staite E, Wu H, Gu Y, Dalton N et al (2005) Hypertension from targeted ablation of chromogranin A can be rescued by the human ortholog. J Clin Invest 115:1942–1952

Mahata SK, O'Connor DT, Mahata M, Yoo SH, Taupenot L, Wu H, Gill BM, Parmer RJ (1997) Novel autocrine feedback control of catecholamine release. A discrete chromogranin a fragment is a noncompetitive nicotinic cholinergic antagonist. J Clin Investig 100:1623–1633

Mahata SK, Mahata M, Parmer RJ, O'Connor DT (1999) Desensitization of catecholamine release: the novel catecholamine release-inhibitory peptide catestatin (chromogranin A344–364) acts at the receptor to prevent nicotinic cholinergic tolerance. J Biol Chem 274:2920–2928

Mahata SK, Mahapatra NR, Mahata M, Wang TC, Kennedy BP, Ziegler MG, O'Connor DT (2003) Catecholamine secretory vesicle stimulustranscription coupling in vivo. Demonstration by a novel transgenic promoter/photoprotein reporter and inhibition of secretion and transcription by the chromogranin A fragment catestatin. J Biol Chem 278:32058–32067

Mahata SK, Mahata M, Wen G, Wong WB, Mahapatra NR, Hamilton BA, O'Connor DT (2004) The catecholamine release-inhibitory 'catestatin' fragment of chromogranin a: naturally occurring human variants with different potencies for multiple chromaffin cell nicotinic cholinergic responses. Mol Pharmacol 66:1180–1191

Mazza R, Mannarino C, Imbrogno S, Barbieri SF, Adamo C, Angelone T, Corti A, Tota B (2007) Crucial role of cytoskeleton reorganization in the negative inotropic effect of chromogranin A-derived peptides in eel and frog hearts. Regul Pept 138:145–151

Mazza R, Gattuso A, Mannarino C, Brar BK, Barbieri SF, Tota B, Mahata SK. Catestatin (chromogranin A344–364 is a novel cardiosuppressive agent, inhibition of isoproterenol and endothelin signaling in the frog heart. Am J Physiol Heart Circ Physiol. 2008;295:H113–H122.

Mazza R, Angelone T, Pasqua T, Gattuso A (2010) Physiological evidence for 3-adrenoceptor in frog (*Rana esculenta*) heart. Gen Comp Endocrinol 169:151–157

Mazza R, Pasqua T, Gattuso A (2012) Cardiac heterometric response: the interplay between Catestatin and nitric oxide deciphered by the frog heart. Nitric Oxide 27:40–49

Mazza R, Pasqua T, Cerra MC, Angelone T, Gattuso A (2013) Akt/eNOS signaling and PLN S-sulfhydration are involved in H_2Sdependent cardiac effects in frog and rat. Am J Physiol Regul Integr Comp Physiol 305:R443–R451

Mazza R, Tota B, Gattuso A (2015) Cardio-vascular activity of catestatin: interlocking the puzzle pieces. Curr Med Chem 22:292–230

Metz-Boutigue MH, Garcia-Sablone P, Hogue-Angeletti R, Aunis D (1993) Intracellular and extracellular processing of chromogranin A. Determination of cleavage sites. Eur J Biochem 217:247–257

O'Connor DT, Kailasam MT, Kennedy BP, Ziegler MG, Yanaihara N, Parmer RJ (2002) Early decline in the catecholamine release-inhibitory peptide catestatin in humans at genetic risk of hypertension. J Hypertens 20:1335–1345

Penna C, Tullio F, Perrelli MG, Mancardi D, Pagliaro P (2012) Cardioprotection against ischemia/reperfusion injury and chromogranin A-derived peptides. Curr Med Chem 19:4074–4085

Peterson JB, Nelson DL, Ling E, Angeletti RH (1987) Chromogranin A-like proteins in the secretory granules of a protozoan, paramecium tetraurelia. J Biol Chem 262:17264–17267

Pieroni M, Corti A, Tota B, Curnis F, Angelone T, Colombo B, Cerra MC, Bellocci F, Crea F, Maseri A (2007) Myocardial production of chromogranin A in human heart: a new regulatory peptide of cardiac function. Eur Heart J 28:1117–1127

Prendergast BD, Sagach VF, Shah AM (1997) Basal release of nitric oxide augments the Frank-Starling response in the isolated heart. Circulation 96:1320–1329

Rao F, Wen G, Gayen JR, Das M, Vaingankar SM, Rana BK, Mahata M, Kennedy BP, Salem RM, Stridsberg M, Abel K, Smith DW, Eskin E, Schork NJ, Hamilton BA, Ziegler MG, Mahata SK, O'Connor DT (2007) Catecholamine release-inhibitory peptide catestatin (chromograninA

352– 372): naturally occurring amino acid variant Gly364Ser causes profound changes in human autonomic activity and alters risk for hypertension. Circulation 115:2271–2281

Reinecke M, Höög A, Ostenson CG, Efendic S, Grimelius L, Falkmer S (1991) Phylogenetic aspects of pancreastatin- and chromogranin-like immunoreactive cells in the gastro-enteropancreatic neuroendocrine system of vertebrates. Gen Comp Endocrinol 83:167–182

Seternes T, Oynebraten I, Sorensen K, Smedsrod B (2001) Specific endocytosis and catabolism in the scavenger endothelial cells of cod (*Gadus morhua* L.) generate high-energy metabolites. J Exp Biol 204:1537–1546

Shiels HA, White E (2008) The Frank–Starling mechanism in vertebrate cardiac myocytes. J Exp Biol 211:2005–2011

Smart D, Johnston CF, Curry WJ, Shaw C, Halton DW, Fairweather I, Buchanan KD (1992) Immunoreactivity to two specific regions of chromogranin A in the nervous system of *Ascaris suum*: an immunocytochemical study. Parasitol Res 78:329–335

Steiner HJ, Weiler R, Ludescher C, Schmid KW, Winkler H (1990) Chromogranins A and B are co-localized with atrial natriuretic peptides in secretory granules of rat heart. J Histochem Cytochem 38:845–850

Strub JM, Goumon Y, Lugardon K, Capon C, Lopez M, Moniatte M, Van Dorsselaer A, Aunis D, Metz-Boutigue MH (1996) Antibacterial activity of glycosylated and phosphorylated chromogranin A-derived peptide 173–194 from bovine adrenal medullary chromaffin granules. J Biol Chem 271:28533–28540

Sys SU, Pellegrino D, Mazza R, Gattuso A, Andries LJ, Tota B (1997) Endocardial endothelium in the avascular heart of the frog: morphology and role of nitric oxide. J Exp Biol 200:3109–3118

Tatemoto K, Efendić S, Mutt V, Makk G, Feistner GJ, Barchas JD (1986) Pancreastatin, a novel pancreatic peptide that inhibits insulin secretion. Nature 324:476–478

Tota B, Mazza R, Angelone T, Nullans G, Metz-Boutigue MH, Aunis D, Helle KB (2003) Peptides from the N-terminal domain of chromogranin A (vasostatins) exert negative inotropic effects in the isolated frog heart. Regul Pept 114:91–99

Tota B, Imbrogno S, Mannarino C, Mazza R (2004) Vasostatins and negative inotropy in vertebrate hearts. Curr Med Chem Immun Endoc Metab Agents 4:195–201

Tota B, Quintieri AM, Di Felice V, Cerra MC (2007) New biological aspects of chromogranin A-derived peptides: focus on vasostatins. Comp Biochem Physiol A Physiol 147:11–18

Trandaburu T, Ali SS, Trandaburu I (1999) Granin proteins (chromogranin A and secretogranin II C23–3 and C26–3) in the intestine of reptiles. Ann Anat 81:261–268

Vesely DL (2006) Which of the cardiac natriuretic peptides is most effective for the treatment of congestive heart failure, renal failure and cancer? Clin Exp Pharmacol Physiol 33:169–176

Vesely DL, Douglass MA, Dietz JR, Gower WR Jr, McCormick MT, Rodriguez-Paz G, Schocken DD (1994) Three peptides from the atrial natriuretic factor prohormone amino terminus lower blood pressure and produce diuresis, natriuresis, and/or kaliuresis in humans. Circulation 90:1129–1140

Weiergraber M, Pereverzev A, Vajna R, Henry M, Schramm M, Nastainczyk W, Grabsch H, Schneider T (2000) Immunodetection of alpha1E voltagegated Ca(2+) channel in chromograninpositive muscle cells of rat heart, and in distal tubules of human kidney. J Histochem Cytochem 48:807–819

Wen G, Mahata SK, Cadman P, Mahata M, Ghosh S, Mahapatra NR, Rao F, Stridsberg M, Smith DW, Mahboubi P, Schork NJ, O'Connor DT, Hamilton BA (2004) Both rare and common polymorphisms contribute functional variation at CHGA, a regulator of catecholamine physiology. Am J Hum Genet 74:197–207

Molecular and Cellular Mechanisms of Action of CgA-Derived Peptides in Cardiomyocytes and Endothelial Cells

Giuseppe Alloatti and Maria Pia Gallo

Abstract Several studies indicate that Chromogranin A (CgA)-derived peptides, in particular Vasostatin-1 (VS-1), Catestatin (CST), Chromofungin and Serpinin, exert important regulatory effects in numerous organs/systems, including the cardiovascular system. This chapter focuses on the recently discovered signalling pathways activated by CgA-derived peptides in cardiomyocytes and endothelial cells, giving insights into the mechanisms at the basis of their inotropic and cardioprotective effects. Several evidences provided convincing support for VS-1 and CST as cardiac inotropic peptides, indirectly acting on cardiomyocytes through a Ca^{2+}-independent/ PI3K-dependent NO release from endothelial cells. This pathway appears to be triggered by the interaction of these peptides with the plasma membrane, as suggested by the biochemical features of VS-1 and CST, structurally characterized by amphipathic properties, and their ability to interact with mammalian and microbial membranes. However, recent data suggest that both VS-1 and CST are also able to exert direct cardioprotective effects in isolated cardiomyocytes, independently from the presence of endothelial cells. Interestingly, both direct and indirect effects seem to be characterized by the absence of specific membrane receptors on target cells, highlighting intriguing novelties in the topic of cell signalling, in particular respect to an hypothetical receptor-independent eNOS activation.

Abbreviations

ANP Atrial natriuretic peptide
AR   Adrenergic receptors

G. Alloatti (✉)
Department of Life Sciences and Systems Biology, University of Turin, Turin, Italy

National Institute for Cardiovascular Research (INRC), Bologna, Italy
e-mail: giuseppe.alloatti@unito.it

M.P. Gallo
Department of Life Sciences and Systems Biology, University of Turin, Turin, Italy
e-mail: mariapia.gallo@unito.it

BAE-1	Bovine Aortic Endothelial cells
BDM	Butanedione monoxime
bFGF	Basic fibroblast growth factor
BNP	Brain natriuretic peptide
cAMP	Cyclic Adenosine monophosphate
Cav1	Caveolin 1
CD	Circular dichroism
CgA	Chromogranin A
cGMP	Cyclic GMP
Chr	Chromofungin
CPPs	Cell penetrating peptides
CST	Catestatin
EE	Endocardial endothelium
eNOS	Endothelial nitric oxide synthase
ET-1	Endothelin-1
GC	Guanylate cyclase
GSK3β	Glycogen synthase kinase 3β
hrVS-1	Human recombinant Vasostatin-1
HSPGs	Heparan Sulfate Proteoglycans
I/R	Ischemia and reperfusion
$I_{Ca,L}$	L-type calcium current
Iso	Isoproterenol
L-NAME	NG-nitro-L-arginine methyl ester
L-NMMA	L-N^G-monomethyl Arginine
MPP	Mitochondrial membrane potential
NMR	Nuclear magnetic resonance
NO	Nitric oxide
PDE	Phosphodiesterase
PI3K	Phosphatidylinositol 3-Kinase
PKG	Protein kinase G
PLN	Phospholamban
PTX	Pertussis toxin
TNFα	Tumour necrosis factor α
VS	Vasostatin
Wm	Wortmannin

1 CgA-Derived Peptides: Novel Regulators of Cardiovascular System

The NH$_2$-terminal fragments of Chromogranin-A (CgA) generated by cleavage at the first and second pair of basic amino acid residues of NH$_2$-terminal domain of CgA, have been termed Vasostatins (VSs) for their vasoinhibitory action in conduit and resistance vessels (Brekke et al. 2002). Initially identified as a potent endogenous nicotinic–cholinergic antagonist, Catestatin (CgA344–364; CST) has subsequently been shown to play a role as a novel regulator of cardiac function and blood pressure. Several findings indeed suggest that CST may act as an endogenous vascular tone regulator, possibly involved in predisposition to hypertension, and a modulator of cardiorespiratory control in the brain stem (Fung et al. 2010; Mahapatra et al. 2005; Mahata et al. 2010; Rao et al. 2007). Besides VSs and CST, Chromofungin (Chr: CgA47–66) and Serpinin, two other CgA-derived peptides displaying cardiovascular modulatory activities, have been recently isolated and characterized (Filice et al. 2015; Tota et al. 2012). In addition, full-length CgA itself exerts negative inotropic and lusitropic effects on mammalian heart (Pasqua et al. 2013). In light of the fact that CgA is produced by human myocardium (Pieroni et al. 2007), and the broad spectrum of cardiovascular effects of CgA and derived peptides, these mediators may represent key players in neuroendocrine regulation of cardiac function and potential therapeutic targets in cardiovascular diseases. This chapter focuses on the recently discovered signalling pathways activated by CgA-derived peptides in cardiomyocytes and endothelial cells, giving insights into the mechanisms at the basis of their inotropic and cardioprotective effects. The effects exerted by CgA-derived peptides on the cardiovascular system have been reviewed in several recent papers (Angelone et al. 2012a; Di Comite and Morganti 2011; Fornero et al. 2012; Helle 2010; Helle and Corti 2015; Mazza et al. 2015; Tota et al. 2014).

2 Cardiac Effects of CgA-Derived Peptides

Taken together, the abovementioned findings indicate a role for CgA-derived peptides as endogenous regulators of cardiovascular system, mainly acting on vascular tone. Moreover, in vivo studies also suggested a physiological role of VSs and CST as inotropic regulators. However, it is difficult to establish whether the reduction of blood pressure observed in vivo experiments is due to a direct effect on cardiac contractility, rather than on vasodilation and reduced afterload, or secondary to reduced catecholamine secretion. To investigate whether CgA-derived peptides are able to directly modulate cardiac inotropism, their effects were tested on ex-vivo and in vitro cardiac preparations such as the isolated perfused heart, the papillary muscle and isolated cardiac cells.

2.1 Vasostatins

The cardiac effects of VSs were initially tested on the frog (Corti et al. 2002, 2004) and the eel heart (Imbrogno et al. 2004). On both preparations, VSs exerted an inhibitory effect on basal cardiac performance and counteracted the positive inotropism induced by adrenergic stimulation. Interestingly, while in the frog heart the effects of VSs were independent from endocardial endothelium (EE) and nitric oxide–cGMP mechanism (Corti et al. 2004), in the eel heart the VS-1-mediated negative inotropism required the presence of an intact EE and the activation of NO-cGMP-PKG pathway (Imbrogno et al. 2004). These findings provided the first evidence that vasostatins exert cardiotropic action in amphibian and fish heart, thus suggesting their long evolutionary history, as well as their species-specific mechanisms of action. Further studies on isolated working eel and frog heart preparations have been performed to study the role of the cytoskeleton in the VSs-mediated inotropic response (Mazza et al. 2007). In both eel and frog hearts, VSs-mediated-negative inotropy was abolished by treatment with inhibitors of cytoskeleton reorganization, such as cytochalasin-D, suggesting that changes in cytoskeletal dynamics play a crucial role in the negative inotropic influence of VSs on these preparations.

The negative inotropic and lusitropic effects of VSs have been further demonstrated in ex-vivo studies on the isolated rat heart. In particular, VS-1 and VS-1-derived peptides containing the disulfide bridged loop reduced cardiac contractility, both under basal conditions and after β-adrenergic stimulation (Cerra et al. 2006, 2008). The action of VS-1 involved both Gi/o protein, as suggested by the blocking effect of pertussis toxin (PTX), and NO-cGMP-PKG pathway (Cerra et al. 2008). Moreover, it has been shown that, like in the eel and frog heart (Mazza et al. 2007), cytoskeleton integrity is involved in the modulation of contractility exerted by human recombinant Vasostatin-1 (hrVS-1) and rat chromogranin A 1–64 (rCgA1–64) in the rat heart (Angelone et al. 2010). Indeed, cytoskeleton impairment by either cytochalasin-D, Butanedione monoxime (BDM), wortmannin or W-7 abolished VSs-induced inotropic response. Moreover, hrVS-1 stimulated actin polymerization in rat cardiac H9C2 cells, supporting the hypothesis that the actin cytoskeletal network strongly contributes to the cardiotropic action of CgA-derived peptides.

The fact that in the isolated rat heart the negative inotropic effect of VSs was not accompanied by significant alterations in heart rate and coronary resistances, strongly suggested a direct action of these peptides. In vitro studies on the isolated rat papillary muscle confirmed the ability of VSs to exert direct inotropic effects. Indeed, the isolated papillary muscle is driven at constant frequency and perfused at constant flow, to avoid any possible alteration of contractility due to variations of heart rate and coronary flow. In these experimental conditions, VS-1 induced dose and time-dependent effects, under both basal conditions and after β-adrenergic stimulation (Gallo et al. 2007).

In agreement with the results obtained on the isolated rat heart, hrVS-1 reduced in a concentration-dependent (5–100 nM) manner the inotropic effect of Iso in papillary muscle. The minimal effective concentration of hrVS-1 was 5 nM, while the higher concentration of hrVS-1 reduced to about 70% the inotropic response to β-adrenergic stimulation (Gallo et al. 2007).

To investigate the structure-function relationship of different VSs, the effects of two modified peptides, i.e., rCgA1–64 without the S-S bridge (rCgA1–64SH) and rCgA1–64 oxidized (rCgA1– 64OX), were compared to those exerted by N-terminal fragment of CgA, reproducing the native rat sequence (rCgA1–64 with S-S bridge:

rCgA1–64S-S). These experiments revealed that the presence of the disulfide bridge is required for the cardiotropic action of VSs. In accordance with the experiments performed on the Langendorff isolated heart (Cerra et al. 2008), rCgA1–64S-S dose-dependently reduced papillary muscle contractility both under basal conditions and after β-adrenergic stimulation. However, neither rCgA1–64SH nor CgA1–64OX affected papillary muscle inotropism.

On the basis of previous observations suggesting that the effects of hrVS-1 mainly depend on the activation of the PI3K-Akt-NO pathway and NO release from endothelial cells (Gallo et al. 2007) (see also below for further details), the effect of rCgA1–64S-S were studied in papillary muscles treated with NG-nitro-L-arginine methyl ester (L-NAME) or with wortmannin (Wm). Pharmacological blockade of both NO synthesis or PI3K activation abrogated the negative inotropic effect of rCgA1–64S-S, suggesting that the PI3K-Akt-NO pathway plays an important role in the inotropic effect induced by VSs.

2.2 Catestatin

The experiments performed on the isolated rat heart showed that, in contrast to VS-1, in basal conditions CST increased heart rate and coronary resistances in a dose-dependent manner, suggesting that these two peptides may display specific different activities in this preparation. On the other hand, CST caused a vasorelaxant influence when coronary arteries were pre-contracted by endothelin-1 (ET-1), through a Gi/o protein–NO–cGMP-dependent mechanism (Angelone et al. 2008). CST (10–50 nM) induced a biphasic effect on isolated rat papillary muscle, characterized by an early transient increase in contractile force, followed by a negative, antiadrenergic effect, similar to that previously reported for the isolated heart (Bassino et al. 2011). Under basal conditions, while a low concentration (5 nM) of CST had no significant effect on myocardial contractility, higher concentrations (10–50 nM) induced a transient positive inotropic effect, reaching near to 50% over control, which was completely reverted within a few minutes. The early positive effect of CST was probably caused by histamine release from cardiac mast cells and H_1 receptors activation (for further details, see below). At concentrations between 5–50 nM, CST exerted a significant anti-adrenergic effect, which was blocked by inhibition of PI3K, NO synthesis or cGMP, thus suggesting that, as in the case of VSs, PI3K, NO and cGMP play an important role in the cardiac effects induced by CST. Additional experiments have been performed to study the effects of two naturally occurring variants of CST (G364S-CST and P370L-CST). In basal conditions, the effects of both G364S-CST and P370L-CST were comparable to that induced by WT-CST, being ineffective at a low concentration (5 nM), while higher concentrations (10–50 nM) induced a transient positive inotropic effect. However, only P370L-CST was able to reduce the positive inotropic effect exerted by β-adrenergic stimulation, while G364S-CST failed to modulate the effect of Isoproterenol (Iso) (Bassino et al. 2011).

The influence exerted by CST on the Frank-Starling response of both normotensive Wistar Kyoto and hypertensive rat hearts was evaluated in a recent study (Angelone et al. 2015). In both rat strains, CST administration improved myocardial mechanical response to increased end-diastolic pressures, shifting to left the Frank–Starling curve. This effect of CST involved the vascular endothelium and required the activation of the AKT/NOS/NO/cGMP/PKG cascade. As suggested by the Authors of this study, the parallel increase of myocardial protein S-Nitrosylation may explain the apparent contradiction of this finding with the previously described negative inotropic effect of CST. Indeed, it has been shown that, in the mammalian heart, among several targets, S-Nitrosylation modulates RyR channels in the sarcoplasmic reticulum, as well as potassium channels and L-type calcium channels at the plasma membrane level (Hess et al. 2005), thus importantly influencing calcium cycling and contractility. The finding that CST is able to modulate the stretch-induced intrinsic regulation of the heart, suggests that this peptide may play an important role in the aged hypertrophic heart, whose function is impaired because of a reduced systolic performance accompanied by delayed relaxation and increased diastolic stiffness.

Further studies showed that, besides mammalian heart, CST also functions as an important negative modulator of heart performance in frog (Mazza et al. 2008) and eel heart (Imbrogno et al. 2010). In the frog heart, CST dose-dependently decreased stroke volume and stroke work, and reduced the positive inotropic effect induced by Iso or ET-1; the threshold concentration was 11 nM, a value comparable to the circulating levels of this peptide (~2–4 nM). By using pharmacological blockers, Mazza et al. (2008) showed that CST effects were due to NOS, guanylate cyclase (GC) and ET(B) receptor activation, as well as phospholamban (PLN) phosphorylation.

In the eel heart, CST was able to reduce both basal contractility and the positive inotropic response induced by β-adrenergic stimulation. In addition, CST induced a significant increase of the Frank–Starling response, which was blocked by L-NMMA and thapsigargin, but independent from GC (Imbrogno et al. 2010). Taken together, these reports indicated that in both frog and fish, CST is able to modulate myocardial performance under basal, as well as under increased preload, conditions, and counteracts the adrenergic-mediated positive inotropism, thus strikingly supporting the evolutionary significance and establishing the cardiomodulatory role of this peptide.

2.3 Chromofungin

Chromofungin (CgA47–66) is a CgA-derived peptide implicated in inflammation and innate immunity, displaying antimicrobial activities and activating neutrophils. The effects of Chr have been recently tested on the isolated Langendorff perfused rat heart (Filice et al. 2015). Under basal conditions, Chr induced dose-dependent negative inotropic effects, while coronary pressure was unaffected. The negative effect of Chr was mediated by the AKT/eNOS/cGMP/PKG pathway. These results suggest that, among CgA-derived peptides, also Chr may considered as a new physiological neuroendocrine modulator of cardiac function.

2.4 Serpinin

Recent data suggest that serpinin peptides act as novel β-adrenergic-like cardiac modulators (Tota et al. 2012). Three forms of serpinin peptides, serpinin (Ala26Leu), pyroglutaminated (pGlu)-serpinin (pGlu23Leu) and serpinin-Arg-Arg-Gly (Ala29Gly) derive from cleavage at pairs of basic residues in the highly conserved C terminus of CgA. Serpinin and pGlu-serpinin exert dose-dependent positive inotropic and lusitropic effects, while Ala29Gly was unable to affect myocardial performance. Moreover, pGlu-serpinin was able to induce positive inotropism also on the isolated rat papillary muscle preparation. Both pGlu-serpinin and serpinin act through a β1-AR/AC/cAMP/PKA pathway, indicating that, contrary to the β-blocking profile of the other CgA-derived cardiosuppressive peptides, VS-1, CST and Chr, these two C-terminal peptides act as β-adrenergic-like agonists, suggesting that CgA derived peptides can play a key role on the modulation of myocardial performance.

Taken together, the experiments performed on isolated perfused hearts of different animal species (amphibian, fish and mammal) as well as on isolated rat papillary muscle, suggested a long evolutionary history of CgA-derived peptides as cardiotropic agents. In particular, while VS-1, CST and Chr displayed negative inotropic and lusitropic effects, opposing to β-adrenergic stimulation, Serpinin and pGlu-serpinin were able to potentiate cardiac performance, eliciting positive inotropic and lusitropic effects. In general, apart from their species-specific mechanisms of action, the negative effects of VS-1, CST and Chr appear to be mediated by endocardial/vascular endothelial cells, and activation of the AKT/eNOS/cGMP/PKG pathway. In contrast, the cardio-stimulatory action of Serpinin and pGlu-serpinin requires a β1-adrenergic receptor/adenylate cyclase/cAMP/PKA pathway. As suggested by Angelone et al. (2012a), all these so far obtained findings indicate that *"the prohormone CgA appears to possess two cardioactive limbs, i.e. an inhibitory N-terminal region and a C-terminal stimulatory domain. These two limbs may function according to a ying/yang strategy whose spatial and temporal traits remain the goal for future research."*

The main effects exerted by CgA derived peptides on cardiac function and related pathways are summarized in Table 1.

Table 1 Main effects exerted by CgA-derived peptides on cardiac contractility and related pathways

Peptide	Model	Force of contraction - Basal condition	β-adrenergic stimulation	Endothelium	NO/cGMP/PKG	Cytoskeleton
Full length CgA	Isolated heart	Reduction Rat (Pasqua et al. 2013)	ND	YES Rat (Pasqua et al. 2013)	YES Rat (Pasqua et al. 2013)	ND
VS-1	Isolated heart	Reduction Frog (Corti et al. 2002, 2004; Mazza et al. 2007) Eel (Imbrogno et al. 2004; Mazza et al. 2007) Rat (Angelone et al. 2010, Cerra et al. 2006, 2008)	Reduction Frog (Corti et al. 2002, 2004; Mazza et al. 2007) Eel (Imbrogno et al. 2004; Mazza et al. 2007) Rat (Angelone et al. 2010; Cerra et al. 2006, 2008)	NO Frog (Corti et al. 2002, 2004) YES Eel (Imbrogno et al. 2004) Rat (Angelone et al. 2010; Cerra et al. 2006, 2008)	NO Frog (Corti et al. 2002, 2004) YES Eel (Imbrogno et al. 2004) Rat (Angelone et al. 2010; Cerra et al. 2006, 2008)	YES Frog (Mazza et al. 2007) Eel (Mazza et al. 2007) Rat (Angelone et al. 2010)
	Papillary muscle	Reduction Rat (Gallo et al. 2007)	Reduction Rat (Gallo et al. 2007)	YES Rat (Gallo et al. 2007)	YES Rat (Gallo et al. 2007)	ND
CST	Isolated heart	Reduction Frog (Mazza et al. 2008) Eel (Imbrogno et al. 2010) Rat (Angelone et al. 2008, 2012b)	Reduction Frog (Mazza et al. 2008) Eel (Imbrogno et al. 2010) Rat (Angelone et al. 2008, 2012b)	YES Rat (Angelone et al. 2012b)	YES Frog (Mazza et al. 2008) Eel (Imbrogno et al. 2010) Rat (Angelone et al. 2008, 2012b)	ND
	Papillary muscle	Increase Rat (Bassino et al. 2011)	Reduction Rat (Bassino et al. 2011)	YES Rat (Bassino et al. 2011)	YES Rat (Bassino et al. 2011)	ND
Chromofungin	Isolated heart	Reduction Rat (Filice et al. 2015)	ND	ND	YES Rat (Filice et al. 2015)	ND
Serpinin	Isolated heart	Increase Rat (Tota et al. 2012)	ND	ND	ND	ND
	Papillary muscle	Increase Rat (Tota et al. 2012)	ND	ND	ND	ND

The table summarizes the main effects exerted by CgA and its derived peptides VS-1, CST, Chromofungin and Serpinin on different cardiac experimental models, and the involvement of endothelial cells, NO/cGMP/PKG pathway and cytoskeleton in their action. The role for a particular mechanism is indicated by YES or NO; *ND* not determined

3 Endothelial-Mediated Effects of VS-1 and CST

To investigate the possible role of endothelial-derived mediators in the action of CgA-derived peptides, the effects of VSs and CST were initially tested on the frog heart. In contrast with the mammalian heart, in which both coronary vascular and endocardial endothelium are present, in the avascular frog heart the endocardial endothelium represents the only barrier between blood and cardiac cells. The fact that both VSs (Corti et al. 2002) and CST (Mazza et al. 2008) maintained their ability to reduce cardiac contractility also in the frog heart, firstly suggested a direct cardiac action of these peptides.

Studies performed on the isolated rat heart confirmed the involvement of the vascular endothelium in CST-induced negative inotropism also in mammals (Angelone et al. 2012b). CST-induced negative inotropism and lusitropism involved β2/β3-adrenergic receptors (AR). In particular, CST interaction with β2-AR activated PI3K/eNOS pathway, increased cGMP levels, and induced activation of type 2 phosphodiesterases (PDE2), leading to a decrease of cAMP levels. The abrogation of CST-dependent negative effect following functional denudation of the endothelium with Triton X-100 strongly suggested that the action of CST is due to stimulation of the vascular endothelium. This assumption was further confirmed by the ability of CST to stimulate eNOS phosphorylation in both cardiac tissue and cultured human umbilical vein endothelial cells. CST also increased S-nitrosylation of both phospholamban and β-arrestin in ventricular extracts. Taken together, these data suggest that endothelium derived NO, PDE2 and S-nitrosylation play crucial roles in the CST regulation of cardiac function (Angelone et al. 2012b).

To test the involvement of NO released from endocardial endothelial cells, rat papillary muscles were treated with VS-1 or CST after endothelium has been removed with Triton X-100. While this protocol did not modify the inotropic effect induced by Iso, the anti-adrenergic effect of the two CgA-derived peptides was completely abolished, suggesting the endothelial origin of NO (Bassino et al. 2011; Gallo et al. 2007). Studies from our laboratory provided novel information on the mechanisms of the cardiac antiadrenergic action of CgA-derived peptides, highlighting the crucial involvement of a Ca^{2+}-independent/PI3K-dependent NO release from endothelial cells. To characterize the mechanisms responsible for the antiadrenergic effect of VS-1 and CST, we studied the effect of these peptides on L-type calcium current ($I_{Ca,L}$) or Ca^{2+} transients in isolated rat ventricular cardiomyocytes. In agreement with the results obtained in isolated papillary muscles treated with Triton X-100, L-type calcium current or Ca^{2+} transients measurements confirmed that, in the absence of endothelial cells, VS-1 and CST are ineffective both in basal conditions and after Iso stimulation (Bassino et al. 2011; Gallo et al. 2007). The lack of effect of VS-1 and CST on basal and Iso-stimulated Ca^{2+} transients amplitude represents an indirect demonstration that the ionic currents involved in the action potential are not affected by these peptides. In addition, the lack of any inhibitory effect of the two CgA-derived peptides on isolated ventricular cells confirmed that, at least in physiological conditions and with acute stimulation by the peptides, they do not act directly on cardiomyocytes but, rather, on other cell types present in cardiac tissue, presumably endothelial cells, as previously suggested by the experiments performed on papillary muscles. Further experiments performed on Bovine Aortic Endothelial cells (BAE-1) support the role of NO released by endocardial endothelium in the effects exerted by VS-1 and CST. In particular, they indicated that both VS-1 and CST promote the release of NO from endothelial cells by means of a Ca^{2+}-independent, PI3-K-dependent mechanism. Indeed, in contrast with ATP, VS-1 and CST enhanced NO production with a mechanism that was independent of intracellular Ca^{2+} concentration, and PI3-K blockade abolished the VS-1/CTS-dependent NO increase in BAE-1 cells. Additional experiments performed on BAE-1 cells reinforced the evidence of an endothelial NO production through a Ca^{2+}-independent, Akt-dependent eNOS phosphorylation (Bassino et al. 2011; Gallo et al. 2007), a pathway previously reported in endothelial cells for insulin, insulin-like growth factor-1, and oestrogens (Dimmeler et al. 1999; Hartell et al. 2005; Maniatis et al. 2006; Shaul 2002). Afterwards, we tested the hypothesis that both peptides could induce a caveolae dependent endocytosis, resulting in Akt-eNOS activation, by interacting with membrane heparan sulphate proteoglycans. First we investigated this pathway for VS-1, grounding on its amphipathic properties and interactions with mammalian and microbial membranes (Kang and Yoo 1997; Maget-Dana et al. 2002). In fact, it has been shown that endocytosis plays a major role in the signaling of different basic and amphipathic exogenous peptides, that is, Antp, R9, and Tat (Duchardt et al. 2007). Our experiments performed on BAE-1 cells clearly showed that VS-1 strongly increases endocytotic vesicles trafficking, thus supporting the hypothesis that this peptide acts through a similar mechanism (Ramella et al. 2010). Moreover, given the critical requirement for the surface Heparan Sulfate Proteoglycans (HSPGs) for endocytosis of cationic peptides (Poon and Gariepy 2007), we supposed a receptorial-like role for HSPGs in the VS-1 pathway. Along with this knowledge, we observed that HSPGs removal by treatment of BAE-1 cells with heparinase completely abolished the VS-1-dependent endocytosis. Moreover, heparinase also reverted the VS-1-induced displacement of caveolin 1 (Cav1) from plasma membrane to cytoplasm and the VS-1-dependent increase in eNOS Ser^{1179} phosphorylation. Since HSPGs and extracellular matrix also seem to participate in the mechanosensing that mediates NO production in response to shear stress (Florian et al. 2003), our results enhance the relevance of the HSPGs–NO axis in the control of the vasomotor tone.

We also observed that VS-1-induced vesicles trafficking and Cav1 displacement were both suppressed by Wortmannin, suggesting that the PI3K pathway plays a central role in the VS-1-activated cellular signaling, by regulating the endocytotic process, the Cav1 trafficking, and the eNOS phosphorylation mechanism. Our findings are in agreement with the important role played by PI3K in membrane budding and fission in endothelial cells (Li et al. 1995; Niles and Malik 1999). In addition, both membrane remodeling and actin filament dynamics during endocytotic traffic are strictly related to the PI3K/eNOS pathway, as suggested by the ability of guanosine triphosphatase dynamin to regulate vesicle scission and to interact with both PI3K and eNOS, causing its activation (Schafer 2004).

In a further work from our laboratory, we investigated this pathway also for CST (Fornero et al. 2014). To this purpose, we studied CST colocalization with heparan sulphate proteoglycans, the effect of CST on endocytotic vesicles trafficking and caveolin 1 internalization, and the modulation of CST-dependent eNOS activation on bovine aortic endothelial cells. Our results demonstrated that CST (5 nM) colocalizes with heparan sulphate proteoglycans and induces a marked increase in the caveolae-dependent endocytosis and Cav 1 internalization; the effects of CST were significantly reduced by pretreatment with heparinase or wortmannin. Our conclusion was that, similarly to VS-1 (Ramella et al. 2010), the intracellular cascade activated by CST in endothelial cells depends on proteoglycans/PI3K-dependent caveolae endocytosis acting as the initiating factor.

Moreover, CTS was unable to induce Ser^{1179} eNOS phosphorylation after pretreatments with heparinase and methyl-β-cyclodextrin. These results are consistent with the biochemical reports on CST, suggesting that this peptide, like other members of the cell penetrating peptides (CPPs) family, exhibits membrane-interaction properties because of both its amphipathic structure and extended hydrophobic region. In particular, circular dichroism (CD) and nuclear magnetic resonance (NMR) spectroscopy data indicated that CST folds into a short helical conformation that interacts with membranes and causes considerable disordering at the level of the phospholipid head groups. Moreover, two of the five residues of the helical region of CST are arginines, an amino acid that has been proposed to form hydrogen bond interactions with phospholipids (Sugawara et al. 2010).

Our experiments also show that CST activated endocytosis required the presence of HSPGs on the surface of endothelial cells and that CST colocalizes with HSPGs. The strong anionic charge present in proteoglycans makes them favorable binding sites for cationic polymers, lipids, and polypeptides, which are used for drug and gene delivery (Belting 2003; Rabenstein 2002). Negatively charged carbohydrates, like HSPGs, located on the plasma membrane may serve as electrostatic traps for the cationic CPPs (Jones and Howl 2012). Interestingly, the most prominent glycosaminoglycans on the surface of endothelial cells are precisely heparan sulphates and one of the major protein core families of HSPGs is the membrane-bound glypicans, that are enriched in caveolae, where a series of molecules involved with eNOS signalling are localized (Fleming 2010; Tarbell 2010). Furthermore, glypican-1 has been hypothesized to be the mechanosensor for eNOS phosphorylation and activation in the shear stress induced response (Lopez-Quintero et al. 2009). It could be speculated that the CST mediated membrane perturbation through HSPGs binding and phospholipid interactions could resemble the acute membrane perturbation involved in shear stress.

This matter, together with our previous finding of CST dependent eNOS activation (Bassino et al. 2011), led us to propose a CST induced mechanism of caveolae endocytosis and consequent eNOS activation. These assumptions are supported by our immunofluorescence experiments on Cav1 transfection and Cav1/eNOS and colocalization. In particular, to test whether the endocytotic process triggered by CST is caveolae-dependent, BAE-1 cells were transfected with GFP-Cav1 plasmid. We observed that in transfected live cells GFP-Cav1 signal was confined in plasma membranes, while in the presence of 5 nM CST green fluorescence appeared diffused in the cytosol, as a consequence of Cav1 internalization. Of note, CST-induced Cav1 internalization was significantly reduced by pretreatment with heparinase.

As CST stimulates endocytosis, induces Cav1 internalization and enhances NO production in endothelial cells (Bassino et al. 2011), we hypothesized that eNOS activation is mediated by the displacement of the protein from Cav1 binding. Previous reports have indeed proposed a mechanism of eNOS activation coupled with caveolae internalization (Maniatis et al. 2006; Sanchez et al. 2009) and dissociation of eNOS from Cav1 has been shown as a marker of eNOS activation (Fleming 2010; Minshall et al. 2003). To verify this hypothesis, we studied cellular colocalization of Cav1 and eNOS by immunofluorescence experiments. We observed that, in comparison with control conditions, CST strongly reduced eNOS/Cav1 colocalization at plasma membrane. The fact that Wm was able to restore this colocalization confirmed the role of PI3K in mediating CST intracellular signaling. Our observation that PI3K activity was required in both endocytosis and eNOS/Cav1 trafficking suggest that PI3K represents the essential key for the CST-activated intracellular signalling. Our results further confirm the notion that PI3K/Akt mediated Ser^{1179} phosphorylation of eNOS represents a common pathway among the multiple regulatory mechanisms affecting the activity of this enzyme (Fleming 2010).

Moreover, PI3K is widely reported to have an important role in membrane budding and fission in endothelial cells (Mellor et al. 2012). Finally, with the last experiments we confirmed our proposed pathway showing that caveolae disruption and HSPGs removal both abolished the CST-induced eNOS phosphorylation.

Taken together, these results highlight the obligatory role for proteoglycans and caveolae internalization in the VS-1/CTS-dependent eNOS activation in endothelial cells. Our results could clarify the mechanism responsible for the physiological properties of VS-1 and CST on endothelial cells, in particular with respect to their ability to activate the intracellular PI3K–eNOS pathways in the absence of a typical high-affinity membrane receptor.

4 Other Mediators Involved in the Action of CgA-Derived Peptides

Several data suggest that, at least in certain tissues and organs, the effects of CgA-derived peptides are due to the release or the interaction with other cell-to-cell messengers, particularly inflammatory mediators. Among these we consider histamine, ANP, BNP, TNF and bFGF.

4.1 Histamine

The potent vasodilator action of CST in rats has been explained as due to the release of the vasodilator histamine from mast cells and H_1 receptors stimulation (Kennedy et al. 1998). The most active N-terminal domain of CST (bCgA344–358: RSMRLSFRARGYGFR) caused a concentration-dependent (0.01–5 µM) release of histamine from peritoneal and pleural mast cells. CST is a very potent activator of histamine release, more active even than the wasp venom mastoparan and the neuropeptide substance P. Interestingly, CST stimulation of histamine release from rat mast cell appears to be due to the same mechanism shared by mastoparan and other cationic charged neuropeptides. The blocking effect of PTX suggested the involvement of a Gi subunit in CST- evoked histamine release (Kruger et al. 2003). To investigate the mechanism involved in the early positive inotropic effect of CST, we performed experiments in which rat papillary muscles were pretreated with mepyramine, a pharmacological blocker of H_1 histamine receptors. Indeed, it has been shown that the positive inotropic effect of histamine on rat ventricular myocardium (Hattori 1999) is due to the activation of H_1 histamine receptors, which are abundantly expressed in this tissue (Matsuda et al. 2004). The fact that H_1 histamine receptors blockade with mepyramine completely abrogated the early transient increase of contractile force induced by CST, strongly suggests that this effect is due to histamine release, possibly from mast cells present within cardiac tissue and cardiac H_1 receptors activation (Bassino et al. 2011).

4.2 ANP, BNP and TNFα

In the rat heart, immunohistochemical evidences showed the co-localization of CgA and atrial natriuretic peptide (ANP) in nonadrenergic myoendocrine atrial cells (Steiner et al. 1990). In patients with chronic heart failure, while CgA plasma level does not correlate with hormones such as catecholamines, vasopressin, endothelins and components of the renin–angiotensin system (Nicholls et al. 1996), it correlates with the levels of tumour necrosis factor (TNF)α and TNFα receptors (Corti et al. 2000), and with the levels of brain natriuretic peptide (BNP) (Pieroni et al. 2007). It has been shown that CgA colocalizes with BNP in biopsies from patients with dilated cardiomyopathy and hypertrophic myopathy (Pieroni et al. 2007). It seems therefore that CgA circulating levels in chronic heart failure reflect myocardial inflammation and distension, more than neuroendocrine autonomic activation.

The fact that CgA-derived peptides, natriuretic peptides (Costa et al. 2000) and TNFα (Alloatti et al. 1999) can stimulate endothelial cells to produce NO, which in turn diffuses to cardiac cells, reinforces the hypothesis that NO represents a key signal molecule on which CgA-derived and other peptide mediators converge, and the key role of endothelial cells in mediating the myocardial actions of CgA-derived peptides. In view of the anti-adrenergic effect of NO, CgA presumably exerts a protective effect on the myocardium, preventing excessive work in stressful conditions.

4.3 Basic Fibroblast Growth Factor

CST induced migration, proliferation and antiapoptotic effect in endothelial cells, and promoted capillary tube formation in vitro in a matrigel assay; all these effects were mediated through the activation PI3K/Akt pathway. The blockade of CST effects by a neutralizing Basic fibroblast growth factor (bFGF) antibody and the ability of this peptide to induce bFGF release strongly suggest that CST regulates endothelial cells functions by stimulating fibroblast growth factor signalling (Teurl et al. 2010).

5 The Cardioprotective Effect of CST on Isolated Cardiomyocytes Suggests a Direct Cardiac Effect

Recent data suggest that CgA-derived peptides may exert important protective effect on the heart undergoing ischemia and reperfusion (I/R). Indeed, infusion of VS-1 before ischemia significantly reduced the development of infarct size and contractile alterations during the reperfusion in the isolated rat heart (Cappello et al. 2007). The protective effect of VS-1 was abolished by either NOS inhibition or PKC blockade and was attenuated, but not suppressed, by the blockade of Adenosine (A_1) receptors, suggesting that VS-1 may trigger two different pathways, the first one mediated by A_1 receptors activation, and the other by NO release. Moreover, Yu et al. (2011) recently showed that overexpression of VS-1 in neonatal cardiomyocytes could limit I/R injury, with a mechanism independent from endothelial cells.

Data regarding the ability of CST to induce cardioprotection are, at present, conflicting. As shown by Brar et al. (2010), both the wild type and Pro^{370} Leu variants increased infarct size and decreased the cardiac levels of phosphorylated Akt and two of its downstream targets, FoxO1 and BAD, when these agents were administered during reperfusion in the Langendorff perfused rat hearts subjected to regional ischemia. However, no significant alteration was present when reperfusion occurred in the presence of the Gly^{364} Ser variant.

In contrast with these findings, it has been also shown that CST induced a cardioprotective effect when infused only in the early phases of reperfusion, thus simulating a post-conditioning effect (Penna et al. 2010; Perrelli et al. 2013). In these conditions, indeed, CST limited the extension of infarct size, reduced post-ischemic development of diastolic contracture, and significantly improved post-ischemic recovery of developed left ventricular pressure during reperfusion. These results suggest a novel cardioprotective role for CST, which appears mainly due to a direct reduction of post-ischemic myocardial damages and dysfunction, rather than to an involvement of adrenergic terminals and/or endothelium. These different results may be tentatively explained on the basis of the different models investigated (regional *vs* global ischemia) and the modality and the dose of administration of CST (the entire reperfusion period or the first 20 min of reperfusion).

It must be remembered, however, that the protective effect of CST already observed on the isolated rat heart, has been confirmed on isolated rat cardiomyocytes exposed to a protocol of simulated I/R injury (Penna et al. 2010). Interestingly, the cardioprotective effect of CST in isolated cardiomyocytes was attained a very low concentration, in comparison with that needed in the isolated heart (5 nM *vs* 75 nM), that is comparable to the circulating concentrations of this peptide found in healthy humans (O'Connor et al. 2002). Although our results do not rule out an additional role for the anti-adrenergic and/or endothelium-dependent mechanisms in the in situ heart, they strongly suggest that CST is able to attain such protection also via a direct effect on cardiomyocytes, independent from the presence of catecholamine in the extracellular milieu or of endothelial cells.

Recent experiments were performed in our laboratory to define the cardioprotective signalling pathways activated by CST on isolated adult rat cardiomyocytes (Bassino et al. 2015). To this purpose, besides cell viability rate, we also evaluated mitochondrial membrane potential (MMP). The involvement of Akt, glycogen synthase kinase 3β (GSK3β) and phospholamban (PLN) cascade was studied by immunofluorescence, while the role of PI3K-Akt pathway was investigated by using the pharmacological blocker Wm. We observed that in isolated cardiomyocytes undergoing simulated I/R, CST increased cell viability rate by 65%. The protective effect of CST was related to its ability to maintain mitochondrial membrane potential (MMP) and to increase Akt, PLN and GSK3β phosphorylation. The cardioprotective effect of CST was abolished by Wm. These results give new insights into the molecular mechanisms involved in the protective role of CST, suggesting that the cardioprotective effect of CST depends on PI3K-Akt-GSK3β cascade activation and is related to MMP stabilization. Moreover, as suggested by PLN phosphorylation, CST may enhance calcium recovery in sarcoplasmatic reticulum, thus reducing calcium overload in cardiomyocytes. Interestingly, similarly to cardiomyocytes, anti-apoptotic properties on endothelial cells were described for CST (Teurl et al. 2010).

6 Conclusions and Perspectives

Accumulating evidences point to a significant role for the CgA-derived peptides VS-1 and CST in the protective modulation of the cardiovascular activity, mainly because of their ability to counteract the adrenergic signal. Indeed, these peptides are able to control catecholamine release from chromaffin cells and noradrenergic neurons, to exert in vivo and in vitro vasodilatory effects, and to limit the inotropic and lusitropic responses to β-adrenergic stimulation of the heart. The cardio-suppressive and vasodilator properties of VS-1 and CST have been recently explained as due to a PI3K-dependent-NO release by endothelial cells. At present, the signalling pathways involved in the cardiovascular responses to VS-I and CST are only fragmentary. As typical high-affinity receptors have not been identified, the cellular processes upstream the eNOS activation exerted by these peptides are still partially unknown. In recent works performed in our laboratory, we showed that, in endothelial cells, on the basis of their cationic and amphipathic properties, both VS-1 and CST act as cell penetrating peptides, binding to heparan sulfate proteoglycans and activating eNOS phosphorylation through a PI3K-dependent, endocytosis-coupled mechanism. These results suggest a novel signal transduction pathway for endogenous cationic and amphipathic peptides in endothelial cells: HSPGs interaction and caveolae endocytosis, coupled with a PI3K-dependent eNOS phosphorylation. Besides to be advantageous to an organism under stress, being able to reduce the adrenergic signal response and to cause a vasodilatatory effect, both VS-1 and CST exert a protective effect against cell death and cardiac alterations induced by ischemia and reperfusion. Interestingly, CST is able to promote cardiomyocyte survival also in the case of isolated ventricular cells undergoing simulated I/R. This effect is attained at a very low concentration, comparable to the circulating concentrations of this peptide found in healthy humans. These results reopen the question concerning the presence of specific receptors for CTS on cardiac cells, suggesting that CST is able to attain such protection also via a direct effect on cardiomyocytes, independent from endothelial cells. In conclusion, CgA-derived peptides, in particular CST, are emerging as very important mediators regulating cardiovascular functions in stress situations, and bear all the potentials to be therapeutic agents to treat several diseases affecting the cardiovascular system, like hypertension or ischemic heart disease.

References

Alloatti G, Penna C, De Martino A, Montrucchio G, Camussi G (1999) Role of nitric oxide and platelet-activating factor in cardiac alterations induced by tumor necrosis factor-alpha in the guinea-pig papillary muscle. Cardiovasc Res 41:611–619

Angelone T, Quintieri AM, Brar BK, Limchaiyawat PT, Tota B, Mahata SK et al (2008) The antihypertensive chromogranin A peptide catestatin acts as a novel endocrine/paracrine modulator of cardiac inotropism and lusitropism. Endocrinology 149:4780–4793

Angelone T, Quintieri AM, Goumon Y, Di Felice V, Filice E, Gattuso A et al (2010) Cytoskeleton mediates negative inotropism and lusitropism of chromogranin A-derived peptides (human vasostatin1-78 and rat CgA(1-64)) in the rat heart. Regul Pept 165:78–85

Angelone T, Mazza R, Cerra MC (2012a) Chromogranin-A: a multifaceted cardiovascular role in health and disease. Curr Med Chem 19:4042–4050

Angelone T, Quintieri AM, Pasqua T, Gentile S, Tota B, Mahata SK et al (2012b) Phosphodiesterase type-2 and NO-dependent S-nitrosylation mediate the cardioinhibition of the antihypertensive catestatin. Am J Phys Heart Circ Phys 302:H431–H442

Angelone T, Quintieri AM, Pasqua T, Filice E, Cantafio P, Scavello F et al (2015) The NO stimulator, Catestatin, improves the Frank-Starling response in normotensive and hypertensive rat hearts. Nitric Oxide 50:10–19

Bassino E, Fornero S, Gallo MP, Ramella R, Mahata SK, Tota B et al (2011) A novel catestatin-induced antiadrenergic mechanism triggered by the endothelial PI3K-eNOS pathway in the myocardium. Cardiovasc Res 91:617–624

Bassino E, Fornero S, Gallo MP, Gallina C, Femmino S, Levi R et al (2015) Catestatin exerts direct protective effects on rat cardiomyocytes undergoing ischemia/reperfusion by stimulating PI3K-Akt-GSK3β pathway and preserving mitochondrial membrane potential. PlosOne. doi:10.1371/journal.pone.0119790. eCollection 2015

Belting M (2003) Heparan sulfate proteoglycan as a plasma membrane carrier. Trends Biochem Sci 28:145–151

Brar BK, Helgeland E, Mahata SK, Zhang K, O'Connor DT, Helle KB et al (2010) Human catestatin peptides differentially regulate infarct size in the ischemic-reperfused rat heart. Regul Pept 165:63–70

Brekke JF, Osol GJ, Helle KB (2002) N-terminal chromogranin-derived peptides as dilators of bovine coronary resistance arteries. Regul Pept 105:93–100

Cappello S, Angelone T, Tota B, Pagliaro P, Penna C, Rastaldo R et al (2007) Human recombinant chromogranin A-derived vasostatin-1 mimics preconditioning via an adenosine/nitric oxide signaling mechanism. Am J Physiol Heart Circ Physiol 293:H719–H727

Cerra MC, De Iuri L, Angelone T, Corti A, Tota B (2006) Recombinant N-terminal fragments of chromogranin-A modulate cardiac function of the Langendorff-perfused rat heart. Basic Res Cardiol 101:43–52

Cerra MC, Gallo MP, Angelone T, Quintieri AM, Pulera E, Filice E et al (2008) The homologous rat chromogranin A1-64 (rCGA1-64) modulates myocardial and coronary function in rat heart to counteract adrenergic stimulation indirectly via endothelium-derived nitric oxide. FASEB J 22:3992–4004

Corti A, Ferrari R, Ceconi C (2000) Chromogranin A and tumor necrosis factor-alpha (TNF) in chronic heart failure. Adv Exp Med Biol 482:351–359

Corti A, Mannarino C, Mazza R, Colombo B, Longhi R, Tota B (2002) Vasostatins exert negative inotropism in the working heart of the frog. Ann N Y Acad Sci 971:362–365

Corti A, Mannarino C, Mazza R, Angelone T, Longhi R, Tota B (2004) Chromogranin A N-terminal fragments vasostatin-1 and the synthetic CGA 7-57 peptide act as cardiostatins on the isolated working frog heart. Gen Comp Endocrinol 136:217–224

Costa MD, Bosc LV, Majowicz MP, Vidal NA, Balaszczuk AM, Arranz CT (2000) Atrial natriuretic peptide modifies arterial blood pressure through nitric oxide pathway in rats. Hypertension 35:1119–1123

Di Comite G, Morganti A (2011) Chromogranin A: a novel factor acting at the cross road between the neuroendocrine and the cardiovascular systems. J Hypertens 29:409–414

Dimmeler S, Fleming I, Fisslthaler B, Hermann C, Busse R, Zeiher AM (1999) Activation of nitric oxide synthase in endothelial cells by Akt-dependent phosphorylation. Nature 399:601–605

Duchardt F, Fotin-Mleczek M, Schwarz H, Fischer R, Brock R (2007) A comprehensive model for the cellular uptake of cationic cell-penetrating peptides. Traffic 8:848–866

Filice E, Pasqua T, Quintieri AM, Cantafio P, Scavello F, Amodio N et al (2015) Chromofungin, CgA47-66-derived peptide, produces basal cardiac effects and postconditioning cardioprotective action during ischemia/reperfusion injury. Peptides 71:40–48

Fleming I (2010) Molecular mechanisms underlying the activation of eNOS. Pflugers Arch 459:793–806

Florian JA, Kosky JR, Ainslie K, Pang Z, Dull RO, Tarbell JM (2003) Heparan sulfate proteoglycan is a mechanosensor on endothelial cells. Circ Res 93:136–142

Fornero S, Bassino E, Gallo MP, Ramella R, Levi R, Alloatti G (2012) Endothelium dependent cardiovascular effects of the Chromogranin A-derived peptides Vasostatin-1 and Catestatin. Curr Med Chem 19:4059–4067

Fornero S, Bassino E, Ramella R, Gallina C, Mahata SK, Tota B et al (2014) Obligatory role for endothelial heparan sulphate proteoglycans and caveolae internalization in Catestatin-dependent eNOS activation. Biomed Res Int 2014:783623

Fung MM, Salem RM, Mehtani P, Thomas B, Lu CF, Perez B et al (2010) Direct vasoactive effects of the chromogranin A (CHGA) peptide catestatin in humans in vivo. Clin Exp Hypertens 32:278–287

Gallo MP, Levi R, Ramella R, Brero A, Boero O, Tota B et al (2007) Endothelium-derived nitric oxide mediates the antiadrenergic effect of human vasostatin-1 in rat ventricular myocardium. Am J Physiol Heart Circ Physiol 292:H2906–H2912

Hartell NA, Archer HE, Bailey CJ (2005) Insulin-stimulated endothelial nitric oxide release is calcium independent and mediated via protein kinase B. Biochem Pharmacol 69:781–790

Hattori Y (1999) Cardiac histamine receptors: their pharmacological consequences and signal transduction pathways. Methods Find Exp Clin Pharmacol 21:123–131

Helle KB (2010) The chromogranin A-derived peptides vasostatin-I and catestatin as regulatory peptides for cardiovascular functions. Cardiovasc Res 85:9–16

Helle KB, Corti A (2015) Chromogranin A: a paradoxical player in angiogenesis and vascular biology. Cell Mol Life Sci 72:339–348

Hess DT, Matsumoto A, Kim SO, Marshall HE, Stamler JS (2005) Protein S-nitrosylation: purview and parameters. Nat Rev Mol Cell Biol 6:150–166

Imbrogno S, Angelone T, Corti A, Adamo C, Helle KB, Tota B (2004) Influence of vasostatins, the chromogranin A-derived peptides, on the working heart of the eel (Anguilla anguilla): negative inotropy and mechanism of action. Gen Comp Endocrinol 139:20–28

Imbrogno S, Garofalo F, Cerra MC, Mahata SK, Tota B (2010) The catecholamine release-inhibitory peptide catestatin (chromogranin A344-364) modulates myocardial function in fish. Exp Biol 213:3636–3643

Jones S, Howl J (2012) Enantiomer-specific bioactivities of peptidomimetic analogues of mastoparan and mitoparan: characterization of inverso mastoparan as a highly efficient cell penetrating peptide. Bioconjug Chem 23:47–56

Kang YK, Yoo SH (1997) Identification of the secretory vesicle membrane binding region of chromogranin A. FEBS Lett 404:87–90

Kennedy BP, Mahata SK, O'Connor DT, Ziegler MG (1998) Mechanism of cardiovascular actions of the chromogranin A fragment catestatin in vivo. Peptides 19:1241–1248

Kruger PG, Mahata SK, Helle KB (2003) Catestatin (CgA344-364) stimulates rat mast cell release of histamine in a manner comparable to mastoparan and other cationic charged neuropeptides. Regul Pept 114:29–35

Li S, Okamoto T, Chun M, Sargiacomo M, Casanova JE, Hansen SH et al (1995) Evidence for a regulated interaction between heterotrimeric G proteins and caveolin. J Biol Chem 270:15693–15701

Lopez-Quintero SV, Amaya R, Pahakis M, Tarbell JM (2009) The endothelial glycocalyx mediates shear-induced changes in hydraulic conductivity. Am J Physiol Heart Circ Physiol 296:H1451–H1456

Maget-Dana R, Metz-Boutigue MH, Helle KB (2002) The N-terminal domain of chromogranin A (CgA1-40) interacts with monolayers of membrane lipids of fungal and mammalian compositions. Ann N Y Acad Sci 971:352–354

Mahapatra NR, O'Connor DT, Vaingankar SM, Hikim AP, Mahata M, Ray S et al (2005) Hypertension from targeted ablation of chromogranin A can be rescued by the human ortholog. J Clin Invest 115:1942–1952

Mahata SK, Mahata M, Fung MM, O'Connor DT (2010) Catestatin: a multifunctional peptide from chromogranin A. Regul Pept 162:33–43

Maniatis NA, Brovkovych V, Allen SE, John TA, Shajahan AN, Tiruppathi C et al (2006) Novel mechanism of endothelial nitric oxide synthase activation mediated by caveolae internalization in endothelial cells. Circ Res 99:870–877

Matsuda N, Jesmin S, Takahashi Y, Hatta E, Kobayashi M, Matsuyama K et al (2004) Histamine H1 and H2 receptor gene and protein levels are differentially expressed in the hearts of rodents and humans. J Pharmacol Exp Ther 309:786–795

Mazza R, Mannarino C, Imbrogno S, Barbieri SF, Adamo C, Angelone T et al (2007) Crucial role of cytoskeleton reorganization in the negative inotropic effect of chromogranin A-derived peptides in eel and frog hearts. Regul Pept 138:145–151

Mazza R, Gattuso A, Mannarino C, Brar BK, Barbieri SF, Tota B et al (2008) Catestatin (chromogranin A344-364) is a novel cardiosuppressive agent: inhibition of isoproterenol and endothelin signaling in the frog heart. Am J Physiol Heart Circ Physiol 295:H113–H122

Mazza R, Tota B, Gattuso A (2015) Cardio-vascular activity of catestatin: interlocking the puzzle pieces. Curr Med Chem 22:292–304

Mellor P, Furber LA, Nyarko JN, Anderson DH (2012) Multiple roles for the p85α isoform in the regulation and function of PI3K signalling and receptor trafficking. Biochem J 441:23–37

Minshall RD, Sessa WC, Stan RV, Anderson RG, Malik AB (2003) Caveolin regulation of endothelial function. Am J Physiol Lung Cell Mol Physiol 285:L1179–L1183

Nicholls DP, Onuoha GN, McDowell G, Elborn JS, Riley MS, Nugent AM et al (1996) Neuroendocrine changes in chronic cardiac failure. Basic Res Cardiol 91(Suppl 1):13–20

Niles WD, Malik AB (1999) Endocytosis and exocytosis events regulate vesicle traffic in endothelial cells. J Membr Biol 167:85–101

O'Connor DT, Kailasam MT, Kennedy BP, Ziegler MG, Yanaihara N, Parmer RJ (2002) Early decline in the catecholamine release-inhibitory peptide catestatin in humans at genetic risk of hypertension. J Hypertens 20:1335–1345

Pasqua T, Corti A, Gentile S, Pochini L, Bianco M, Metz-Boutigue MH et al (2013) Full-length human chromogranin-A cardioactivity: myocardial, coronary, and stimulus-induced processing evidence in normotensive and hypertensive male rat hearts. Endocrinology 154:3353–3365

Penna C, Alloatti G, Gallo MP, Cerra MC, Levi R, Tullio F et al (2010) Catestatin improves postischemic left ventricular function and decreases ischemia/reperfusion injury in heart. Cell Mol Neurobiol 30:1171–1179

Perrelli MG, Tullio F, Angotti C, Cerra MC, Angelone T, Tota B et al (2013) Catestatin reduces myocardial ischaemia/reperfusion injury: involvement of PI3K/Akt, PKCs, mitochondrial KATP channels and ROS signalling. Pflugers Arch 465:1031–1040

Pieroni M, Corti A, Tota B, Curnis F, Angelone T, Colombo B et al (2007) Myocardial production of chromogranin A in human heart: a new regulatory peptide of cardiac function. Eur Heart J 28:1117–1127

Poon GM, Gariepy J (2007) Cell-surface proteoglycans as molecular portals for cationic peptide and polymer entry into cells. Biochem Soc Trans 35:788–793

Rabenstein DL (2002) Heparin and heparan sulfate: structure and function. Nat Prod Rep 19:312–331

Ramella R, Boero O, Alloatti G, Angelone T, Levi R, Gallo MP (2010) Vasostatin 1 activates eNOS in endothelial cells through a proteoglycan-dependent mechanism. J Cell Biochem 110:70–79

Rao F, Wen G, Gayen JR, Das M, Vaingankar SM, Rana BK et al (2007) Catecholamine release-inhibitory peptide catestatin (chromogranin A(352-372)): naturally occurring amino acid variant Gly364Ser causes profound changes in human autonomic activity and alters risk for hypertension. Circulation 115:2271–2281

Sanchez FA, Rana R, Kim DD, Iwahashi T, Zheng R, Lal BK et al (2009) Internalization of eNOS and NO delivery to subcellular targets determine agonist-induced hyperpermeability. Proc Natl Acad Sci U S A 106:6849–6853

Schafer DA (2004) Regulating actin dynamics at membranes: a focus on dynamin. Traffic 5:463–469

Shaul PW (2002) Regulation of endothelial nitric oxide synthase: location, location, location. Annu Rev Physiol 64:749–774

Steiner HJ, Weiler R, Ludescher C, Schmid KW, Winkler H (1990) Chromogranins A and B are co-localized with atrial natriuretic peptides in secretory granules of rat heart. J Histochem Cytochem 38:845–850

Sugawara M, Resende JM, Moraes CM, Marquette A, Chich JF, Metz-Boutigue et al (2010) Membrane structure and interactions of human catestatin by multidimensional solution and solid-state NMR spectroscopy. FASEB J 24:1737–1746

Tarbell JM (2010) Shear stress and the endothelial transport barrier. Cardiovasc Res 87:320–330

Teurl M, Schgoer W, Albrecht K, Jeschke J, Egger M, Beer AG et al (2010) The neuropeptide catestatin acts as a novel angiogenic cytokine via a basic fibroblast growth factor-dependent mechanism. Circ Res 107:1326–1335

Tota B, Gentile S, Pasqua T, Bassino E, Koshimizu H, Cawley NX et al (2012) The novel Chromogranin A-derived serpinin and pyroglutaminated serpinin peptides are positive cardiac ß-adrenergic-like inotropes. FASEB J 26:2888–2898

Tota B, Angelone T, Cerra MC (2014) The surging role of Chromogranin A in cardiovascular homeostasis. Front Chem 2:64

Yu M, Wang Z, Fang Y, Xiao MD, Yuan ZX, Lu CB et al (2011) Overexpression of Vasostatin-1 protects hypoxia/reoxygenation injuries in cardiomyocytes independent of endothelial cells. Cardiovasc Ther 30:145–151

Chromogranin A-Derived Peptides in Cardiac Pre- and Post-conditioning

Claudia Penna and Pasquale Pagliaro

Abstract *Chromogranin A* (CgA, also known as secretory protein I) is produced by cells of sympathoadrenal system, and by mammalian ventricular myocardium. In clinical settings CgA was primarily used as a marker of neuroendocrine tumors. However, in the last 10 years, many data have been published on the role of the CgA and its derived peptides, especially Catestatin and Vasostatin, in the regulation of cardiovascular function and cardiovascular disease, including heart failure and hypertension. Several CgA-derived fragments, *e.g.* Catestatin, Chromofungin, Serpinin and Vasostatin, may affect several physiological features of cardiovascular system, including inotropic and lusitropic properties of the heart. As a matter of fact, CgA processing, leading to derived peptide formation, has been proposed as a part of a complex feedback system involved in the regulation and modulation of catecholamine release and effects. The CgA system can also be regarded as a cardioprotective tool against ischemic myocardial injury that can be active before, during and/or after an ischemic insult. In fact, it has been shown that Vasostatin can trigger cardioprotective effects similar to those achieved with ischemic preconditioning (a cardioprotective phenomenon activated before ischemia). Yet, while Catestatin and Chromofungin resulted to be potent cardioprotective agents in the post-ischemic early stage, that is they are postconditioning agents (protection is activated at the onset of reperfusion), Serpinin displayed the ability to act as both pre- and post-conditioning agent. All these peptides have proven to be able to activate multiple cardioprotective pathways, and each of them displayed similar and unique properties. For instance, while both Catestatin and Vasostatin can induce nitric oxide dependent pathway, Serpinin acts *via* adenylate cyclase and cAMP/PKA pathway, and all of them can play key roles in cardioprotection against ischemia/reperfusion injury. Clearly, the exact cardioprotective mechanism of the CgA system is far from being fully understood. Here, before to consider the cardioprotective effects of CgA-derived peptides, we describe the main mechanisms of cardiac ischemic injury and protection.

C. Penna (✉) • P. Pagliaro
Department of Clinical and Biological Sciences, University of Turin,
Regione Gonzole, 10, 10043 Orbassano (TO), Italy
e-mail: claudia.penna@unito.it; pasquale.pagliaro@unito.it

Abbreviations

A/R	Anoxia/Reoxygenation
Akt	Serine/threonine protein kinase
AMI	Acute Myocardial Infarction
BNP	Brain Natriuretic Peptide
CgA	Chromogranin A
cGMP/PKG	Cyclic guanosin monophosphate/protein kinase G
Chr	Chromofungin
CST	Catestatin
eNOS	Endothelial NO synthase
ERK1/2	Extracellular regulated kinase 1/2
GSK3β	Glycogen synthase kinase 3 β
H_2O_2	Hydrogen peroxide
IP	Ischemic Preconditioning
JAK	Janus Kinase
LVDP	end Diastolic Left Ventricular Pressure
MPG	Mercaptopropionyl Glycine
MEK	Mitogen-activated protein kinase kinase
mitoK$_{ATP}$	mitochondrial ATP-dependent K$^+$ channels
mPTP	Mitochondrial permeability transition pore
NO	Nitric oxide
O_2^-	Superoxide Anion
ONOO$^-$	Peroxynitrite
P70S6K	p70 ribosomal S6 protein kinase
PI3K	Phosphoinositide 3-kinase
PKC	Protein kinase C
PKG	Protein kinase G
PLC	Phospholipase C
PostC	Postconditioning
RISK	Reperfusion Injury Salvage Kinase
RNS	Reactive Nitrogen Species
ROS	Reactive Oxygen Species
SAFE	Survivor Activating Factor Enhancement
PIP3	Phosphatidylinositol Triphosphate
Serp	Serpinin
SNO	S-nitrosylation
STAT3	Signal Transducer and Activator of Transcription 3
VS-1	Vasostatin 1
VS-2	Vasostatin 2

1 Introduction

Several endogenous cardioprotective factors, including gasotransmitters nitric oxide, hydrogen sulphide and carbon monoxide, as well as microvesicles and exosomes, may induce cardioprotection. Moreover, the cardioprotective effects may be elicited by growth factors, cytokines and many endogenous peptides, like natriuretic peptides, bradykinin, opioids and ghrelin-associated peptides (Garcia-Dorado et al. 2009; Penna et al. 2015). Among emerging peptides in the cardiovascular system, chromogranin A (CgA) derived peptides are occupying a role of paramount importance.

CgA is a key player in neuroendocrine regulation of cardiac function. (Aardal and Helle 1992; Angeletti et al. 1994; Helle et al. 2001, 2007; Pieroni et al. 2007). Intriguingly, human ventricular myocardium produces and releases CgA and brain natriuretic peptide (BNP), and, in fact, there is strong correlation between BNP and CgA circulating levels in heart failure patients. Therefore, CgA may be a potential therapeutic target in heart failure (Helle 2004) and CgA derived peptides may play a role in regulating cardiovascular function. Moreover, CgA is emerging as a prognostic marker. For instance, in a cohort of elderly patients with typical symptoms of heart failure, it was demonstrated that the plasma level of CgA is a good marker of death; in fact, it identifies those patients at increased risk of short- and long-term mortality (Goetze et al. 2014). This aspect of CgA as new marker may have a clinical relevance when the natriuretic peptide, a classical marker, is below the diagnostic cutoff values ($\leq$400 ng/l) proposed by European Society of Cardiology Guidelines 2008 (Goetze et al. 2014). The role of CgA as Heart Failure marker has also been proposed by a recent echocardiographic study conducted on 112 patients ($\geq$60 years old) with normal Ejection Fraction (18 controls and 94 with hypertension). In this study, the CgA levels resulted increased in subject with diastolic dysfunction respect to controls (Szelényi et al. 2015).

As said, CgA is a precursor of several active peptides. In fact, a proteolytic processing gives rise to several peptides of biological importance (Aardal et al. 1993; Filice et al. 2015; Hou et al. 2016; Mahata et al. 1997, 1999, 2000, 2003, 2004; Pasqua et al. 2015; Tatemoto et al. 1986) (For more details, see other chapters in this book). Catestatin (CST), Vasostatin 1 (VS-1), Vasostatin 2 (VS-2) and Serpinin (Serp) are CgA derivatives involved in the control of cardiovascular homeostasis. These fragments of CgA present different and significative cardiac effects, ranging from negative to positive inotropic effects, and their levels may be indicative of pathological conditions such as left ventricular hypertrophy or metabolic syndrome. Actually, Meng et al. (2011) have observed that in hypertensive patients the ratio of Catestatin to Norepinephrine was lower in patients with left ventricular hypertrophy respect to patients without hypertrophy; these results suggest that CST modulates the cardiac hypertrophic response to high blood pressure. However, for instance, the levels of VS-2 are significantly reduced in patients with important atherosclerosis lesions (Cappello et al. 2007). Clearly, CgA and its derived peptides have a role in the pathogenesis of hypertension, being a complex system able to modulate sympatho-adrenal tone and cardiovascular functions. Importantly, CST, VS-1 and Serp also act as cardioprotective agents against ischemia/reperfusion injury in both pre- and post-conditioning mechanisms, through specific effects, which include NO-dependent mechanisms, the activation of Reperfusion Injury Salvage Kinase signaling and the modulation of mitochondrial activity (Cappello et al. 2007; Penna et al. 2010a, b, 2014; Perrelli et al. 2013). In this chapter, we describe the studies that elucidated the cardioprotective role of CgA derivatives.

Before proceeding with this description, we treat briefly some aspects of cardiac ischemia/reperfusion (I/R) damage. Indeed, in different animal models the levels of CgA and CgA derivatives were evaluated in the presence of I/R and have been suggested as independent predictors of mortality after acute myocardial infarction. For example, in pigs after 1 h of regional myocardial ischemia followed by 3 h of reperfusion, the plasma level of N-terminal CgA (VS-1) revealed a 30% increase 1 h after the re-establishment of coronary perfusion, whereas the level of pancreastatin, another CgA fragment, did not increase in response to I/R, but decreased during the entire experiment. These results suggest a differentiated CgA processing in myocardial I/R and can reflect tissue-specific post-translational modifications and release of these peptides (Frydland et al. 2013).

We now consider the ischemia/reperfusion injury, the cardioprotective strategies and pathways and, finally, the cardioprotective effects of CgA derived peptides.

2 Ischemia/Reperfusion Injury and Cardioprotective Strategies: Preconditioning and PostConditioning (Fig. 1)

2.1 Ischemia/Reperfusion Injury

Ischemic heart disease is one of the leading causes of death in the industrialized countries. The only way to treat coronary occlusion leading to acute myocardial infarction (AMI) is based on rapid restoration of blood flow to the ischemic zone, *i.e.,* reperfusion therapy. The rapid reperfusion, however, has the potential to induce additional lethal injury, in fact it may lead to further myocardial cell death, termed lethal myocardial reperfusion injury (Jordan et al. 1999; Pagliaro et al. 2011; Piper et al. 2003). Complex biochemical and mechanical mechanisms are involved in the reperfusion injury.

In the myocardium, reperfusion injury includes cellular death, which can occur for different types of death (*e.g.* necrosis, autophagy and apoptosis). It has been proposed that necrosis can be caused mainly by ischemia as well as by reperfusion, whereas the apoptosis is typically induced by reperfusion (Zhao and Vinten-Johansen 2002). Autophagy may be both deleterious and beneficial, depending on a number of circumstances (for review see Gottlieb et al. 2009). Reperfusion injury also includes myocardial stunning, endothelial dysfunction and no-reflow phenomenon (Pagliaro et al. 2011).

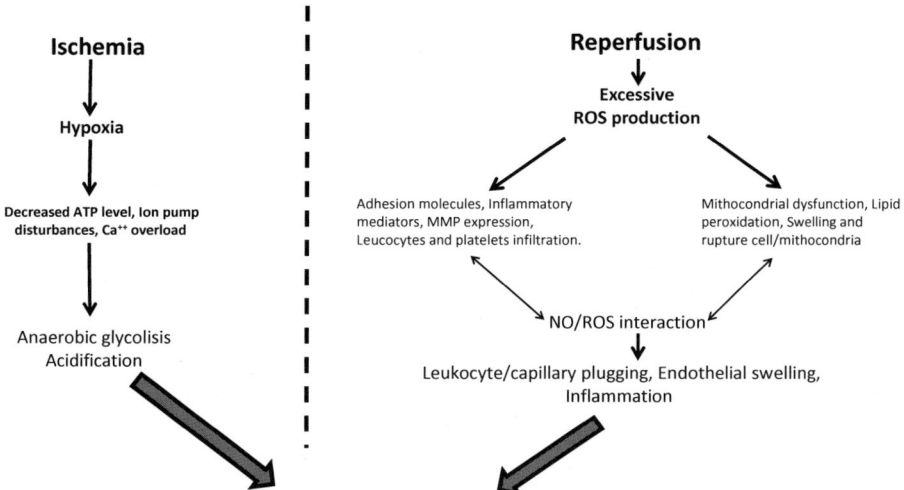

Fig. 1 Flowchart depicting the main mechanisms of myocardial ischemia/reperfusion injury. During ischemia, several mechanisms related to absence of oxygen may lead to tissues damage. Also during the reperfusion several mechanisms can be responsible for the negative effects observed after an ischemic insult. In particular, the reactive oxygen species (*ROS*) production increases with many deleterious action at vascular and cardiac levels, including No-reflow phenomenon. Of course, the final results of the processes of ischemia and reperfusion are loss of cardiac tissues (see text for more explanation)

The effects of *reperfusion injury* include (a) vascular and endothelial dysfunction, reduction of *nitric oxide* (NO) production, and consequently "no-reflow phenomenon"; (b) contractile and metabolic dysfunction; (c) arrhythmias; and (d) cell death, by apoptosis, swelling and contraction band-necrosis.

It is known that myocardial damages during reperfusion among others is due to different agents: the production/formation of *reactive oxygen species* (ROS) and *reactive nitrogen species* (RNS), the cellular/mitochondrial overload of Ca^{2+}, the activation of mitochondrial permeability transition pore (mPTP), the reduced availability of NO and to the activation of the nuclear factor kappa B (NFκB), and many other known and unknown mechanisms (For extensive reviews see Dan Dunn et al. 2015; Pagliaro and Penna 2015; Tullio et al. 2013).

The production or the formation of ROS induces an oxidative stress, which may play an important role in determining the extension of damage (Pagliaro et al. 2011; Pagliaro and Penna 2015; Tritto and Ambrosio 2001, 2013; Tullio et al. 2013). In fact, oxidative stress is responsible of direct and indirect damages of molecular components, *e.g.* oxidation of membrane components.

During myocardial ischemia, due to the occlusion of a coronary branch, the production of ROS, in particular the superoxide anion (O_2^-), increases as a result of the activation of various enzymatic complexes. In the event of reperfusion the production and formation of various ROS/RNS strongly oxidize the myocardial fibers already damaged by the ischemia, thus favoring the cell death and in particular the contraction band-necrosis and apoptosis (Ambrosio et al. 1991, 1993; Gottlieb et al. 2009; Hoffman et al. 2004; Pagliaro et al. 2011; Tritto and Ambrosio 2001; Zhao 2004; Zhao and Vinten-Johansen 2002). Superoxide anion may react with NO possibly present, forming peroxynitrite ($ONOO^-$). Actually, the scarcity of NO is correlated to the production of $ONOO^-$ (Beauchamp et al. 1999; Kaeffer et al. 1997) that takes part with O_2^- to the myocardial injury (Ferdinandy and Schulz 2003; Lefer and Lefer 1991; Ronson et al. 1999). The preserved production of NO may be also protective *via* peroxynitrite reduction, in the so-called *secondary reaction*, which is due to the reaction of $ONOO^-$ with NO and which in turn will lead to protein S-nitrosylation (SNO of proteins) (Penna et al. 2011a). This S-nitrosylation of proteins is a phenomenon involved in the cardiac effects induced by Catestatin, including modulation of cardiac force of contraction and cardioprotection in normotensive and hypertensive rat hearts (Angelone et al. 2015, 2012; Penna et al. 2011b; Perrelli et al. 2013) (see below). On the other hand, the dismutation of O_2^- in hydrogen peroxide (H_2O_2) mediated by superoxide-dismutase can also reduce significantly the injury; yet in the presence of Fe^{2+} or Cu^{2+}, H_2O_2 is transformed in hydroxyl anion (OH^-), thus resulting in more toxic effects than O_2^- and H_2O_2. This brief description of ROS/RNS production/formation during I/R may lead to the misconception that radicals are prevalently deleterious and some radical species are good and other are bad. This is not always the case, as we can see in the following description of cardioprotective pathways, some reactive species, including OH^- may exert protective effects, depending on several factors, including compartmentalization and flux velocity of reactions.

Nevertheless, the *oxidative stress* in the context of I/R may result in acute inflammatory response with activation of vascular endothelial cells and leukocytes and with the expression on cell surface of adhesion molecules, leading to *leukocyte/capillary plugging*, release of cytokines, and pro-inflammatory agents which determine the onset and maintenance of post-ischemic inflammation (Zhao and Vinten-Johansen 2002).

Other deleterious factor of reperfusion injury is the cellular Ca^{2+}*overload;* this phenomenon starts during ischemia, for depletion of ATP and consequent inhibition of ionic pump, and it may be further increased during reperfusion. During I/R, the altered cytosolic Ca^{2+} handling may induce structural fragility and excessive contractile activation, with a band-necrosis and progressive increase of diastolic contracture (Hoffman et al. 2004; Piper et al. 2003; Siegmund et al. 1993). The overload of Ca^{2+} contributes to the augmentation of cellular osmolarity, which will be responsible of *explosive swelling* and necrosis of cardiomyocytes. Also, the mitochondria undergo rapid changes in matrix Ca^{2+} concentration; in fact, while cytosolic Ca^{2+} overload is responsible of expression/release of proapoptotic elements, Ca^{2+} overload within mitochondria leads to the release of pro-apoptotic cofactors, and to the opening of *mitochondrial transition pore* (mPTP) (Zhao 2004).

Actually, mPTP are kept closed during the ischemia by the acidic environment. During reperfusion, the pore formation is favored by several factors, including the pH recovery, the oxidative stress, the ATP depletion and, as said, the high levels of intramitochondrial Ca^{2+} concentration (Gateau-Roesch et al. 2006). The opening of mPTP, besides the release of pro-apoptotic factors, comports disruption of mitochondria (*mitochondrial swelling*) and consequently cell death for necrosis or for apoptosis. In fact, the formation of mPTPs abolishes the electrochemical gradients between the cytoplasm and mitochondrial matrix and consequently blocks ATP production, thus determining cell necrosis. The release of *cytochrome c* will be the major responsible of cell apoptosis. It is, thus, likely that in reperfusion the majority of cells die by means of these two processes (Halestrap 2006, 2009; Juhaszova et al. 2009; Zorov et al. 2009).

In the reperfusion, the activation of NFκB is also possible. This activation may be induced by several agents, including ROS/RNS. When activated, NFκB contributes to the exacerbation of the myocardium injury by sustaining inflammatory processes, and determining an upregulation of the genes involved in the production of cell adhesion molecules (Baldwin and Thurston 2001; Lefer and Lefer 1996; Marczin et al. 2003; Schreck et al. 1992). These molecules may favor the adhesion of leukocytes to the endothelium and then infiltration into the myocardial wall (Baldwin and Thurston 2001). Therefore, during reperfusion neutrophils are the main source of ROS (Jordan et al. 1999). This production together with the deficient production of NO induced by I/R, may favor a vicious cycle leading to the activation/transcription of genes for the production of further cell adhesion molecules (Beauchamp et al. 1999; Lefer and Lefer 1996).

The deficit of NO causes vasoconstriction and formation of micro-thrombi in small vessels (Radomski et al. 1987; Schulz et al. 2004). These mechanisms, combined with the adhesion of leucocytes to the endothelium, can lead to the so-called "*no-reflow phenomenon*" (Reffelmann and Kloner 2002). Other pathological manifestations of I/R injury induced by oxidative stress are represented by arrhythmias and myocardial stunning. In fact, increased ROS production and depletion of energy may also contribute to alterations in excitation-contraction coupling, thus sustaining myocardial stunning and arrhythmias (Pagliaro et al. 2011; Penna et al. 2008; Vinten-Johansen et al. 2011).

In summary, reperfusion injury is due to several mechanisms in which mPTP play a central role being primed by ischemia and opening upon reperfusion, because of the sudden recovery of pH and simultaneous presence of several damaging factors, such as Ca^{2+} overload, ROS generation, and reduced NO bioavailability. All these factors contribute to the activation of NFκB, which leads to the augmented expression of cellular adhesion molecules, leukocyte infiltration and no-reflow phenomenon. Therefore, necrotic and apoptotic cell death contributes to reperfusion-injury which is exacerbated by inflammatory processes (see Fig. 1).

2.2 Ischemic Preconditioning

Murry et al. (1986, 1991) described in 1986 the ***ischemic preconditioning*** (IP) phenomenon. In this groundbreaking study the cardioprotective maneuvers consisted in 4 cycles of 5-min ischemia/5-min reperfusion just prior to a 40-min coronary occlusion. This protocol in dogs induced a marked (75%) reduction of infarct size. Thus, the brief periods of ischemia had "preconditioned" the myocardium to make it more resistant to the stress of a longer ischemic interval followed by a full reperfusion. IP was then studied in a variety of animal models and in many independent laboratories, and it became clear that this intervention was a robust cardioprotective strategy that greatly diminished myocardial infarction after a sustained period of coronary occlusion followed by reperfusion. In all animal models, brief periods (a few minutes) of ischemia, separated from one another by brief periods (a few minutes) of reperfusion just prior to a prolonged period of ischemia followed by reperfusion induce the IP protection (see Fig. 2). The severity of the I/R injury are limited by IP and the cardioprotective effects of IP include infarct size reduction, limitation of apoptosis, reduction of stunning, anti-arrhythmic effects, and vascular preconditioning consisting in reduction of endothelial dysfunction and limitation of endothelial activation with reduction of neutrophil adherence and platelet aggregation.

During the years, it has been reported that many molecules released during the short periods of preconditioning ischemia are responsible of IP protection triggering. One of the first molecule demonstrated to be responsible of triggering IP has been adenosine, which can be released during the brief periods of preconditioning maneuvers (Liu et al. 1991). Several other autacoids were identified as IP triggers, *e.g.* bradykinin, platelet activating factor and opioids, produced during the brief periods of ischemia of IP protocols (Bolli 2001; Dawn and Bolli 2002; Hausenloy and Yellon 2007a, 2008; Ludman et al. 2010; Yellon and Hausenloy 2007; Pagliaro et al. 2001; Penna et al. 2008; Wink et al. 2003). Several studies have identified many of IP signaling steps. These steps are represented as complex protective pathways, which can be divided at least into three phases: (1) a pre-ischemic trigger phase, (2) a memory phase and (3) a mediation phase, which occurs in early reperfusion after the infarcting ischemia.

In *the trigger phase* the released cardioprotective substances induce the activation of signal transduction pathways with the final point converging on mitochondria (Fig. 2) (Gomez et al. 2007; Hausenloy and Yellon 2007b; Juhaszova et al. 2004). Actually, the opening of mitochondrial ATP-sensitive potassium channels (mitoK$_{ATP}$) is an important step for the cardioprotection by IP achieved by a complex signaling cascade (Carroll et al. 2001; Cohen et al. 2001; Forbes et al. 2001; Garlid et al. 2003; Juhaszova et al. 2004; Lim et al. 2007; O'Rourke 2000; Weiss et al. 2003). In fact, the cardioprotective autacoids via their specific receptor may activate a molecular pathway which includes the activation of PI3K/Akt, nitric oxide synthase (NOS), guanylyl cyclase (GC) and Protein Kinase G (PKG), Nevertheless, this complex cascade can be by-passed by the mitoK$_{ATP}$ opener, Diazoxide, which pharmacologically preconditions the heart (Juhaszova et al. 2004; Weiss et al. 2003). The cardioprotective effect of Diazoxide can be completely inhibited by the infusion of a free radical scavenger, such as mercaptopropionyl glycine (MPG) or N-acetyl-cysteine. These results are in line with the observation that the antioxidant compounds infused during preconditioning ischemia avoid the protective effects of IP (Juhaszova et al. 2004; Weiss et al. 2003). Therefore, ROS step is downstream to mitoK$_{ATP}$ channels.

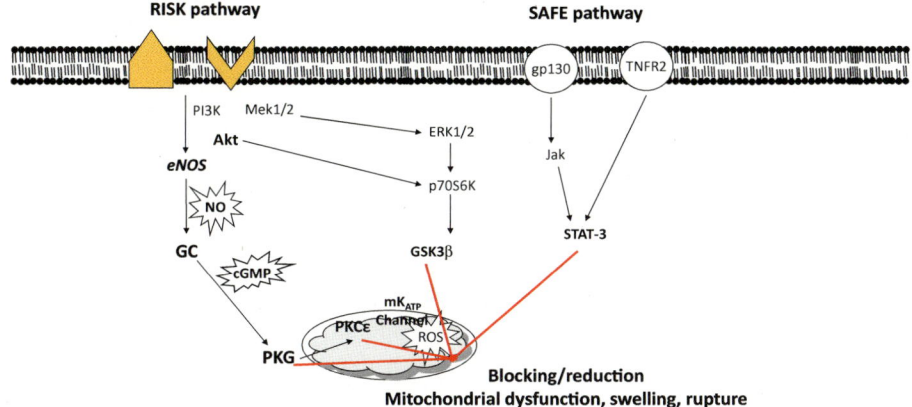

Fig. 2 Scheme depicting the principal factors involved in cardioprotective pathways triggered by preconditioning. The activation of cellular surface receptors in response to an ischemic conditioning stimulus recruits *SAFE* and *RISK* pathways. The end-point of these signal transduction pathways are the salvage of mitochondrial structure and activity with activation of mitochondria-dependent protective pathways (*Akt* Serine/threonine protein kinase, *cGMP/PKG* Cyclic guanosin monophosphate/protein kinase G, *eNOS* Endothelial NO synthase, *ERK1/2* Extracellular regulated kinase 1/2, *GSK3β* Glycogen synthase kinase 3 β, *MEK* Mitogen-activated protein kinase kinase, *mPTP* Mitochondrial permeability transition pore, *mitoK$_{ATP}$* mitochondrial ATP-dependent K$^+$ channels, *NO* Nitric oxide, *P70S6K* p70 ribosomal S6 protein kinase, *PLC* phospholipase C, *PI3K* Phosphoinositide 3-kinase, *PKG* Protein kinase G, *RISK* Reperfusion injury salvage kinases, *ROS* reactive oxygen species)

Pivotal role is assigned to protein kinase C (PKC) as point of convergence of the protective triggering pathway. In fact, the opening of mitoK$_{ATP}$ channels and release of ROS may activate PKC (Oldenburg et al. 2002, 2004) (Fig. 2). PKC may also be directly activated by adenosine via activation of phospholipase C (PLC) mechanism. Activation of PKC and other kinases represent *the memory phase*.

When these above described pathways are activated, the heart displays a protected phenotype which persists for a couple of hours even after the triggering autacoids have been washed out.

In *the mediation phase* the adenosine A$_{2b}$ receptors are activated by PKC and during the reperfusion phase, which follows the infarcting ischemia, these receptors may be responsible of the re-activation of PKC (Oldenburg et al. 2002; Yue et al. 2002). The downstream signaling to A$_{2b}$ receptors recapitulates somehow those seen in the trigger pathway. The final point of these signaling is to prevent the formation of mPTP (Tissier et al. 2007; Zorov et al. 2009). Several investigations of the signaling pathways, underlying IP mediation phase, have identified a number of different signal transduction pathways conveying the cardioprotective signal from the sarcolemma to the mitochondria, some of which overlap with postconditioning (see below). These reperfusion signaling pathways include the *Reperfusion Injury Salvage Kinase* (RISK) pathway and the more recently described *Survivor Activating Factor Enhancement* (SAFE) pathway, two apparently distinct signal cascades which may actually interact to mediate IP and Postconditioning (PostC) cardioprotection (Philipp et al. 2006; Lecour 2009) (Fig. 2).

The fundamental role of activation of Akt in the cardioprotection, has been demonstrated by numerous studies. The cardioprotective effect induce by Akt activation is correlated to signaling mediated G protein–coupled receptors (Kuno et al. 2008; Means et al. 2008; Perrelli et al. 2011; Sarbassov et al. 2005), receptors of tyrosine kinases and glycoprotein 130–linked receptors (Fujio et al. 2000; Hausenloy et al. 2011). The activation of these receptors induces the PI3K system activation with an increase of phosphatidylinositol triphosphate (PIP3) levels (Fujio et al. 2000; Hausenloy et al. 2011; Matsui et al. 1999). The translocation of Akt to the plasma membrane is favored by PIP3. Activation/phosphorylation of Akt occurs through phosphorylation at Thr308 by phosphoinositide-dependent kinase 1 (PDK1) and through phosphorylation at Ser473 *via* both TORC2 mechanism and the intrinsic catalytic activity of Akt (Craig et al. 2001; Miyamoto et al. 2008; Negoro et al. 2001). Recently Goodman et al. (2008) and Gross et al. (2006) reported the possibility that Akt is also phosphorylated by signal transducer and activator of transcription 3 (STAT3) or Janus Kinase (JAK). It is well known that STAT3 leads to changes in gene transcription, transducing stress signals from the plasma membrane to the nucleus. The cardiac role of STAT3 is very intriguingly, in fact it is involved in hypertrophy and apoptosis and development of infarct size after I/R protocol. In addition, it has been recently reported that STAT3 plays a cardioprotective role in the SAFE pathway which includes JAK/STAT signal transduction (Newton 2003; Sarbassov et al. 2005). In particular, during cardioprotective protocols, STAT3 has been localized with molecular biology and confocal laser scan microscopy in isolated mitochondria (Boengler et al. 2008a; Goodman et al. 2008; Gross et al. 2006). During ischemia/reperfusion STAT3 is activated by phosphorylation (Lacerda et al. 2009; Wegrzyn et al. 2009), different cardioprotective maneuvers induce its phosphorylation/activation (Boengler et al. 2010, 2011; Fuglesteg et al. 2008; Gross et al. 2006; McCormick et al. 2006; Negoro et al. 2000), with reduction of cardiomyocyte death and adverse cardiac remodeling after I/R injury (Kelly et al. 2010). Both IP and PostC comprises STAT3 phosphorylation with closure of mPTP (Boengler et al. 2008a, b; Goodman et al. 2008; Lecour et al. 2005; Sarbassov et al. 2005).

Other important IP mechanism is dependent on NO/nitroxyl (HNO) effects (Pagliaro 2003; Penna et al. 2008). NO-dependent, PKG-independent mechanism is also described (Inserte and Garcia-Dorado 2015; Penna et al. 2008). The latter includes the intervention of ONOO$^-$ and other RNS, which in concert with ROS can activate PKC (Cohen and Downey 2008). NO/cGMP/PKG signaling is cardioprotective also in PostC (see below).

The activation of adenosine receptors induces the activation of PKC, and this pathway is independent of the NO/cGMP pathway (Yang et al. 2010), but, as said, PKC can also be activated *via* the activation of a NO-cGMP-PKG pathway (Lacerda et al. 2009; Pagliaro et al. 2003; Simkhovich et al. 2013). Therefore, adenosine may have much more direct and strong coupling to PKC (Lacerda et al. 2009; Cohen and Downey 2011; Cohen et al. 2000, 2006; Penna et al. 2005; Peart and Gross 2003).

2.3 Postconditioning

Postconditioning consists in rapid intermittent interruptions of blood flow in the early reperfusion applied just after the prolonged ischemic insult (Zhao et al. 2003). While preconditioning is applied before an ischemic insult, the postconditioning (PostC) is applied at the end of ischemia: these differences are fundamental for the clinical application of these cardioprotective maneuvers.

The first observation of PostC as protective against infarct size was made by the Vinten-Johansen group's in 2003 (Zhao et al. 2003). The authors demonstrated that this cardioprotective protocol reduced myocardial inflammation, Ca^{+2} overload, oxidative stress, infarct size and apoptotic cell death. In this study PostC preserved endothelial function and the injured tissue produced less edema. Several endogenous pro-survival signaling pathways are activated during PostC, and their activation is fundamental for the protection of the heart against lethal myocardial reperfusion injury (Penna et al. 2006a, b; Tsang et al. 2004; Zhao et al. 2003; Yang et al. 2004).

The protection of the heart at the time of reperfusion has been previously investigated using several pharmacological approaches (*e.g.* Ca^{+2} channel blockers, anti-inflammatory agents, anti-oxidant therapy, and so forth with mixed outcomes). Actually, the cardioprotective reperfusion was first suggested in the late 1990s when it was discovery that several growth factors and drugs are capable of reducing myocardial infarct size if administered at the onset of coronary artery reopening (Penna et al. 2008). However, only when Zhao et al. (2003) reported the paradoxical observation that several brief coronary occlusions after an infarcting coronary occlusion significantly reduced infarct size in vivo model, the interest of researchers to this phase of reperfusion greatly increased. This cardioprotective approach with brief ischemia applied at the beginning of reperfusion was called postconditioning and it has been demonstrated in different animal models (see review Iliodromitis et al. 2009; Penna et al. 2008; Skyschally et al. 2009). The signaling pathways leading to protection resulted similar to those already demonstrated for preconditioning and for many pharmacologic agents that protect when infused at reperfusion.

In clinical practice, PostC is a promising adjunctive technique to reperfusion since it can improve post-infarction outcome and limit left ventricle dilatation and maladaptive remodeling. Whether postconditioning can attenuate acute contractile dysfunction is a matter of controversy.

With no doubt postconditioning is appealing for clinician as it can be under their control. However, ischemic PostC cannot be applied to all patients with acute myocardial infarction and this makes pharmacological PostC an intriguing, but unmet, clinical objective. For example, ischemic PostC cannot be applied in case of thrombolytic intervention, rather than angioplasty. It is, thus necessary to deeply study the procedure and the mechanisms of protection.

As said, the molecular pathways responsible of PostC are similar to those of IP protection; in fact, also in PostC there is the activation of RISK and SAFE (Fig. 2). However, the mechanisms of cardioprotection involved in PostC are not exactly the same as those involved in preconditioning (for reviews see Hausenloy et al. 2011; Kaur et al. 2009; Lacerda et al. 2009; Pagliaro and Penna 2015; Philipp et al. 2006; Tullio et al. 2013; Vinten-Johansen et al. 2011).

The RISK pathway comprises the activation of several enzymes, such as PI3K-Akt and MEK-ERK-1/2 as above described. Also, the end point are mitochondria and sarcoplasmic reticulum which in concert limit cytoplasmic and mitochondrial Ca^{+2} overload, thus limiting cell damages. Subsequently, the attention of researchers was attracted by STAT3 in SAFE pathway: the deletion of STAT3 abolished the reduction of infarct size usually observed after PostC maneuvers (Negoro et al. 2000). Moreover, the pharmacological postconditioning was obtained by exogenous tumor necrosis factor α (TNFα) in wildtype, but not in STAT3 knockout mice (Lacerda et al. 2009; Negoro et al. 2000; Tsang et al. 2004). Of note, mitochondrial translocation of STAT3 has been observed in these models of PostC and TNFα induced cardioprotection. Of note, only the mechanical maneuvers of ischemic PostC induced this translocation, while this mechanism is absent when PostC is induced by Diazoxide (Penna et al. 2013). Since Diazoxide acts *via* a redox-mechanism, these data suggest a redox-independent mechanism in the SAFE pathway. Whether or not mitochondrial STAT3 contributes to the cardioprotection by PostC should be confirmed in further studies (Boengler et al. 2011; Lacerda et al. 2009).

3 Chromogranin Fragments and Cardioprotection

3.1 Vasostatin 1

It has been reported that the N-terminal CgA-derived Vasostatin 1 (VS-1) counteracts the effects of adrenergic stimulation and mediates NO-dependent vasodilation, acting on both the endothelial and endocardial cells (Cerra et al. 2006). VS-1 thus contributes to protection against excessive excitatory sympathetic changes (Cerra et al. 2006, 2008; Gallo et al. 2007; Ramella et al. 2010). We have shown that in the isolated rat heart VS-1 induces cardioprotection when infused before of I/R: this protection recapitulates many aspects of IP. We have shown that low concentration of VS-1 (80 nM), administered before I/R, may activate adenosine/NO/PKC signaling (Cappello et al. 2007). In basal condition, VS-1 at 33 nM in isolated rat heart, induces the reduction of the left ventricular pressure (LVP), the maximal values of the first derivative of LVP (dP/dt_{max}) and of the rate-pressure. These effects of VS-1 were abolished blocking the $G_{i/o}$ proteins/NOS/NO/solubleGC/PKG pathway. These data demonstrate the involvement of molecular mechanisms in the VS-1-dependent negative inotropic effects similar to those involved in cardioprotection. In fact, the infusion of VS-1 before ischemia reduced the infarct size, which was by about 50% of the risk area. This protection was NO/PKC dependent, in fact it was abolished by either NOS or PKC inhibition. However, cardioprotection was only attenuated by the blockade of Adenosine type 1 (A_1) receptors, thus suggesting a peculiarity in VS-1 mechanism of action. VS-1 activity results mediated by two different pathways that converge on PKC: one is mediated by NO release, and the other is mediated by A_1 receptors. Similarly, to preconditioning ischemia, VS-1 may be considered a stimulus strong enough to trigger the two pathways, which may converge on PKC. The lack of a cellular receptor for VS-1 (Gallo et al. 2007; Ramella et al. 2010) can induce the infused VS-1 to interfere with other membrane receptors (Cappello et al. 2007). Importantly, between VS-1 infusion and 30-min ischemia there was a wash-out period which allowed the recovery from inotropic effect. Therefore, cardioprotection was not due to a reduction of oxygen consumption because of cardiac inotropic depression.

Recently in an in vitro trans-well co-colture model of cardiomyocyte-endothelial cells was demonstrated that VS-1 exerts protective effects directly on the cardiomyocytes or indirectly by cardiomyocyte-endothelial cells interaction (Liu et al. 2014).

These data emphasize the potential importance of the release of CgA as a mechanism of the cardiac system to protect itself against I/R injury and, eventually, against sympathetic overstimulation. Actually, increased plasma levels of CgA are present in patients after myocardial infarction (Wang et al. 2011).

Importantly, in acute myocardial infarction (AMI) patients an initial reduction with a subsequent increase in catestatin plasma levels has been recently reported (Wang et al. 2011). The increased levels of catestatin and its precursor, CgA, are more supportive of the potential importance of the release of CgA and derivatives as attempt to protect the heart against I/R injury. Since the majority of I/R damage occurs in the early reperfusion, we wondered whether a supplementation of VS-1 or catestatin in the early reperfusion phase may protect the heart against reperfusion injury. In our laboratory, we observed that VS-1 does not protect when given in postischemic phase (unpublished observations). On the contrary, catestatin only given in reperfusion resulted highly protective against I/R damages (Penna et al. 2010a).

3.2 Catestatin

In humans, Catestatin (CST), was initially studied as an important factor in blood pressure control (Angelone et al. 2008; O'Connor et al. 2002). However, a comparative study has recently demonstrated that the plasma levels of CST in healthy humans and in individuals with coronary heart disease are significantly different and correlated somehow with norepinephrine levels. In fact, a positive and significant correlation between the plasma catestatin level and norepinephrine level was observed in different myocardial ischemia states. However, in patients plasma catestatin on admission was not associated with adverse cardiac events. Moreover, there was not an appreciable relationship between plasma catestatin and onset of new cardiovascular events (Liu et al. 2014). More recently, the cardioprotective effect of CST has been described. In particular, this peptide seems to be cardioprotective when added to the coronary perfusate during reperfusion, unlike VS-1, which resulted cardioprotective when given before ischemia, but not in reperfusion. Actually, we recently suggested that catestatin induces a sort of *pharmacological postconditioning* (Penna et al. 2010a). We have reported that in an isolated and perfused heart model, CST at a concentration of 75 nM is able to reduce I/R injuries when given during the early reperfusion. The CST concentration we used in the isolated rat heart is within the same range of concentrations of the precursor CgA detected in plasma of patients suffering IMA (about 1 nM) or cardiac heart failure (about 10 nM) (O'Connor et al. 2002; Wang et al. 2011). It is also similar to the peptide concentration (IC50 ~ 100 nM) which depresses myocardial force of contraction in normal hearts (O'Connor et al. 2002; Wang et al. 2011), and seems slightly lower than the IC50 value for the inhibition of the nicotinic cholinergic receptor-mediated catecholamine release in bovine adrenal chromaffin cells induced by CST (Mahata et al. 1997; Mahata 2004). In the isolated rat heart model, CST not only reduces infarct size, but also limits post-ischemic contracture and improves post-ischemic systolic function. Moreover, we reported that CST is protective in a model of isolated cardiomyocytes (Penna et al. 2010a). In this model CST increases viability rate of cells exposed to simulated ischemia. This direct protective effect on cardiomyocytes may explain why CST applied in the reperfusion is protective especially in terms of improvement of post-ischemic cardiac function. Since protection was observed in both isolated heart and isolated cardiomyocytes, we suggested that the protective effect is primarily due to a direct effect on the myocardium and does not necessarily depend on the antiadrenergic and/or endothelial effects of CST (Penna et al. 2010a). However, endothelial effects could be additive. In fact, Alloatti and coworkers (Bassino et al. 2015) have shown that catestatin may also act on endothelial cells.

About the mechanisms of action of CST, we have shown that CST given in early reperfusion facilitates the phosphorylation of Akt, PKCε and GSK3β which may regulate mitochondrial function (Bassino et al. 2015; Penna et al. 2014; Perrelli et al. 2013). The mechanisms seem similar to those described in ischemic PostC. However, the protective pathways partially diverge, as mitoK$_{ATP}$ channel blockade, by 5-hydroxydecanoate, 5-HD, or ROS scavenger does not avoid CST-dependent contracture limitation, whereas PKC inhibition abolishes infarct size, antiapoptotic activity, contracture limitation and systolic function recovery. Since 5-HD attenuates the PKCε activation due to CST, a reverberant circuit (PKCε-dependent/mitoK$_{ATP}$ channel activation/ROS formation/PKCε re-activation, Fig. 3) has been hypothesized (Perrelli et al. 2013). We also observed that the anti-infarct effect of CST is abolished by scavenging ROS with a sulfhydryl donor specific for mitochondrial activity (Perrelli et al. 2013), namely N-(2-mercaptopropionyl) glycine (MPG, 300 μM); whereas the contracture limitation is not affected by MPG. Recently these results have been confirmed in isolated cardiac cells, in fact in this in vitro model, catestatin exerts a direct action on cardiomyocytes with activation of PI3K/NO/cGMP pathway as trigger and mitochondrial membrane potential preservation as the end point of its action (Bassino et al. 2015; Perrelli et al. 2013). The protection is confirmed in H9c2

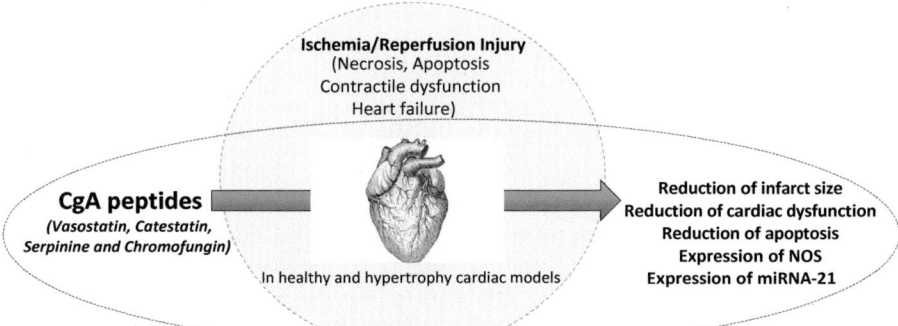

Fig. 3 Scheme depicting the main outcomes of cardiac ischemia/reperfusion injury (*IRI*) and the main cardioprotective results obtained with CgA-derived peptides in this context. The IRI leads to heart failure, the CgA-derived peptides limits IRI and increase the expression of protective factors in healthy and hypertrophy cardiac model. See text for further explanation

cardiomyoblasts with Anoxia/Reoxygenation (A/R) protocol. In this model the siRNA-mediated knockdown of catestatin exaggerated endoplasmic reticulum stress induced apoptosis. This protective effect is dependent on ERK1/2/PI3K/Akt pathway. In particular, the protective effect of catestatin may depend on the activation of type 2 Muscarinic receptor, and seems independent of type 1 Muscarinic receptor activation (Liao et al. 2015). Despite the aforementioned studies report a cardioprotective effects for CST in different models, Brar et al. (2010) described deleterious effects against I/R injury for both the wild type and Pro^{370} Leu variants of CST, but not for Gly^{364} Ser variant. These authors studied these three different variants of CST in the isolated perfused rat hearts subjected to regional ischemia and the CTSs were administered during the entire period of reperfusion. These different results might be explained on the basis of the different models investigated (isolated rat heart perfused at constant pressure and subjected to regional ischemia *vs* hearts perfused at constant flow and subjected to global ischemia), the modality of administration of CST (the entire reperfusion period or the first 20 min of reperfusion, when the majority of I/R injury occurs) and the dose used (75 nM for 20 min *vs* 100 nM for 120 min). Finally, it is noteworthy that in the setting of isolated hearts, when studying I/R injury, ventricular volume adjustment is crucial part of maintaining a good heart preparation and the subsequent interpretation of data obtained. We usually pierce the left ventricle and set the volume of the balloon to obtain an end diastolic left ventricular pressure (LVDP) below 5 mmHg (Penna et al. 2006a, 2007, 2014; Perrelli et al. 2013), whereas Brar et al. (2010) imposed an LVDP of 5–10 mm Hg. This higher range of LVDP may have affected the amount of myocardial necrosis. "The volume of the balloon should be such that it is able to fill the ventricular cavity; … too large and the potential for inducing endocardial necrosis once a pre-load diastolic pressure is set" (Bell et al. 2011).

The cardioprotective effects of CST is also preserved in a mode of cardiac hypertrophy, a model usually not protected by ischemic conditioning protocols (Ma et al. 2013; Penna et al. 2010a, 2011a, b; Wagner et al. 2013). CST, in fact, resulted cardioprotective in isolated heart obtained from Spontaneously Hypertensive Rat. In particular, in this model CST displayed an anti-apoptotic effect and induced the expression of pro-angiogenetic factors (*i.e.*, HIF-1α and eNOS expression) after 2 h of reperfusion (Penna et al. 2014; Perrelli et al. 2013). Importantly, the chronic administration of catestatin results to be protective after myocardial ischemia in rats. In this chronic model the protective effect of CST may derive from its ability to decrease the cardiac sympathetic drive and to improve autonomic function (Wang et al. 2016).

3.3 Chromofungin

Recently Angelone and coworkers have demonstrated that another fragment of CgA, chromofungin (Chr: CgA47-66) displays cardioprotective effects. Chr presents an immediate protective anti-microbical effect (Metz-Boutigue et al. 2003), and the ability to penetrate the cell membrane, thus inducing extracellular calcium entry by a Calmodulin-regulated iPLA2 pathway (Zhang et al. 2009). Chr induces cardioprotective effects acting as a post-conditioning agent through the activation of the RISK pathways and via the involvement of mitoK$_{ATP}$ channels. In particular, the protective effect of Chr was cGMP-dependent (Filice et al. 2015). Of note, this study has been the first one that demonstrated a correlation between this cyclic nucleotide and cardioprotection by CgA fragments. This is important because cGMP plays a central role in cardioprotection. In fact, PostC depends on GC activation via either NOS-dependent or NOS-independent pathways (Penna et al. 2006a, b).

Another interesting aspect of the cardioprotective effects mediated by Chr is the capacity to increase miRNA-21 expression. This effect has been correlated to the reduction of infarct size, and the inhibition of pro-apoptotic genes and the increase of anti-apoptotic genes (Cheng et al. 2009; Dong et al. 2009).

3.4 Serpinin

Recently, another CgA fragment, serpinin (Serp), has been studied on its ability to induce cardioprotective effects. Interestingly, this fragment has been shown to exert cardioprotective effects in both healthy and hypertensive rat heart models (Pasqua et al. 2015). Perhaps more importantly, Serp, differently from other CgA fragments, has been highly protective both in pre- and in post-conditioning protocols (Pasqua et al. 2015). The cardioprotective effects of serpinin is mediated by an intracellular signaling which, starting from β1-adrenergic receptor, includes adenylate cyclase and cAMP/PKA pathway, with a consequent phosphorylation of ERK1/2 and GSK3β, which are components of RISK cardioprotective cascade. This promising cardioprotective factor deserves to be studied more thoroughly.

4 Conclusions

It is conceivable that the effects of CgA and its derived fragments may be multifunctional. In fact, on cardiovascular system they act not only *via* the nervous and sympathoadrenal systems, but also via direct timely protective mechanisms on endothelium and cardiomyocytes.

Interestingly, all studied fragments are cardioprotective, some are more effective as preconditioning mediator (VS-1), other as postconditioning agent (CST) and other as both pre- and post-conditioning inducer (Serp). As a matter of fact, delayed increased levels of CST after infarction (O'Connor et al. 2002; Wang et al. 2011) are in line with an attempt of compensatory protective response against cardiac injury (Ceconi et al. 2002; Helle et al. 2007; Omland et al. 2003). Early interventions which target the first few minutes of reperfusion, such as pharmacological postconditioning, may be clinically useful at the time of angioplasty, thrombolysis or cardiac surgery. Importantly, VS-1 (Cappello et al. 2007; Cerra et al. 2008; Gallo et al. 2007; Ramella et al. 2010), CST (Angelone et al. 2008; Bassino et al. 2015; Penna et al. 2010a, 2014; Perrelli et al. 2013) and Serp (Pasqua et al. 2015) positively influence endothelial function, and this may be of pivotal importance in organ protection. Our studies on cardioprotection also provide insights into the importance of the stimulus-secretion associating CgA and its processing as an attempt of the cardiovascular system to protect itself against I/R injury and associated pathophysiological events. Altogether, our results suggest that CgA derived peptides might represent a class of compounds dedicated to reduce cardiac reperfusion injury in a time dependent fashion.

Acknowledgments The authors were funded by: National Institutes of Cardiovascular Research (INRC) - Italy; Regione Piemonte, PRIN, ex-60% and Compagnia di San Paolo. We thank Prof. Donatella Gattullo for her invaluable support.

References

Aardal S, Helle KB (1992) The vasoinhibitory activity of bovine Chromogranin A fragment (vasostatin) and its independence of extracellular calcium in isolated segments of human blood vessels. Regul Pept 41:9–18

Aardal S, Helle KB, Elsayed S, Reed RK, Serck-Hanssen G (1993) Vasostatins, comprising the N-terminal domain of chromogranin A, suppress tension in isolated human blood vessel segments. J Neuroendocrinol 5:405–412

Ambrosio G, Flaherty JT, Duilio C, Tritto I, Santoro G, Elia PP et al (1991) Oxygen radicals generated at reflow induce peroxidation of membrane lipids in reperfused hearts. J Clin Invest 87:2056–2066

Ambrosio G, Zweierj JL, Duilio C, Kuppusamyj P, Santoro G, Elia PP et al (1993) Evidence that mitochondrial respiration is a source of potentially toxic oxygen free radicals in intact rabbit hearts subjected to ischemia and reflow. J Biol Chem 268:18532–18541

Angeletti RH, Aardal S, Serck-Hanssen G, Gee P, Helle KB (1994) Vasoinhibitory activity of synthetic peptides from the amino terminus of chromogranin A. Acta Physiol Scand 152:11–19

Angelone T, Quintieri AM, Brar BK, Limchaiyawat PT, Tota B, Mahata SK et al (2008) The antihypertensive chromogranin a peptide catestatin acts as a novel endocrine/paracrine modulator of cardiac inotropism and lusitropism. Endocrinology 149:4780–4793

Angelone T, Quintieri AM, Pasqua T, Gentile S, Tota B, Mahata SK et al (2012) Phosphodiesterase type-2 and NO-dependent S-nitrosylation mediate the cardioinhibition of the antihypertensive catestatin. Am J Physiol Heart Circ Physiol 302:H431–H442

Angelone T, Quintieri AM, Pasqua T, Filice E, Cantafio P, Scavello F et al (2015) The NO stimulator, Catestatin, improves the Frank-Starling response in normotensive and hypertensive rat hearts. Nitric Oxide 50:10–19

Baldwin AL, Thurston G (2001) Mechanics of endothelial cell architecture and vascular permeability. Crit Rev Biomed Eng 29:247–278

Bassino E, Fornero S, Gallo MP, Gallina C, Femminò S, Levi R et al (2015) Catestatin exerts direct protective effects on rat cardiomyocytes undergoing ischemia/reperfusion by stimulating PI3K-Akt-GSK3β pathway and preserving mitochondrial membrane potential. PLoS One 10:e0119790

Beauchamp P, Richard V, Tamion F, Lallemand F, Lebreton JP, Vaudry H et al (1999) Protective effects of preconditioning in cultured rat endothelial cells: effects on neutrophil adhesion and expression of ICAM-1 after anoxia and reoxygenation. Circulation 100:541–546

Bell RM, Mocanu MM, Yellon DM (2011) Retrograde heart perfusion: the Langendorff technique of isolated heart perfusion. J Mol Cell Cardiol 50:940–950

Boengler K, Buechert A, Heinen Y, Roeskes C, Hilfiker-Kleiner D, Heusch G, Schulz R (2008a) Cardioprotection by ischemic postconditioning is lost in aged and STAT3- deficient mice. Circ Res 102:131–135

Boengler K, Hilfiker-Kleiner D, Drexler H, Heusch G, Schulz R (2008b) The myocardial JAK/STAT pathway: from protection to failure. Pharmacol Therap 120:172–185

Boengler K, Hilfiker-Kleiner D, Heusch G, Schulz R (2010) Inhibition of permeability transition pore opening by mitochondrial STAT3 and its role in myocardial ischemia/reperfusion. Basic Res Cardiol 105:771–785

Boengler K, Heusch G, Schulz R (2011) Mitochondria in postconditioning. Antioxid Redox Signal 14:863–880

Bolli R (2001) Cardioprotective function of inducible nitric oxide synthase and role of nitric oxide in myocardial ischemia and preconditioning: an overview of a decade of research. J Mol Cell Cardiol 33:1897–1918

Brar BK, Helgeland E, Mahata SK, Zhang K, O'Connor DT, Helle KB et al (2010) Human catestatin peptides differentially regulate infarct size in the ischemic-reperfused rat heart. Regul Pept 165:63–70

Cappello S, Angelone T, Tota B, Pagliaro P, Penna C, Rastaldo R et al (2007) Human recombinant chromogranin A-derived vasostatin-1 mimics preconditioning via an adenosine/nitric oxide signaling mechanism. Am J Physiol Heart Circ Physiol 293:H719–H727

Carroll R, Gant VA, Yellon DM (2001) Mitochondrial K_{ATP} channels protects a human atrial-derived cell line by a mechanism involving free radical generation. Cardiovasc Res 51:691–700

Ceconi C, Ferrari R, Bachetti T, Opasich C, Volterrani M, Colombo B et al (2002) Chromogranin A in heart failure; a novel neurohumoral factor and a predictor for mortality. Eur Heart J 23:967–974

Cerra MC, De Iuri L, Angelone T, Corti A, Tota B (2006) Recombinant N-terminal fragments of chromogranin-A modulate cardiac function of the Langendorff-perfused rat heart. Basic Res Cardiol 101:43–52

Cerra MC, Gallo MP, Angelone T, Quintieri AM, Pulerà E, Filice E et al (2008) The homologous rat chromogranin A1-64 (rCGA1-64) modulates myocardial and coronary function in rat heart to counteract adrenergic stimulation indirectly via endothelium-derived nitric oxide. FASEB J 22:3992–4004

Cheng Y, Liu X, Zhang S, Lin Y, Yang J, Zhang C (2009) MicroRNA-21 protects against the H2O2-induced injury on cardiac myocytes via its target gene PDCD4. J Mol Cell Cardiol 47:5–14

Cohen MV, Downey JM (2008) Adenosine: trigger and mediator of cardioprotection. Basic Res Cardiol 103:203–215

Cohen MV, Downey JM (2011) Ischemic postconditioning: from receptor to end-effector. Antioxid Redox Signal 14:821–831

Cohen MV, Baines CP, Downey JM (2000) Ischemic preconditioning: from adenosine receptor to K_{ATP} channel. Annu Rev Physiol 62:79–109

Cohen MV, Yang XM, Liu GS, Heusch G, Downey JM (2001) Acetylcholine, bradykinin, opioids, and phenylephrine, but not adenosine, trigger preconditioning by generating free radicals and opening mitochondrial KATP channels. Circ Res 89:273–278

Cohen MV, Yang XM, Downey JM (2006) Nitric oxide is a preconditioning mimetic and cardioprotectant and is the basis of many available infarct sparing strategies. Cardiovasc Res 70:231–239

Craig R, Wagner M, McCardle T, Craig AG, Glembotski CC (2001) The cytoprotective effects of the glycoprotein 130 receptor-coupled cytokine, cardiotrophin-1, require activation of NF-kappa B. J Biol Chem 276:37621–37629

Dan Dunn J, Alvarez LA, Zhang X, Soldati T (2015) Reactive oxygen species and mitochondria: a nexus of cellular homeostasis. Redox Biol 6:472–485

Dawn B, Bolli R (2002) Role of nitric oxide in myocardial preconditioning. Ann N Y Acad Sci 962:18–41

Dong S, Cheng Y, Yang J, Li J, Liu X, Wang X et al (2009) MicroRNA expression signature and the role of microRNA-21 in the early phase of acute myocardial infarction. J Biol Chem 284:29514–29525

Ferdinandy P, Schulz R (2003) Nitric oxide, superoxide, and peroxynitrite in myocardial ischaemia-reperfusion injury and preconditioning. Br J Pharmacol 138:532–543

Filice E, Pasqua T, Quintieri AM, Cantafio P, Scavello F, Amodio N et al (2015) Chromofungin, CgA47-66-derived peptide, produces basal cardiac effects and postconditioning cardioprotective action during ischemia/reperfusion injury. Peptides 71:40–48

Forbes RA, Steenbergen C, Murphy E (2001) Diazoxide induced cardioprotection requires signaling through a redox-sensitive mechanism. Circ Res 88:802–809

Frydland M, Kousholt B, Larsen JR, Burnettr JC Jr, Hilsted L, Hasenkam JM et al (2013) Increased N-terminal CgA in circulation associated with cardiac reperfusion in pigs. Biomark Med 7:959–967

Fuglesteg BN, Suleman N, Tiron C, Kanhema T, Lacerda L, Andreasen TV et al (2008) Signal transducer and activator of transcription 3 is involved in the cardioprotective signalling pathway activated by insulin therapy at reperfusion. Basic Res Cardiol 103:444–453

Fujio Y, Nguyen T, Wencker D, Kitsis RN, Walsh K (2000) Akt promotes survival of cardiomyocytes in vitro and protects against ischemia-reperfusion injury in mouse heart. Circulation 101:660–667

Gallo MP, Levi R, Ramella R, Brero A, Boero O, Tota B et al (2007) Endothelium-derived nitric oxide mediates the antiadrenergic effect of human vasostatin-1 in rat ventricular myocardium. Am J Physiol Heart Circ Physiol 292:H2906–H2912

Garcia-Dorado D, Agulló L, Sartorio CL, Ruiz-Meana M (2009) Myocardial protection against reperfusion injury: the cGMP pathway. Thromb Haemost 101:635–642

Garlid KD, Dos Santos P, Xie ZJ, Costa AD, Paucek P (2003) Mitochondrial potassium transport: the role of the mitochondrial ATP-sensitive K$^+$ channel in cardiac function and cardioprotection. Biochim Biophys Acta 1606:1–21

Gateau-Roesch O, Argaud L, Ovize M (2006) Mitochondrial permeability transition pore and postconditioning. Cardiovasc Res 70:264–273

Goetze JP, Hilsted LM, Rehfeld JF, Alehagen U (2014) Plasma chromogranin A is a marker of death in elderly patients presenting with symptoms of heart failure. Endocr Connect 3:47–56

Gomez L, Thibault HB, Gharib A, Dumont JM, Vuagniaux G, Scalfaro P et al (2007) Inhibition of mitochondrial permeability transition improves functional recovery and reduces mortality following acute myocardial infarction in mice. Am J Physiol Heart Circ Physiol 293:H1654–H1661

Goodman MD, Koch SE, Fuller-Bicer GA, Butler KL (2008) Regulating RISK: a role for JAK-STAT signaling in postconditioning? Am J Physiol Heart Circ Physiol 295:H1649–H1656

Gottlieb RA, Finley KD, Mentzer RM Jr (2009) Cardioprotection requires taking out the trash. Basic Res Cardiol 104:169–180

Gross ER, Hsu AK, Gross GJ (2006) The JAK/STAT pathway is essential for opioid induced cardioprotection: JAK2 as a mediator of STAT3, Akt, and GSK-3beta. Am J Physiol Heart Circ Physiol 291:H827–H834

Halestrap AP (2006) Calcium, mitochondria and reperfusion injury: a pore way to die. Biochem Soc Trans 34:232–237

Halestrap AP (2009) Mitochondria and reperfusion injury of the heart--a holey death but not beyond salvation. J Bioenerg Biomembr 41:113–121

Hausenloy DJ, Yellon DM (2007a) Preconditioning and postconditioning: united at reperfusion. Pharmacol Ther 116:173–191

Hausenloy DJ, Yellon DM (2007b) Reperfusion injury salvage kinase signalling: taking a RISK for cardioprotection. Heart Fail Rev 12:217–234

Hausenloy DJ, Yellon DM (2008) Preconditioning and postconditioning: new strategies for cardioprotection. Diabetes Obes Metab 10:451–459

Hausenloy DJ, Lecour S, Yellon DM (2011) Reperfusion injury salvage kinase and survivor activating factor enhancement prosurvival signaling pathways in ischemic postconditioning: two sides of the same coin. Antioxid Redox Signal 14:893–907

Helle KB (2004) The granin family of uniquely acidic proteins of the diffuse neuroendocrine system: comparative and functional aspects. Biol Rev Camb Philos Soc 79:769–794

Helle KB, Metz-Boutigue MH, Aunis D (2001) Chromogranin A as a calcium binding precursor for multiple regulatory peptides for the immune, endocrine and metabolic systems. Curr Med Chem 1:119–140

Helle KB, Corti A, Metz-Boutigue MH, Tota B (2007) The endocrine role for chromogranin A: a prohormone for peptides with regulatory properties. Cell Mol Life Sci 64:2863–2886

Hoffman JW Jr, Gilbert TB, Poston RS, Silldorff EP (2004) Myocardial reperfusion injury: etiology, mechanisms, and therapies. J Extra Corpor Technol 36:391–411

Hou J, Xue X, Li J (2016) Vasostatin-2 inhibits cell proliferation and adhesion in vascular smooth muscle cells, which are associated with the progression of atherosclerosis. Biochem Biophys Res Commun 469:948–953

Iliodromitis EK, Downey JM, Heusch G, Kremastinos DT (2009) What is the optimal postconditioning algorithm? J Cardiovasc Pharmacol Ther 14:269–273

Inserte J, Garcia-Dorado D (2015) The cGMP/PKG pathway as a common mediator of cardioprotection: translatability and mechanism. Br J Pharmacol 172:1996–2009

Jordan JE, Zhao ZQ, Vinten-Johansen J (1999) The role of neutrophils in myocardial ischemia-reperfusion injury. Cardiovasc Res 43:860–878

Juhaszova M, Zorov DB, Kim SH, Pepe S, Fu Q, Fishbein KW et al (2004) Glycogen synthase kinase-3beta mediates convergence of protection signaling to inhibit the mitochondrial permeability transition pore. J Clin Invest 113:1535–1549

Juhaszova M, Zorov DB, Yaniv Y, Nuss HB, Wang S, Sollott SJ (2009) Role of glycogen synthase kinase-3beta in cardioprotection. Circ Res 104:1240–1252

Kaeffer N, Richard V, Thuillez C (1997) Delayed coronary endothelial protection 24 hours after preconditioning: role of free radicals. Circulation 96(7):2311–2316

Kaur S, Jaggi AS, Singh N (2009) Molecular aspects of ischaemic postconditioning. Fundam Clin Pharmacol 23:521–536

Kelly RF, Lamont KT, Somers S, Hacking D, Lacerda L, Thomas P et al (2010) Ethanolamine is a novel STAT-3 dependent cardioprotective agent. Basic Res Cardiol 105:763–770

Kuno A, Solenkova NV, Solodushko V, Dost T, Liu Y, Yang X-M et al (2008) Infarct limitation by a protein kinase G activator at reperfusion in rabbit hearts is dependent on sensitizing the heart to A2b agonists by protein kinase C. Am J Physiol Heart Circ Physiol 295:H1288–H1295

Lacerda L, Somers S, Opie LH, Lecour S (2009) Ischaemic postconditioning protects against reperfusion injury via the SAFE pathway. Cardiovasc Res 84:201–208

Lecour S (2009) Activation of the protective survivor activating factor enhancement (SAFE) pathway against reperfusion injury: does it go beyond the RISK pathway? J Mol Cell Cardiol 47:32–40

Lecour S, Suleman N, Deuchar GA, Somers S, Lacerda L, Huisamen B et al (2005) Pharmacological preconditioning with tumor necrosis factor-alpha activates signal transducer and activator of transcription-3 at reperfusion without involving classic prosurvival kinases (Akt and extracellular signal-regulated kinase). Circulation 112:3911–3918

Lefer AM, Lefer DJ (1991) Endothelial dysfunction in myocardial ischemia and reperfusion: role of oxygen-derived free radicals. Basic Res Cardiol 86:109–116

Lefer AM, Lefer DJ (1996) The role of nitric oxide and cell adhesion molecules on the microcirculation in ischaemia-reperfusion. Cardiovasc Res 32:743–751

Liao F, Zheng Y, Cai J, Fan J, Wang J, Yang J et al (2015) Catestatin attenuates endoplasmic reticulum induced cell apoptosis by activation type 2 muscarinic acetylcholine receptor in cardiac ischemia/reperfusion. Sci Rep 5:16590

Lim SY, Davidson SM, Hausenloy DJ, Yellon DM (2007) Preconditioning and postconditioning: the essential role of the mitochondrial permeability transition pore. Cardiovasc Res 75:530–535

Liu GS, Thornton J, Van Winkle DM, Stanley AW, Olsson RA, Downey JM (1991) Protection against infarction afforded by preconditioning is mediated by A1 adenosine receptors in rabbit heart. Circulation 84:350–356

Liu J, Yang D, Shi S, Lin L, Xiao M, Yuan Z et al (2014) Overexpression of vasostatin-1 protects hypoxia/reoxygenation injuries in cardiomyocytes-endothelial cells transwell co-culture system. Cell Biol Int 38:26–31

Ludman AJ, Yellon DM, Hausenloy DJ (2010) Cardiac preconditioning for ischaemia: lost in translation. Dis Model Mech 3:35–38

Ma LL, Zhang FJ, Kong FJ, Qian LB, Ma H, Wang JA et al (2013) Hypertrophied myocardium is refractory to sevoflurane-induced protection with alteration of reperfusion injury salvage kinase/glycogen synthase kinase 3β signals. Shock 40:217–221

Mahata SK (2004) Catestatin: the catecholamine release inhibitory peptide: a structural and functional overview. Curr Med Chem Immun Endoc Metab Agents 4:221–234

Mahata SK, O'Connor DT, Mahata M, Yoo SH, Taupenot L, Wu H et al (1997) Novel autocrine feedback control of catecholamine release. A discrete chromogranin a fragment is a noncompetitive nicotinic cholinergic antagonist. J Clin Invest 100:1623–1633

Mahata SK, Mahata M, Parmer RJ, O'Connor DT (1999) Desensitization of catecholamine release. The novel catecholamine release-inhibitory peptide catestatin (chromogranin a344-364) acts at the receptor to prevent nicotinic cholinergic tolerance. J Biol Chem 274:2920–2928

Mahata SK, Mahata M, Wakade AR, O'Connor DT (2000) Primary structure and function of the catecholamine release inhibitory peptide catestatin (chromogranin A (344-364)): identification of amino acid residues crucial for activity. Mol Endocrinol 14:1525–1535

Mahata SK, Mahapatra NR, Mahata M, Wang TC, Kennedy BP, Ziegler MG et al (2003) Catecholamine secretory vesicle stimulus-transcription coupling in vivo. Demonstration by a novel transgenic promoter/photoprotein reporter and inhibition of secretion and transcription by the chromogranin A fragment catestatin. J Biol Chem 278:32058–32067

Mahata SK, Mahata M, Wen G, Wong WB, Mahapatra NR, Hamilton BA et al (2004) The catecholamine release-inhibitory "catestatin" fragment of chromogranin a: naturally occurring human variants with different potencies for multiple chromaffin cell nicotinic cholinergic responses. Mol Pharmacol 66:1180–1191

Marczin N, El-Habashi N, Hoare GS, Bundy RE, Yacoub M (2003) Antioxidants in myocardial ischemia-reperfusion injury: therapeutic potential and basic mechanisms. Arch Biochem Biophys 420:222–236

Matsui T, Li L, del Monte F, Fukui Y, Franke TF, Hajjar RJ et al (1999) Adenoviral gene transfer of activated phosphatidylinositol 3′-kinase and Akt inhibits apoptosis of hypoxic cardiomyocytes in vitro. Circulation 100:2373–2379

McCormick J, Barry SP, Sivarajah A, Stefanutti G, Townsend PA, Lawrence KM et al (2006) Free radical scavenging inhibits STAT phosphorylation following in vivo ischemia/reperfusion injury. FASEB J 20:2115–2117

Means CK, Miyamoto S, Chun J, Brown JH (2008) S1P1 receptor localization confers selectivity for Gi-mediated cAMP and contractile responses. J Biol Chem 283:11954–11963

Meng L, Ye XJ, Ding WH, Yang Y, Di BB, Liu L, Huo Y (2011) Plasma catecholamine release-inhibitory peptide catestatin in patients with essential hypertension. J Cardiovasc Med (Hagerstown) 12:643–647

Metz-Boutigue MH, Kieffer AE, Goumon Y, Aunis D (2003) Innate immunity: involvement of new neuropeptides. Trends Microbiol 11:585–592

Miyamoto S, Murphy AN, Brown JH (2008) Akt mediates mitochondrial protection in cardiomyocytes through phosphorylation of mitochondrial hexokinase-II. Cell Death Differ 15:521–529

Murry CE, Jennings RB, Reimer KA (1986). Preconditioning with ischemia: a delay of lethal cell injury in ischemic myocardium. Circulation 74:1124–36

Murry CE, Jennings RB, Reimer KA (1991) New insights into potential mechanisms of ischemic preconditioning. Circulation 84:442–445

Negoro S, Kunisada K, Tone E, Funamoto M, Oh H, Kishimoto T et al (2000) Activation of JAK/STAT pathway transduces cytoprotective signal in rat acute myocardial infarction. Cardiovasc Res 47:797–805

Negoro S, Oh H, Tone E, Kunisada K, Fujio Y, Walsh K et al (2001) Glycoprotein 130 regulates cardiac myocyte survival in doxorubicin-induced apoptosis through phosphatidylinositol 3-kinase/Akt phosphorylation and Bcl-xL/caspase-3 interaction. Circulation 103:555–561

Newton AC (2003) Regulation of the ABC kinases by phosphorylation: protein kinase C as a paradigm. Biochem J 370:361–371

O'Connor DT, Kailasam MT, Kennedy BP, Ziegler MG, Yanaihara N, Parmer RJ (2002) Early decline in the catecholamine release-inhibitory peptide catestatin in humans at genetic risk of hypertension. J Hypertens 20:1335–1345

O'Rourke B (2000) Myocardial $K_{(ATP)}$ channels in preconditioning. Circ Res 87:845–855

Oldenburg O, Cohen MV, Yellon DM, Downey JM (2002) Mitochondrial K_{ATP} channels: role in cardioprotection. Cardiovasc Res 55:429–437

Oldenburg O, Qin Q, Krieg T, Yang XM, Philipp S, Critz SD et al (2004) Bradykinin induces mitochondrial ROS generation via NO, cGMP, PKG, and mitoKATP channel opening and leads to cardioprotection. Am J Physiol Heart Circ Physiol 286:H468–H476

Omland T, Dickstein K, Syversen U (2003) Association between plasma chromogranin A concentration and long-term mortality after MI. Am J Med 114:25–30

Pagliaro P (2003) Differential biological effects of products of nitric oxide (NO) synthase: it is not enough to say NO. Life Sci 73:2137–2149

Pagliaro P, Gattullo D, Rastaldo R, Losano G (2001) Ischemic preconditioning: from the first to the second window of protection. Life Sci 69:1–15

Pagliaro P, Penna C (2015) Redox signalling and cardioprotection: translatability and mechanism. Br J Pharmacol 172:1974–1995

Pagliaro P, Mancardi D, Rastaldo R, Penna C, Gattullo D, Miranda KM, Feelisch M et al (2003) Nitroxyl affords thiol-sensitive myocardial protective effects akin to early preconditioning. Free Radic Biol Med 34:33–43

Pagliaro P, Moro F, Tullio F, Perrelli MG, Penna C (2011) Cardioprotective pathways during reperfusion: focus on redox signaling and other modalities of cell signaling. Antioxid Redox Signal 14:833–850

Pasqua T, Tota B, Penna C, Corti A, Cerra MC, Loh YP et al (2015) pGlu-serpinin protects the normotensive and hypertensive heart from ischemic injury. J Endocrinol 227:167–178

Peart JN, Gross GJ (2003) Adenosine and opioid receptor-mediated cardioprotection in the rat: evidence for cross-talk between receptors. Am J Physiol Heart Circ Physiol 285:H81–H89

Penna C, Alloatti G, Cappello S, Gattullo D, Berta G, Mognetti B et al (2005) Platelet-activating factor induces cardioprotection in isolated rat heart akin to ischemic preconditioning: role of phosphoinositide 3-kinase and protein kinase C activation. Am J Physiol Heart Circ Physiol 88:H2512–H2520

Penna C, Cappello S, Mancardi D, Raimondo S, Rastaldo R, Gattullo D et al (2006a) Postconditioning reduces infarct size in the isolated rat heart: role of coronary flow and pressure and the nitric oxide/cGMP pathway. Basic Res Cardiol 101:168–179

Penna C, Rastaldo R, Mancardi D, Raimondo S, Cappello S, Gattullo D et al (2006b) Post-conditioning induced cardioprotection requires signaling through a redox-sensitive mechanism, mitochondrial ATP-sensitive K+ channel and protein kinase C activation. Basic Res Cardiol 101:180–189

Penna C, Mancardi D, Rastaldo R, Losano G, Pagliaro P (2007) Intermittent activation of bradykinin B2 receptors and mitochondrial KATP channels trigger cardiac postconditioning through redox signaling. Cardiovasc Res 75:168–177

Penna C, Mancardi D, Raimondo S, Geuna S, Pagliaro P (2008) The paradigm of postconditioning to protect the heart. J Cell Mol Med 12:435–458

Penna C, Alloatti G, Gallo MP, Cerra MC, Levi R, Tullio F et al (2010a) Catestatin improves post-ischemic left ventricular function and decreases ischemia/reperfusion injury in heart. Cell Mol Neurobiol 30:1171–1179

Penna C, Tullio F, Moro F, Folino A, Merlino A, Pagliaro P (2010b) Effects of a protocol of ischemic postconditioning and/or captopril in hearts of normotensive and hypertensive rats. Basic Res Cardiol 105:181–192

Penna C, Perrelli MG, Tullio F, Moro F, Parisella ML, Merlino A et al (2011a) Post-ischemic early acidosis in cardiac postconditioning modifies the activity of antioxidant enzymes, reduces nitration, and favors protein S-nitrosylation. Pflugers Arch 462:219–233

Penna C, Tullio F, Perrelli MG, Moro F, Abbadessa G, Piccione F et al (2011b) Ischemia/reperfusion injury is increased and cardioprotection by a postconditioning protocol is lost as cardiac hypertrophy develops in nandrolone treated rats. Basic Res Cardiol 106:409–420

Penna C, Perrelli MG, Tullio F, Angotti C, Camporeale A, Poli V et al (2013) Diazoxide postconditioning induces mitochondrial protein S-nitrosylation and a redox-sensitive mitochondrial phosphorylation/translocation of RISK elements: no role for SAFE. Basic Res Cardiol 108:371

Penna C, Pasqua T, Amelio D, Perrelli MG, Angotti C, Tullio F et al (2014) Catestatin increases the expression of anti-apoptotic and pro-angiogenetic factors in the post-ischemic hypertrophied heart of SHR. PLoS One 9:e102536

Penna C, Granata R, Tocchetti CG, Gallo MP, Alloatti G, Pagliaro P (2015) Endogenous cardioprotective agents: role in pre and postconditioning. Curr Drug Targets 16:843–867

Perrelli MG, Pagliaro P, Penna C (2011) Ischemia/reperfusion injury and cardioprotective mechanisms: role of mitochondria and reactive oxygen species. World J Cardiol 3:186–200

Perrelli MG, Tullio F, Angotti C, Cerra MC, Angelone T, Tota B et al (2013) Catestatin reduces myocardial ischaemia/reperfusion injury: involvement of PI3K/Akt, PKCs, mitochondrial KATP channels and ROS signalling. Pflugers Arch 465:1031–1040

Philipp S, Yang X-M, Cui L, Davis AM, Downey JM, Cohen MV (2006) Postconditioning protects rabbit hearts through a protein kinase C-adenosine A2b receptor cascade. Cardiovasc Res 70:308–314

Pieroni M, Corti A, Tota B, Curnis F, Angelone T, Colombo B et al (2007) Myocardial production of chromogranin A in human heart: a new regulatory peptide of cardiac function. Eur Heart J 28:1117–1127

Piper HM, Meuter K, Schäfer C (2003) Cellular mechanisms of ischemia-reperfusion injury. Ann Thorac Surg 75:S644–S648

Radomski MW, Palmer RM, Moncada S (1987) The role of nitric oxide and cGMP in platelet adhesion to vascular endothelium. Biochem Biophys Res Commun 148:1482–1489

Ramella R, Boero O, Alloatti G, Angelone T, Levi R, Gallo MP (2010) Vasostatin 1 activates eNOS in endothelial cells through a proteoglycan-dependent mechanism. J Cell Biochem 110:70–79

Reffelmann T, Kloner RA (2002) The "no-reflow" phenomenon: basic science and clinical correlates. Heart 87:162–168

Ronson RS, Nakamura M, Vinten-Johansen J (1999) The cardiovascular effects and implications of peroxynitrite. Cardiovasc Res 44:47–59

Sarbassov DD, Guertin DA, Ali SM, Sabatini DM (2005) Phosphorylation and regulation of Akt/PKB by the rictor-mTOR complex. Science 307:1098–1101

Schreck R, Albermann K, Baeuerle PA (1992) Nuclear factor kappa B: an oxidative stress-responsive transcription factor of eukaryotic cells (a review). Free Radic Res Commun 17:221–237

Schulz R, Kelm M, Heusch G (2004) Nitric oxide in myocardial ischemia/reperfusion injury. Cardiovasc Res 61:402–413

Siegmund B, Schlüter KD, Piper HM (1993) Calcium and the oxygen paradox. Cardiovasc Res 27:1778–1783

Simkhovich BZ, Przyklenk K, Kloner RA (2013) Role of protein kinase C in ischemic "conditioning": from first evidence to current perspectives. J Cardiovasc Pharmacol Ther 18:525–532

Skyschally A, van Caster P, Iliodromitis EK, Schulz R, Kremastinos DT, Heusch G (2009) Ischemic postconditioning: experimental models and protocol algorithms. Basic Res Cardiol 104:469–483

Szelényi Z, Fazakas Á, Szénási G, Kiss M, Tegze N, Fekete BC et al (2015) Inflammation and oxidative stress caused by nitric oxide synthase uncoupling might lead to left ventricular diastolic and systolic dysfunction in patients with hypertension. J Geriatr Cardiol 12:1–10

Tatemoto K, Efendić S, Mutt V, Makk G, Feistner GJ, Barchas JD (1986) Pancreastatin, a novel pancreatic peptide that inhibits insulin secretion. Nature 324:476–478

Tissier R, Cohen MV, Downey JM (2007) Protecting the acutely ischemic myocardium beyond reperfusion therapies: are we any closer to realizing the dream of infarct size elimination? Arch Mal Coeur Vaiss 100:794–802

Tritto I, Ambrosio G (2001) Role of oxidants in the signaling pathway of preconditioning. Antioxid Redox Signal 3:3–10

Tsang A, Hausenloy DJ, Mocanu MM, Yellon DM (2004) Postconditioning: a form of "modified reperfusion" protects the myocardium by activating the phosphatidylinositol 3-kinase-Akt pathway. Circ Res 95:230–232

Tullio F, Angotti C, Perrelli MG, Penna C, Pagliaro P (2013) Redox balance and cardioprotection. Basic Res Cardiol 108:392

Vinten-Johansen J, Granfeldt A, Mykytenko J, Undyala VV, Dong Y, Przyklenk K (2011) The multidimensional physiological responses to postconditioning. Antioxid Redox Signal 14:791–810

Wagner C, Ebner B, Tillack D, Strasser RH, Weinbrenner C (2013) Cardioprotection by ischemic postconditioning is abrogated in hypertrophied myocardium of spontaneously hypertensive rats. J Cardiovasc Pharmacol 61:35–41

Wang X, Xu S, Liang Y, Zhu D, Mi L, Wang G et al (2011) Dramatic changes in catestatin are associated with hemodynamics in acute myocardial infarction. Biomarkers 16:372–377

Wang D, Liu T, Shi S, Li R, Shan Y, Huang Y, Hu D, Huang C (2016) Chronic administration of catestatin improves autonomic function and exerts cardioprotective effects in myocardial infarction rats. J Cardiovasc Pharmacol Ther 21(6):526–535

Wegrzyn J, Potla R, Chwae YJ, Sepuri NB, Zhang Q, Koeck T et al (2009) Function of mitochondrial Stat3 in cellular respiration. Science 323:793–797

Weiss JN, Korge P, Honda HM, Ping P (2003) Role of the mitochondrial permeability transition in myocardial disease. Circ Res 93:292–301

Wink DA, Miranda KM, Katori T, Mancardi D, Thomas DD et al (2003) Orthogonal properties of the redox siblings nitroxyl and nitric oxide in the cardiovascular system: a novel redox paradigm. Am J Physiol Heart Circ Physiol 285:H2264–H2276

Yang XM, Proctor JB, Cui L, Krieg T, Downey JM, Cohen MV (2004) Multiple, brief coronary occlusions during early reperfusion protect rabbit hearts by targeting cell signaling pathways. J Am Coll Cardiol 44:1103–1110

Yang X, Cohen MV, Downey JM (2010) Mechanism of cardioprotection by early ischemic preconditioning. Cardiovasc Drugs Ther 24:225–234

Yellon DM, Hausenloy DJ (2007) Myocardial reperfusion injury. N Engl J Med 357:1121–1135

Yue Y, Qin Q, Cohen MV, Downey JM, Critz SD (2002) The relative order of mK_{ATP} channels, free radicals and p38 MAPK in preconditioning's protective pathway in rat heart. Cardiovasc Res 55:681–689

Zhang D, Shooshtarizadeh P, Laventie BJ, Colin DA, Chich JF, Vidic J et al (2009) Two chromogranin a-derived peptides induce calcium entry in human neutrophils by calmodulin-regulated calcium independent phospholipase A2. PLoS One 4:e4501

Zhao ZQ (2004) Oxidative stress-elicited myocardial apoptosis during reperfusion. Curr Opin Pharmacol 4:159–165

Zhao ZQ, Vinten-Johansen J (2002) Myocardial apoptosis and ischemic preconditioning. Cardiovasc Res 55:438–455

Zhao ZQ, Corvera JS, Halkos ME, Kerendi F, Wang NP, Guyton RA et al (2003) Inhibition of myocardial injury by ischemic postconditioning during reperfusion: comparison with ischemic preconditioning. Am J Physiol Heart Circ Physiol 285:H579–H588

Zorov DB, Juhaszova M, Yaniv Y, Nuss HB, Wang S, Sollott SJ (2009) Regulation and pharmacology of the mitochondrial permeability transition pore. Cardiovasc Res 83:213–225

Naturally Occurring Single Nucleotide Polymorphisms in Human Chromogranin A (*CHGA*) Gene: Association with Hypertension and Associated Diseases

Nitish R. Mahapatra, Sajalendu Ghosh, Manjula Mahata, Gautam K. Bandyopadhyay, and Sushil K. Mahata

Abstract Single nucleotide polymorphisms (SNPs) refer to changes of a single DNA base, which accounts for ~90% of human sequence variations. Chromogranin A (CgA) is a secretory protein whose plasma concentration is augmented in established essential (hereditary) hypertension, in heart failure patients and in patients with renal failure. Resequencing of *CHGA* gene in several ethnic groups across the globe led to the discovery of both common (minor allele frequency > 5%) and rare (minor allele frequency < 5%) SNPs in the coding and in the regulatory regions such as the proximal promoter and 3′-UTR (untranslated region). Variants in both the proximal promoter and the 3′-UTR showed statistical associations with hypertension. In contrast, the hypertensive renal disease was best predicted by variants at the 3′-UTR. Non-synonymous (change in amino acid) SNPs in the pancreastatin (PST) and catestatin (CST) domains affected insulin-stimulated glucose uptake and nicotine-evoked catecholamine secretion, respectively. The heterozygote (Gly364Ser) and minor allele homozygote (Ser364Ser) variant of CST influenced both cardiovascular (autonomic function and hypertension) and metabolic (plasma glucose,

---

We dedicate this review article to Daniel T. O'Connor who passed away on August 6, 2014. He took the lead on the discovery and functional characterization of *CHGA* SNPs at UCSD.

N.R. Mahapatra
Department of Biotechnology, Bhupat and Jyoti Mehta School of Biosciences, Indian Institute of Technology Madras, Chennai, India

S. Ghosh
Postgraduate Department of Zoology, Ranchi College, Ranchi, India

M. Mahata • G.K. Bandyopadhyay
Department of Medicine, University of California, San Diego, CA, USA

S.K. Mahata, Ph.D. (✉)
VA San Diego Healthcare System, San Diego, CA, USA

Metabolic Physiology & Ultrastructural Biology Laboratory, Department of Medicine, University of California, San Diego, 9500 Gilman Drive, La Jolla, CA 92093-0732, USA
e-mail: smahata@ucsd.edu

triglycerides, and high-density lipoproteins) functions. The above SNP-associated findings may thus lead to novel approaches to the pathophysiology, diagnosis, and treatments of the autonomic and metabolic dysfunctions.

1 Introduction

The secretory proprotein Chromogranin A (CgA) is co-stored and co-released with catecholamines from dense core vesicles in the adrenal medulla and the post-ganglionic sympathetic axons (Bartolomucci et al. 2011; Winkler and Fischer-Colbrie 1992). CgA gene (*CHGA*) is located on human chromosome 14q32, with eight exons (encoding 439 amino acids) separated by 7 introns (Mouland et al. 1994). Although initially detected in chromaffin granules, this protein was subsequently detected ubiquitously in secretory vesicles of endocrine, neuroendocrine, and neuronal cells (Bartolomucci et al. 2011; Winkler and Fischer-Colbrie 1992). The circulatory concentrations of CgA show a ~7 to ~22-fold (Tramonti et al. 2001; Ziegler et al. 1990) increase in patients with renal failure and a ~2-fold increase in essential hypertensive subjects (O'Connor et al. 2008). Both stimulated and basal CgA concentrations in the circulation are influenced by exocytotic catecholamine secretion. CgA plays an important role in formation of catecholamine storage vesicles and storage of catecholamines in chromaffin cells (Kim et al. 2001; Mahapatra et al. 2005; Pasqua et al. 2016). The extracellular function of CgA includes the generation of bioactive peptides, such as the insulin-regulatory hormone pancreastatin (Bandyopadhyay et al. 2015; Gayen et al. 2009; Sanchez-Margalet et al. 2010; Tatemoto et al. 1986), the vasodilating and cardioprotective vasostatin (Aardal et al. 1993; Tota et al. 2008), the anti-adrenergic (Mahata et al. 1997, 2003, 2010), anti-hypertensive (Mahapatra et al. 2005), anti-obesity (Bandyopadhyay et al. 2012), pro-angiogenic (Theurl et al. 2010) and cardioprotective catestatin (CST) (Angelone et al. 2008; Mahata et al. 2010), and the pro-adrenergic serpinin (Tota et al. 2012).

Single nucleotide polymorphisms (SNPs) refer to changes of a single DNA base, which accounts for ~90% of human sequence variations (Collins et al. 1998). The majority of SNPs are functionally neutral. SNPs in gene regulatory regions such as promoters, enhancers, silencers, and introns affect gene expression level in an allele specific manner (Ponomarenko et al. 2002). SNPs in gene coding regions (cSNPs) may lead to changes in the function of the encoded protein (Bell et al. 1993; Ingram and Stretton 1959).

CgA has been implicated in the pathogenesis of human essential (genetic) hypertension as hypertensive subjects display heritable increase in plasma CgA (O'Connor 1985; Takiyyuddin et al. 1995). Plasma CST in monozygotic and dizygotic twins from two continents (North America and Australia) displayed significant heritability (O'Connor et al. 2008). Because of the heritable nature of CgA, CST, and hypertension it was reasoned that genetic variation at the *CHGA* locus might contribute to hypertension and hypertensive renal disease. Systematic polymorphism discovery at the *CHGA* locus was conducted in several ethnic groups at the University of California, San Diego (UCSD) (Wen et al. 2004), in an Indian population at the Indian Institute of Technology Madras (IITM) (Allu et al. 2014; Sahu et al. 2012), and in a Japanese population at the University of Tsukuba (Choi et al. 2015). Functional genetic variations at the human *CHGA* locus, in both the proximal promoter (Chen et al. 2008a) and 3′-UTR (Chen et al. 2008b) showed associations with essential hypertension and hypertensive end-organ damage (Salem et al. 2008), and non-synonymous variations within the CST and PST regions displayed changes in potencies of these peptide hormones in regulation of catecholamine secretion (Mahata et al. 2004; Sahu et al. 2012; Wen et al. 2004) and glucose uptake (Allu et al. 2014; O'Connor et al. 2005), respectively. The above SNP-associated findings may lead to novel approaches to the pathophysiology, diagnosis, and treatments of the autonomic and metabolic dysfunctions.

2 SNPs in the Gene Regulatory Regions

2.1 *Discovery of SNPs in the Regulatory Region of CHGA*

The reference sequence (RefSeq) for human *CHGA* gene was obtained from the UCSC Genome Browser. To probe for the SNPs at *CHGA* locus that might alter its function, resequencing of *CHGA* was conducted at the University of California at San Diego (UCSD) in all eight exons and adjacent intronic regions from 180 ethnically diverse (88 Asian American, 114 African American, 56 Hispanic American and 102 European American) human subjects (2n = 360 chromosomes) (Wen et al. 2004). Resequencing detected 53 SNPs and 2 single-base insertion/deletions in 5725 bp footprint. Eight of the common SNPs (minor allele frequency ≥ 5%) were found in the promoter region (**G**-1106A, rs9658628; **A**-1018T, rs9658629; **T**-1014C, rs9658630; **T**-988G, rs9658631; **G**-462A, rs9658634; **T**-415C, rs9658635; **C**-89A, rs7159323; **C**-57T, rs9658638) (Figs. 1 and 2). The use of PHASE software for reconstructing haplotypes (variants that are inherited together) revealed 8 haplotypes in *CHGA* proximal promoter: haplotype 1 (GATTGTCC), haplotype 2 (AATTGTCC), haplotype 3 (GACGATAC), haplotype 4 (GATTGCCC), haplotype 5 (GTTTGCCT), haplotype 6 (GACGATCC), haplotype 7 (GATTGCCT), and haplotype 8 (GTTTGCCC) (Table 1). Four promoter SNPs in tight LD with each other (T-1014, T-988, G-462, and C-89) were common in the general population. Haplotype networks were constructed through use of Arlequin to illustrate relationships among the promoter haplotypes (Fig. 3a). Four linked SNPs appeared to divide *CHGA* promoter haplotypes into two

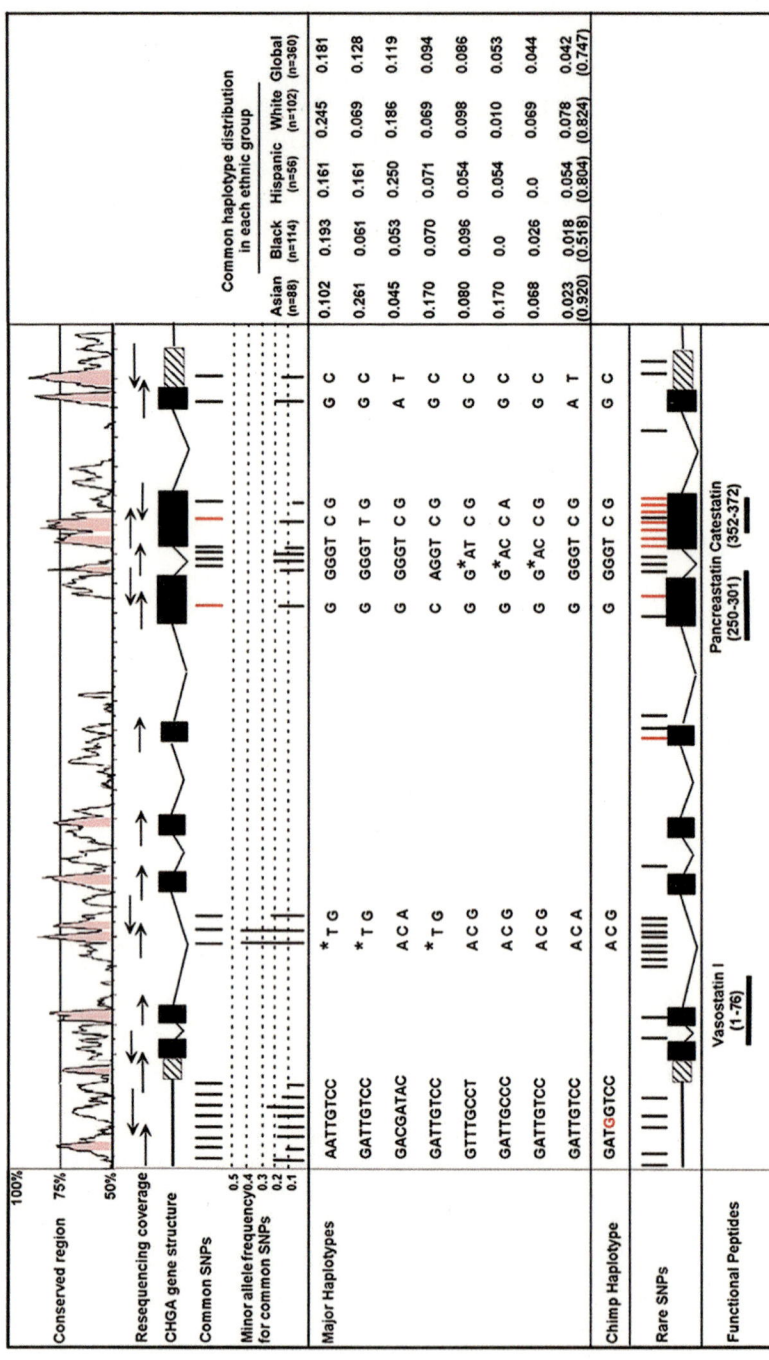

Fig. 1 Resequencing strategy and identified variants in *CHGA* gene. Note conserved sequence homology between mouse and human *CHGA* as deduced by VISTA. Red rods indicate nonsynonymous SNPs, and black rods indicate synonymous SNPs. The nucleotide in red in the chimpanzee haplotype indicates the minor allele in the human sequence. Computationally reconstructed haplotypes along with their relative frequencies in ethnogeographic groups within UCSD population are also shown. *Asterisk* (*) indicates deletions in haplotype sequences (Reprinted with permission from ELSEVIER Publishing Company)

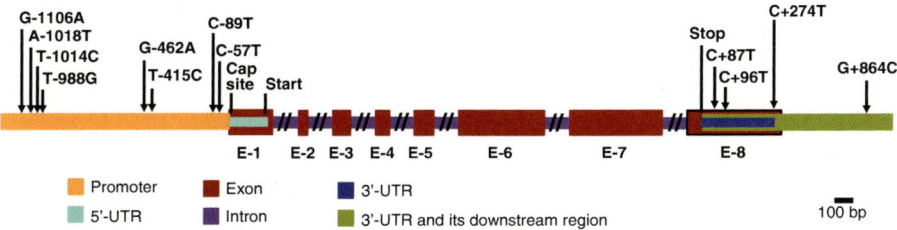

Fig. 2 Schematic diagram showing promoter and 3'UTR SNPs in *CHGA* gene

clusters, which define a deep division in the human lineage. Resequencing also identified four different naturally occurring SNPs in the *CHGA* 3'-UTR (untranslated region): **C+87T**, (**C**11825T; rs7610); **C+96T** (**C**11834T; rs9658672), **C+274T** (**C**12012T; rs9658673), and **G+864C** (**G**12602C) (Fig. 2).

3 Function of SNPs in the Regulatory Region of *CHGA*

3.1 Regulation of *CHGA* Expression by Promoter SNPs

The functional significance of *CHGA* promoter variants was verified at UCSD by examining the influence of individual SNPs on the expression of *CHGA* gene products in vivo. Testing associations of SNPs across *CHGA* with four plasma CgA peptide fragments revealed significant association with plasma CgA for only three SNPs (**T**-1014C, **T**-988G, and **G**-462A), which showed absolute linkage disequilibrium (LD) with each other in the promoter region (Fig. 3b). The promoter haplotype-specific reporter constructs were made by placing eight inferred promoter haplotypes upstream of a luciferase reporter and evaluated *CHGA* expression in PC12 cells (Wen et al. 2004). The lowest *CHGA* expression was detected in two common haplotypes containing minor alleles at the four complete-LD sites (Fig. 3c). In addition, *CHGA* expression was different in haplotypes 3 (GACGAT**A**C) and 6 (GACGAT**C**C), which differ in sequence only at −89 bp position (**bold** letter). Based on these findings, it was suspected that SNP C-89A and at least one of the three SNPs in absolute LD would play a prominent role in expression of *CHGA*. To validate this, each of the three SNPs in absolute-LD was mutated in the promoter reporter constructs for high-expressing haplotype 1 (GATTGTCC) and low-expressing haplotype 6 (GACGATCC). While mutation of **G**-462 to **A**-462 in haplotype 1 caused significant decrease in promoter activity, mutation of **A**-462 to **G**-462 resulted in significant increase in *CHGA* expression (Wen et al. 2004), implicating a prominent role for this SNP in regulation of *CHGA* expression (Fig. 3d).

Table 1 Haplotype distribution in the *CHGA* promoter region among four populations[*]

Promotor haplotype number	Nucleotide at position								Frequency (no. of chromosomes) in population				
	−1106	−1018	−1014	−988	−462	−415	−89	−57	Asian (n = 88)	Black (n = 114)	Hispanic (n = 114)	White (n = 102)	Total (n = 360)
1	G	A	T	T	G	T	C	C	0.446 (41)	0.211 (24)	0.321 (18)	0.265 (27)	0.306 (110)
2	A	A	T	T	G	T	C	C	0.102 (9)	0.272 (31)	0.196 (11)	0.255 (26)	0.214 (77)
3	G	A	C	G	A	T	A	C	0.045 (4)	0.123 (14)	0.268 (15)	0.206 (21)	0.15 (54)
4	G	A	T	T	G	C	C	C	0.26 (23)	0.156 (18)	0.071 (4)	0.078 (8)	0.147 (53)
5	G	T	T	T	G	C	C	T	0.102 (9)	0.105 (12)	0.107 (6)	0.176 (18)	0.125 (45)
6	G	A	C	G	A	T	C	C	0 (0)	0.114 (13)	0.036 (2)	0 (0)	0.042 (15)
7	G	A	T	T	G	C	C	T	0 (0)	0.018 (2)	0 (0)	0.02 (2)	0.011 (4)
8	G	T	T	T	G	C	C	C	0.023 (2)	0 (0)	0 (0)	0 (0)	0.006 (2)

[*]This Table is reprinted with permission from ELSEVIER Publishing Company.

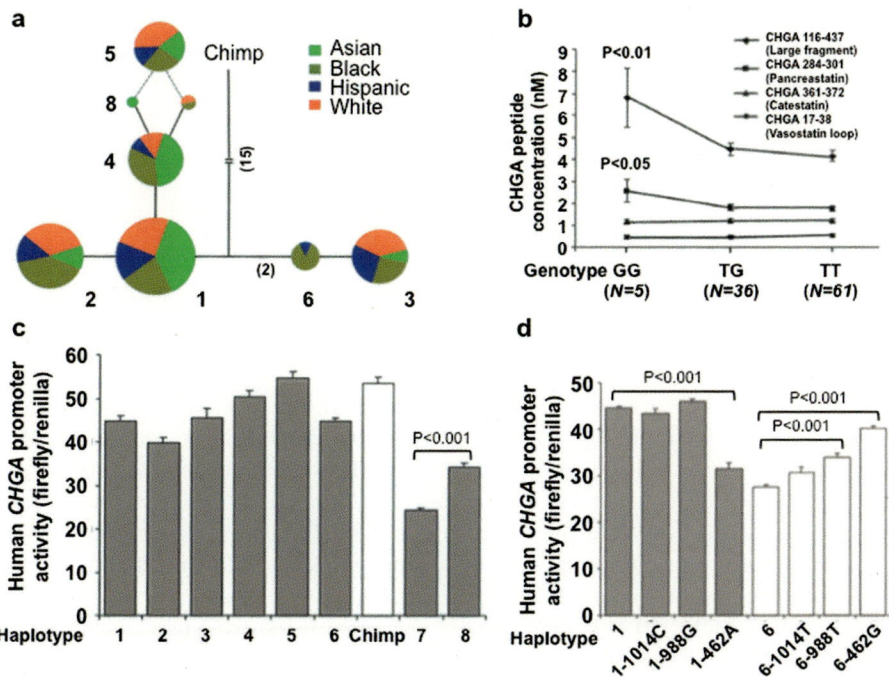

Fig. 3 Functional variation in the *CHGA* promoter and validation by changes in nucleotide in the haplotype. (**a**) Circle represents promoter haplotypes, and the circle area represents overall frequency. The proportion of the individual haplotype frequency in each of the four populations (indicated by different colors) is shown by the subdivision within the circle. *Dashed lines* indicate alternative topologies of equal length. Lines connecting haplotypes indicate nucleotide substitution. (**b**) Association of *CHGA* promoter SNP (GG, TG, and TT) with plasma CgA peptides. (**c**) Haplotype (haplotype 1–8)-specific *CHGA* promoter activity as evaluated by luciferase reporter activity with transfected promoter/reporter plasmids in PC12 cells. (**d**) Role of nucleotide substitution in haplotype 1 (T-1014C, T-988G, and G-462A) or haplotype 6 (C-1014T, G-988T, and A-462G) in *CHGA* promoter activity as evaluated by luciferase reporter activity with transfected promoter/reporter plasmids in PC12 cells (Reprinted with permission from ELSEVIER Publishing Company)

3.2 Regulation of Autonomic Activity and Blood Pressure by CHGA Promoter SNPs

A UCSD study used the HAP algorithm to ascertain physiological significance of promoter SNPs in five tightly linked common promoter variants (at positions **C**-1014T, **G**-988T, **C**-415T, and **A**-89C), which inferred three most common haplotypes: haplotype A (TTGTC, 56.9%), haplotype B (CGATA, 23%), and haplotype C (TTGCC, 16.5%). Of these haplotypes, haplotype B was found to blunt the blood pressure (BP) response to cold stress, which included immersion of the left hand in ice water for 60 s after a 10 min rest upon arrival and continuous recording of BP and heart rate with a calibrated radial artery applanation device. Interestingly, the

BP response showed molecular heterosis with the greatest change in blood pressure in heterozygotes (haplotype A/haplotype B) (Chen et al. 2008a). Besides playing a prominent role in *CHGA* expression, a **G**-462A variant that altered a COUP-TF transcriptional control motif was found to predict resting BP. Molecular heterosis was also evident in this variant because of the presence of high BP in heterozygotes (Chen et al. 2008a).

3.3 Regulation of Autonomic Activity and Blood Pressure by CHGA 3′-UTR SNPs

The association of *CHGA* 3′-UTR with autonomic activity and BP was established at UCSD from genotype (resequencing) and phenotype (plasma CgA, BP, catecholamines) data in four different cohorts. The 1st cohort consisted of initial resequencing of *CHGA* locus (8 exons, intron/exon borders, UTRs, and proximal promoters) in 180 subjects as described above. The 2nd cohort comprised of twin pairs from Southern California were 69% monozygotic (MZ) and 31% dizygotic (DZ): n = 103 monozygotic (MZ) pairs (M/M 21, F/F 82), and n = 45 dizygotic (DZ) pairs (8 M/M, 30 F/F, and 7 M/F). Amongst these twin pairs, 9.9% were hypertensive (8.8% treated with antihypertensive medications). Changes in BP in response to environmental (cold) stress were conducted in 149 twin pairs. The third cohort was from a population cohort with extreme BPs consisted of 470 male and 558 female European American, which were selected based on DBP in the upper or lower most extreme (fifth) percentiles of DBP distribution in 25,599 men and 27,479 women is a primary care practice at Kaiser-Permanente of Southern California medical group. The DBP status was as follows: 189 men with DBP ≥96 mm Hg and 281 men with DBP ≤61 mm Hg; 175 women with DBP ≥92 mm Hg and 383 women with DBP ≤59 mm Hg. The 4th cohort (purely phenotypic) consisted of 724 individuals with normal renal function (serum creatinine ≤1.5 mg/dl), stratified by BP status: normal BP (SBP < 135 and DBP < 85 mm Hg, on no medications), versus a diagnosis of essential hypertension (DBP ≥ 90 mm Hg, 75% were on antihypertensive medications). Consistent with previous studies, hypertensive subjects showed increased plasma CgA. A common (~27% frequency) genetic variant in the *CHGA* 3′-UTR (**C+87T**) was found to be strongly associated with human essential hypertension (Chen et al. 2008b). **C+87T** caused significant effects on both SBP and DBP as well as with genotype-by-sex interactions. Substantial BP differences between homozygote (C/C, T/T) classes were evidenced in men: ~12 mm Hg for SBP and ~9 mm Hg for DBP. While **C+87T** accounted for ~1.9% of the population variance in SBP (or ~13.7 mm Hg), and ~1.2% of the variance for DBP (or ~5.8 mm Hg) in men, **C+87T** did not affect either SBP or DBP in women. Increased numbers of the minor T allele was found to be associated with diminished plasma CgA (by ~10%) (Chen et al. 2008b). Likewise, increased numbers of the minor T allele blunted cold-stress-induced increments in SBP (decreased in men by ~12 mm Hg and in women by ~6 mm Hg) (Chen et al. 2008b). Sex-dependent responses to acute adrenergic stimuli, and the sex-dependent long-term consequences of repeated stressors on resting BP or the late appearance of hypertension may account for the above different consequences in men and women for *CHGA* C+87T 3′-UTR variant.

3.4 Association of CHGA Promoter and 3′-UTR Polymorphisms with Hypertensive Nephrosclerosis

UCSD study revealed that plasma CgA was inversely proportional to renal function as plasma CgA rose systematically (from 3.9 ± 1 nM to >10 nM) with the decline in renal function, from healthy individuals (GFR 101 ± 1 ml/min), to subjects with chronic kidney diseases (CKD, GFR 28 ± 3 ml/min), to subjects with end-stage renal disease (ESRD: GFR <10 ml/min). Like ESRD, plasma CgA level was also high in hypertensive subjects (Salem et al. 2008). In contrast, plasma CST was positively correlated with decreased renal function. In black ESRD subjects, plasma CST decreased by ~34% (from 1.2 ± 0.11 nM in black controls to 0.79 ± 0.03 nM in black ESRD) (Salem et al. 2008). Like ESRD, patients with essential hypertension also showed decreased (by ~14%) plasma CST (from 1.47 ± 0.06 in normotensive to 1.26 ± 0.08 in hypertensive subjects). The demographic description of the ESRD and control population included the following: age, gender, diabetes, hypertension, the family history of hypertension or ESRD, body mass index, and serum creatinine. Because of the important role of CgA in storage and release of catecholamines and its association with hypertension, the role of *CHGA* variants has been explored for their effects on hypertensive nephrosclerosis in three cohorts. The UCSD cohort consisted of 58 African American patients with hypertensive ESRD and 150 control subjects (74 normotensive and 76 hypertensive). The ESRD cases had a glomerular filtration rate (GFR) of essentially zero (all were sustained by chronic hemodialysis) as compared to the control group, which had a mean GFR of ~111 ml/min. The hypertensive control (with or without antihypertensive medication) subjects had a SBP >140 mm Hg or DBP >90 mm Hg. Replication study in North Carolina cohort included 301 hypertension-associated ESRD and 305 controls (normotensive as well as hypertensive). The National Institute of Diabetes and Digestive and Kidney Diseases (NIDDK) African American Study of Kidney Disease and Hypertension (AASK) trial cohort comprised of 830 subjects with progressive renal disease. *CHGA* promoter and 3′-UTR polymorphisms showed significant association with hypertensive (HT)-ESRD. *CHGA* promoter haplotype ATC (**G**-462A, **T**-415C, and **C**-89A) were more common in ESRD cases than controls. While the ATC haplotype was found in only ~14% individuals without renal failure, it was found in ~34% subjects with HT-ESRD. This indicates that ATC haplotype constitute a risk factor for developing ESRD. Likewise, *CHGA* 3′-UTR minor allele haplotype TC (**C**+87T and **G**+864C) were more common in ~40% of

ESRD cases, than ~15% of control subjects, implicating association of TC haplotype with ESRD (Salem et al. 2008). The above findings indicate that common variants in *CHGA* are associated with an increased risk for hypertensive ESRD in blacks. Since CgA induces secretion of endothelin-1 from glomerular endothelial cells and TGF-β1 from mesangial cells cocultured with glomerular endothelial cells, it was believed that CgA acted through the glomerular endothelium to regulate renal function.

4 SNPs in Gene Coding Regions

4.1 Functional Genetic Polymorphisms in the Pancreastatin (PST) Domain

4.1.1 Discovery of PST SNPs

The initial discovery of SNPs on PST domain was made at UCSD after resequencing 180 individuals from ethnic groups as described above (Wen et al. 2004). A UCSD study revealed three naturally occurring non-synonymous (amino acid replacement) variants in the PST region: **Arg**253Trp (rs9658662) (two heterozygotes; minor allele frequency, 0.6%), **Ala**256Gly (rs9658663) (two heterozygotes and one homozygote; minor allele frequency, 1.1%), and **Gly**297Ser (rs9658664) (two heterozygotes; minor allele frequency, 0.6%) (Wen et al. 2004). A subsequent study on PST SNP discovery at the Indian Institute of Technology Madras (IITM) resequenced *CHGA* in an Indian cohort of 410 individuals (= 820 chromosomes), which revealed three non-synonymous SNPs: **Arg**253Trp (minor allele frequency, 0.24%), **Glu**287Arg (minor allele frequency, 0.12%), and **Gly**297Ser (minor allele frequency, 6.7%, ~10-fold higher than the UCSD population) (Allu et al. 2014) (Fig. 4a, c). While Ala256Gly was not detected in the Indian cohort, the study discovered the novel **Glu**287Arg variant. Likewise, **Gly**297Ser variant was not detected in a Japanese cohort of 343 individuals (143 men and 200 women) (Choi et al. 2015).

4.1.2 Functional Consequences of PST SNPs

Effects of PST Variants on Glucose Homeostasis

Both UCSD and IITM studies showed greater inhibition of insulin-stimulated glucose uptake by **Gly**297Ser variant followed by **Glu**287Arg variant compared to WT-PST (O'Connor et al. 2005). The IITM study also revealed increased expression of gluconeogenic genes by PST variant as compared to WT-PST with comparable potencies by **Glu**287Arg and **Gly**297Ser variants (Allu et al. 2014).

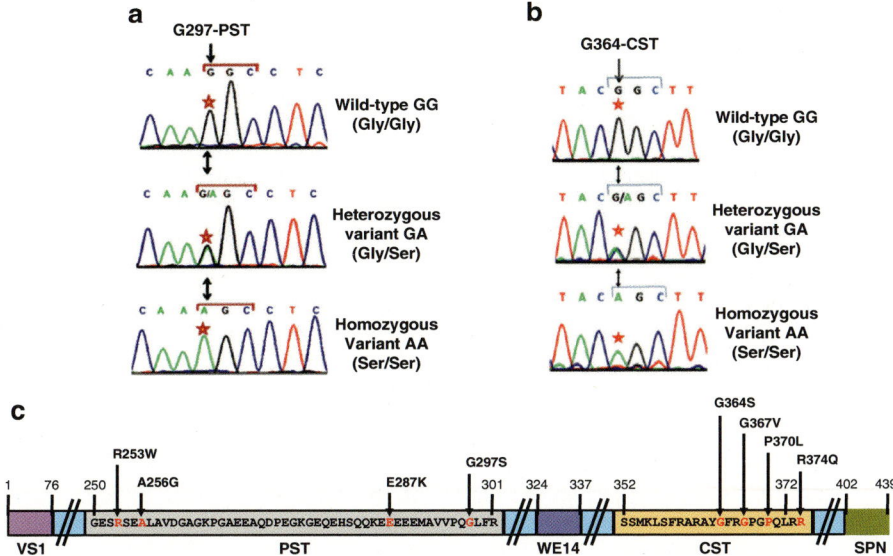

Fig. 4 (**a**) Electropherogram showing changes of G (major allele) to A (minor allele) in the PST domain at 9358 bp. (**b**) Electropherogram showing changes of G (major allele) to A (minor allele) in the CST domain at 9559 bp. (**c**) Schematic diagram showing nonsynonymous SNPs in the PST and CST domains of the *CHGA* gene

Association of PST Gly297Ser Variant with Biochemical Parameters

Because the **Gly**297Ser variant occurred in fairly large section of the Indian population (~13% subjects in the IITM study) it was possible to carry out association of various biochemical parameters with **Gly**297Gly and **Gly**297Ser genotype groups. The **Gly**297Ser subjects displayed markedly elevated plasma glucose (by ~17 mg/dl) and cholesterol (by ~12 mg/dl) compared to the **Gly**297Gly individuals. The higher glucose level in the carriers of PST 297Ser allele was consistent with the greater inhibition of insulin-stimulated glucose uptake and increased expression of gluconeogenic genes (viz. glucose-6-phosphatase and phosphoenolpyruvate carboxykinase-1) by the PST variant peptide as compared to WT-PST.

Interestingly, while the variants of PST in the C-terminal half of the molecule at 287 (Glu287Arg) and at 297 (Gly297Ser) enhance anti-insulin effects and elevate plasma glucose by inhibition of glucose uptake and stimulation of gluconeogenic effects, experimental deletion of N-terminal three amino acids Pro-Glu-Gly of human WT-PST (CgA$_{273-276}$) demonstrated opposite effects by reducing plasma glucose level and hepatic gluconeogenesis in rodent obese model (Bandyopadhyay et al. 2015). Therefore, finding the variants of N-terminal end of PST in human population may lead to discovery of a trait which would confer protection against insulin resistance as shown in mouse obese model with the N-terminal deletion variant of PST (Bandyopadhyay et al. 2015).

4.2 Functional Genetic Polymorphisms in the Catestatin (CST) Domain

4.2.1 Discovery of CST SNPs

Resequencing of all eight exons and adjacent intronic regions from 180 ethnically diverse human subjects (2n = 360 chromosomes) at UCSD as described above revealed two non-synonymous SNPs in the CST region: **Gly**364Ser variation (rs9658667; occurred in 11 of 180 subjects; minor allele frequency, 3.1%) and Pro370Leu (rs9658668; occurred in 2 of 180 subjects; minor allele frequency, 0.6%) (Wen et al. 2004). Although the **Gly**364Ser variant was distributed across other ethnic groups (5 Asian American, 5 European American, and 1 Hispanic American), it was absent in African American sample. In contrast, **Pro**370Leu variant occurred only in African American samples. Resequencing CST region of *CHGA* in an Indian population (n = 1010 subjects) at IITM detected 2 SNPs: **Gly**364Ser (minor allele frequency, 7.92%, ~2.6-fold higher than UCSD population) (Fig. 4b, c), and a new SNP **Gly**367Val (rs200576557) (minor allele frequency, 0.099%), which was not detected in UCSD population (Sahu et al. 2012). Resequencing of 343 Japanese subjects also detected **Gly**364Ser with a minor allele frequency of 6.1%, which is comparable to the Indian population (Choi et al. 2015). Thus, Gly364Ser variant is much more preponderant in Asian populations.

4.2.2 Functional Consequences of CST SNPs

Effects of CST Variants on Catecholamine Secretion

UCSD studies in PC12 cells revealed that the **Gly**364Ser variant of CST was ~4.7-fold less potent than WT-CST and Pro370Leu variant was ~2.3-fold more potent than WT-CST in terms of the inhibition of nicotine-evoked catecholamine secretion (Mahata et al. 2004; Wen et al. 2004) (Fig. 5b). IITM study in PC12 cells also found comparable loss of potency for **Gly**364Ser variant in inhibition of nicotine-evoked catecholamine secretion (Sahu et al. 2012). In addition, UCSD studies revealed less renal catecholamine secretion in **Gly**364Ser heterozygotes (n = 13) compared to Gly364/Gly364 homozygotes (n = 236) (Rao et al. 2007) (Fig. 5c). Furthermore, IITM study found marked decrease in plasma catecholamines in both the Gly364Ser heterozygotes and Ser364Ser homozygotes (Sahu et al. 2012).

Effects of CST Variants on Cardiovascular Function

The initial UCSD study consisted of 166 hypertensive (DBP > 95 mm Hg) individuals (146 men, 20 women) and 353 unmedicated normotensive controls (DBP < 85 mm Hg; 186 men, 167 women) from European descent. The replication study (Kaiser Permanente) included 1361 white (European ancestry; 643 male,

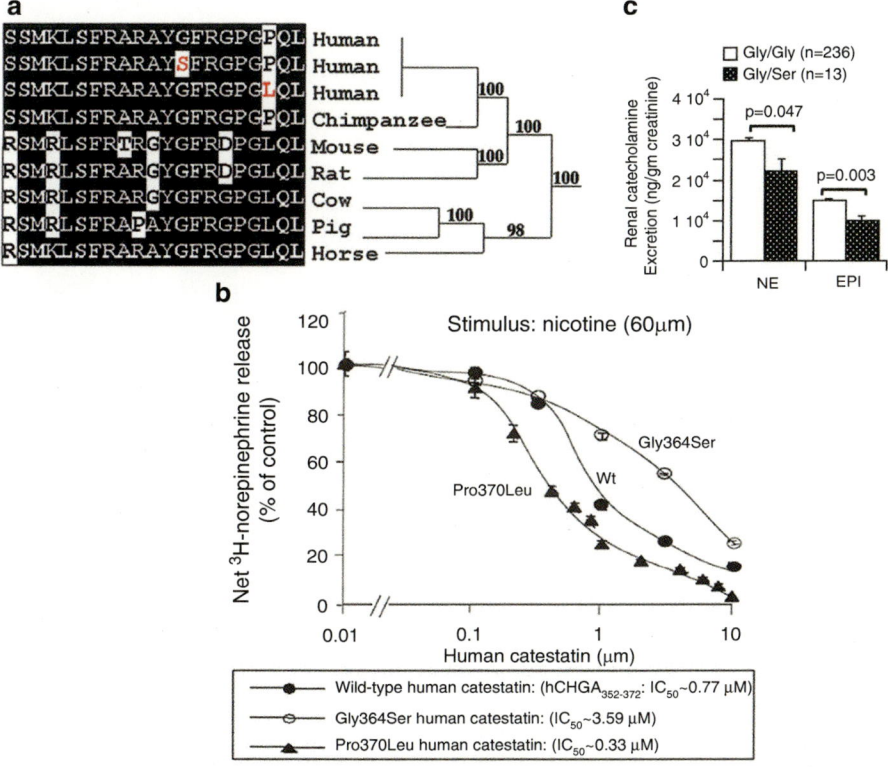

Fig. 5 (**a**) Sequence alignment showing sequence conservation across several species at Gly364Ser and Pro370Leu regions. (**b**) Potencies of CST variants (WT-CST, Gly364Ser-CST, and Pro370Leu-CST) on nicotine-evoked catecholamine secretion from PC12 cells (Reprinted with permission from ELSEVIER Publishing Company). (**c**) Effects of Gly364Ser variants on catecholamine secretion

718 female) subjects drawn from 53,078 individuals (27,475 women, 25,538 men) recruited from a large primary care (Kaiser Permanente) population in San Diego. Of note, Gly364Ser variant was associated with lower BP in the San Diego population (Rao et al. 2007) (Fig. 6a) Seventeen **Gly**364Ser heterozygote (as cases) and 48 **Gly364/Gly**364 homozygote (as controls) individuals underwent cold pressor test, which included immersion of the left hand in ice water for 60 s after a 10 min rest upon arrival and continuous recording of BP and heart rate with a calibrated radial artery applanation device. Although resting BP and HR did not differ significantly between the groups, **Gly**364Ser heterozygotes displayed decreased BP rise after cold pressor test (by ~16/~8 mm Hg) (Rao et al. 2007). **Gly**364Ser heterozygotes differed significantly with the controls in the time domain of autonomic monitoring: increased baroreceptor slope during upward deflections (by ~47%) and downward deflections (by ~44%), increased parasympathetic index by

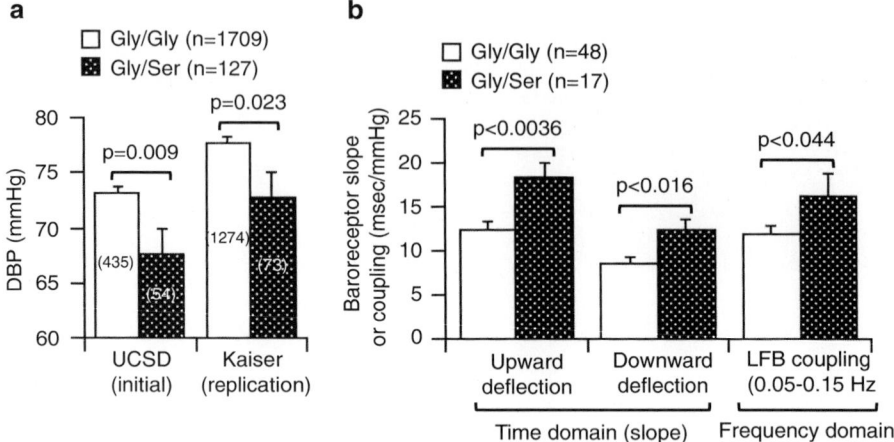

Fig. 6 (**a**) Association of Gly364Ser variant with DBP. (**b**) Effects of Gly364Ser variants on changes in baroreceptor function as evaluated by time and frequency domains

~2.4-fold, and decreased sympathetic index by ~26%. Increased pulse interval variability (by ~96%) and increased baroreflex coupling were noted in **Gly**364Ser heterozygotes in the parasympathetic high frequency band (0.15–0.4 Hz) (Rao et al. 2007) (Fig. 6b). Although the baroreceptor slope in **Gly**364Ser heterozygotes increased significantly in the mixed parasympathetic/sympathetic low frequency band (0.05–0.15 Hz), no difference was detected in pulse interval power. The pulse interval power and baroreceptor coupling were comparable between cases and controls in the sympathetic very low frequency band (0.01–0.05 Hz) (Rao et al. 2007) (Fig. 6b). Interestingly, the BP changes associated with this **Gly**364Ser variant were restricted to men only (Rao et al. 2007). Thus, the **Gly**364Ser variant displayed the most robust phenotype in human autonomic activity, both parasympathetic and sympathetic. In contrast, the Japanese study showed an association of **Gly**364Ser variant with high SBP and pulse pressure compared to **Gly**364**Gly** homozygote (controls), thereby increasing the risk for cardiovascular diseases in Japanese population (Choi et al. 2015). This directionally opposite association of the CST **Gly**364Ser SNP with blood pressure in two different studies suggests that the association may be specific to the particular ethnic population and points towards the genetic heterogeneity in different human populations.

5 Conclusions and Perspectives

In USA, the prevalence of hypertension and uncontrolled hypertension is highest among African Americans, where reconciliation of differences between patient and provider expectations for management of hypertension improves adherence to and acceptance of medical treatments. In addition, African Americans are more present-oriented regarding the daily management (or treatment) of hypertension as compared to whites.

Discovery and analysis of genetic variations at the *CHGA* locus in geographically/ethnically-different human populations revealed the presence of functional SNPs in the regulatory regions (viz., promoter and 3′-UTR) as well as coding regions (viz. within PST and CST peptide domains). A significant extent of heterogeneity in the occurrence (presence/absence/frequency) of several of these SNPs across different populations was observed. Nonetheless, *CHGA* SNPs showed strong associations with a number of cardiovascular phenotypes (viz. blood pressure, cardiac functions, plasma glucose, catecholamines and cholesterol levels). Further studies in this area may lead to utilization of *CHGA* genetic variations for clinical management of cardiovascular/metabolic disease states including hypertension and hypertensive kidney disease.

Acknowledgments Studies at UCSD were supported by grants from the National Institutes of Health and Veterans Affairs Medical Research. Research at IIT Madras was supported by grants from the Department of Biotechnology (BT/PR9546/MED/12/349/2007) and Science and Engineering Research Board (SR/SO/HS-084/2013A), Govt. of India.

References

Aardal S, Helle KB, Elsayed S, Reed RK, Serck-Hanssen G (1993) Vasostatins, comprising the N-terminal domain of chromogranin A, suppress tension in isolated human blood vessel segments. J Neuroendocrinol 5:405–412

Allu PK, Chirasani VR, Ghosh D, Mani A, Bera AK, Maji SK, Senapati S, Mullasari AS, Mahapatra NR (2014) Naturally occurring variants of the dysglycemic peptide pancreastatin: differential potencies for multiple cellular functions and structure-function correlation. J Biol Chem 289:4455–4469

Angelone T, Quintieri AM, Brar BK, Limchaiyawat PT, Tota B, Mahata SK, Cerra MC (2008) The antihypertensive chromogranin a peptide catestatin acts as a novel endocrine/paracrine modulator of cardiac inotropism and lusitropism. Endocrinology 149:4780–4793

Bandyopadhyay GK, Lu M, Avolio E, Siddiqui JA, Gayen JR, Wollam J, Vu CU, Chi NW, O'Connor DT, Mahata SK (2015) Pancreastatin-dependent inflammatory signaling mediates obesity-induced insulin resistance. Diabetes 64:104–116

Bandyopadhyay GK, Vu CU, Gentile S, Lee H, Biswas N, Chi NW, O'Connor DT, Mahata SK (2012) Catestatin (chromogranin A(352-372)) and novel effects on mobilization of fat from adipose tissue through regulation of adrenergic and leptin signaling. J Biol Chem 287:23141–23151

Bartolomucci A, Possenti R, Mahata SK, Fischer-Colbrie R, Loh YP, Salton SR (2011) The extended granin family: structure, function, and biomedical implications. Endocr Rev 32:755–797

Bell DA, Taylor JA, Butler MA, Stephens EA, Wiest J, Brubaker LH, Kadlubar FF, Lucier GW (1993) Genotype/phenotype discordance for human arylamine N-acetyltransferase (NAT2) reveals a new slow-acetylator allele common in African-Americans. Carcinogenesis 14:1689–1692

Chen Y, Rao F, Rodriguez-Flores JL, Mahapatra NR, Mahata M, Wen G, Salem RM, Shih PA, Das M, Schork NJ, Ziegler MG, Hamilton BA, Mahata SK, O'Connor DT (2008a) Common genetic variants in the chromogranin A promoter alter autonomic activity and blood pressure. Kidney Int 74:115–125

Chen Y, Rao F, Rodriguez-Flores JL, Mahata M, Fung MM, Stridsberg M, Vaingankar SM, Wen G, Salem RM, Das M, Cockburn MG, Schork NJ, Ziegler MG, Hamilton BA, Mahata SK, Taupenot

L, O'Connor DT (2008b) Naturally occurring human genetic variation in the 3′-untranslated region of the secretory protein chromogranin A is associated with autonomic blood pressure regulation and hypertension in a sex-dependent fashion. J Am Coll Cardiol 52:1468–1481

Choi Y, Miura M, Nakata Y, Sugasawa T, Nissato S, Otsuki T, Sugawara J, Iemitsu M, Kawakami Y, Shimano H, Iijima Y, Tanaka K, Kuno S, Allu PK, Mahapatra NR, Maeda S, Takekoshi K (2015) A common genetic variant of the chromogranin A-derived peptide catestatin is associated with atherogenesis and hypertension in a Japanese population. Endocr J 62:797–804

Collins FS, Brooks LD, Chakravarti A (1998) A DNA polymorphism discovery resource for research on human genetic variation. Genome Res 8:1229–1231

Gayen JR, Saberi M, Schenk S, Biswas N, Vaingankar SM, Cheung WW, Najjar SM, O'Connor DT, Bandyopadhyay G, Mahata SK (2009) A novel pathway of insulin sensitivity in chromogranin a null mice: a crucial role for pancreastatin in glucose homeostasis. J Biol Chem 284:28498–28509

Ingram VM, Stretton AO (1959) Genetic basis of the thalassaemia diseases. Nature 184:1903–1909

Kim T, Tao-Cheng J, Eiden LE, Loh YP (2001) Chromogranin A, an "On/Off" switch controlling dense-core secretory granule biogenesis. Cell 106:499–509

Mahapatra NR, O'Connor DT, Vaingankar SM, Hikim AP, Mahata M, Ray S, Staite E, Wu H, Gu Y, Dalton N, Kennedy BP, Ziegler MG, Ross J, Mahata SK (2005) Hypertension from targeted ablation of chromogranin A can be rescued by the human ortholog. J Clin Invest 115:1942–1952

Mahata SK, Mahapatra NR, Mahata M, Wang TC, Kennedy BP, Ziegler MG, O'Connor DT (2003) Catecholamine secretory vesicle stimulus-transcription coupling in vivo. Demonstration by a novel transgenic promoter/photoprotein reporter and inhibition of secretion and transcription by the chromogranin A fragment catestatin. J Biol Chem 278:32058–32067

Mahata SK, Mahata M, Fung MM, O'Connor DT (2010) Catestatin: a multifunctional peptide from chromogranin A. Regul Pept 162:33–43

Mahata SK, Mahata M, Wen G, Wong WB, Mahapatra NR, Hamilton BA, O'Connor DT (2004) The catecholamine release-inhibitory "catestatin" fragment of chromogranin a: naturally occurring human variants with different potencies for multiple chromaffin cell nicotinic cholinergic responses. Mol Pharmacol 66:1180–1191

Mahata SK, O'Connor DT, Mahata M, Yoo SH, Taupenot L, Wu H, Gill BM, Parmer RJ (1997) Novel autocrine feedback control of catecholamine release. A discrete chromogranin A fragment is a noncompetitive nicotinic cholinergic antagonist. J Clin Invest 100:1623–1633

Mouland AJ, Bevan S, White JH, Hendy GN (1994) Human chromogranin A gene. Molecular cloning, structural analysis, and neuroendocrine cell-specific expression. J Biol Chem 269:6918–6926

O'Connor DT (1985) Plasma chromogranin A. Initial studies in human hypertension. Hypertension 7:I76–I79

O'Connor DT, Cadman PE, Smiley C, Salem RM, Rao F, Smith J, Funk SD, Mahata SK, Mahata M, Wen G, Taupenot L, Gonzalez-Yanes C, Harper KL, Henry RR, Sanchez-Margalet V (2005) Pancreastatin: multiple actions on human intermediary metabolism in vivo, variation in disease, and naturally occurring functional genetic polymorphism. J Clin Endocrinol Metab 90:5414–5425

O'Connor DT, Zhu G, Rao F, Taupenot L, Fung MM, Das M, Mahata SK, Mahata M, Wang L, Zhang K, Greenwood TA, Shih PA, Cockburn MG, Ziegler MG, Stridsberg M, Martin NG, Whitfield JB (2008) Heritability and genome-wide linkage in US and australian twins identify novel genomic regions controlling chromogranin a: implications for secretion and blood pressure. Circulation 118:247–257

Pasqua T, Mahata S, Bandyopadhyay GK, Biswas A, Perkins GA, Sinha Hikim AP, Goldstein DS, Eiden LE, Mahata SK (2016) Impact of Chromogranin A deficiency on catecholamine storage, catecholamine granule morphology, and chromaffin cell energy metabolism in vivo. Cell Tissue Res 363:693–712

Ponomarenko JV, Orlova GV, Merkulova TI, Gorshkova EV, Fokin ON, Vasiliev GV, Frolov AS, Ponomarenko MP (2002) rSNP_Guide: an integrated database-tools system for studying SNPs and site-directed mutations in transcription factor binding sites. Hum Mutat 20:239–248

Rao F, Wen G, Gayen JR, Das M, Vaingankar SM, Rana BK, Mahata M, Kennedy BP, Salem RM, Stridsberg M, Abel K, Smith DW, Eskin E, Schork NJ, Hamilton BA, Ziegler MG, Mahata

SK, O'Connor DT (2007) Catecholamine release-inhibitory peptide catestatin (chromogranin A(352-372)): naturally occurring amino acid variant Gly364Ser causes profound changes in human autonomic activity and alters risk for hypertension. Circulation 115:2271–2281

Sahu BS, Obbineni JM, Sahu G, Allu PK, Subramanian L, Sonawane PJ, Singh PK, Sasi BK, Senapati S, Maji SK, Bera AK, Gomathi BS, Mullasari AS, Mahapatra NR (2012) Functional genetic variants of the catecholamine-release-inhibitory peptide catestatin in an Indian population: allele-specific effects on metabolic traits. J Biol Chem 287:43840–43852

Salem RM, Cadman PE, Chen Y, Rao F, Wen G, Hamilton BA, Rana BK, Smith DW, Stridsberg M, Ward HJ, Mahata M, Mahata SK, Bowden DW, Hicks PJ, Freedman BI, Schork NJ, O'Connor DT (2008) Chromogranin A polymorphisms are associated with hypertensive renal disease. J Am Soc Nephrol 19:600–614

Sanchez-Margalet V, Gonzalez-Yanes C, Najib S, Santos-Alvarez J (2010) Metabolic effects and mechanism of action of the chromogranin A-derived peptide pancreastatin. Regul Pept 161:8–14

Takiyyuddin MA, Parmer RJ, Kailasam MT, Cervenka JH, Kennedy B, Ziegler MG, Lin MC, Li J, Grim CE, Wright FA et al (1995) Chromogranin A in human hypertension. Influence of heredity. Hypertension 26:213–220

Tatemoto K, Efendic S, Mutt V, Makk G, Feistner GJ, Barchas JD (1986) Pancreastatin, a novel pancreatic peptide that inhibits insulin secretion. Nature 324:476–478

Theurl M, Schgoer W, Albrecht K, Jeschke J, Egger M, Beer AG, Vasiljevic D, Rong S, Wolf AM, Bahlmann FH, Patsch JR, Wolf D, Schratzberger P, Mahata SK, Kirchmair R (2010) The neuropeptide catestatin acts as a novel angiogenic cytokine via a basic fibroblast growth factor-dependent mechanism. Circ Res 107:1326–1335

Tota B, Angelone T, Mazza R, Cerra MC (2008) The chromogranin A-derived vasostatins: new players in the endocrine heart. Curr Med Chem 15:1444–1451

Tota B, Gentile S, Pasqua T, Bassino E, Koshimizu H, Cawley NX, Cerra MC, Loh YP, Angelone T (2012) The novel chromogranin A-derived serpinin and pyroglutaminated serpinin peptides are positive cardiac beta-adrenergic-like inotropes. FASEB J 26:2888–2898

Tramonti G, Ferdeghini M, Annichiarico C, Norpoth M, Donadio C, Bianchi R, Bianchi C (2001) Relationship between renal function and blood level of chromogranin A. Ren Fail 23:449–457

Wen G, Mahata SK, Cadman P, Mahata M, Ghosh S, Mahapatra NR, Rao F, Stridsberg M, Smith DW, Mahboubi P, Schork NJ, O'Connor DT, Hamilton BA (2004) Both rare and common polymorphisms contribute functional variation at CHGA, a regulator of catecholamine physiology. Am J Hum Genet 74:197–207

Winkler H, Fischer-Colbrie R (1992) The chromogranins A and B: the first 25 years and future perspectives. Neuroscience 49:497–528

Ziegler MG, Kennedy B, Morrissey E, O'Connor DT (1990) Norepinephrine clearance, chromogranin A and dopamine beta hydroxylase in renal failure. Kidney Int 37:1357–1362

Serpinin Peptides: Tissue Distribution and Functions

Y. Peng Loh, Niamh Cawley, Alicja Woronowicz, and Josef Troger

Abstract Serpinins are a family of peptides derived from proteolytic processing at paired basic residues at the C-terminus of chromogranin A, followed by aminopeptidase activity to trim the N-terminus of the liberated peptides and pyroglutamination. Three serpinin peptides have been identified that are released from the mouse endocrine pituitary cell line, AtT20. These include serpinin, pyroglutaminated serpinin (pGlu-serpinin) and C-terminal extended serpinin, serpinin-RRG. Each of these peptides have been found in different amounts in various tissues such as, adrenal medulla, heart, retina and brain. Cellular localization and secretion studies of these peptides indicate that they are packaged in secretory granules and secreted in a regulated (stimulated) manner and therefore function extracellularly as signaling molecules. Serpinin and pGlu-serpinin play an important role in up-regulating secretory granule biogenesis in endocrine cells. In the heart, endocrine cells and neurons, serpinin and pGlu-serpinin have been found to protect these cells against cell death under oxidative or ischemic stress. Serpinin and pGlu-serpinin are also positive cardiac β-adrenergic-like inotropes with a powerful effect on enhancing myocardial contractility. Additionally, serpinin, pGlu-serpinin and serpinin-RRG have been shown to have anti-angiogenesis effects with the C-terminal extended peptide being the most potent. The many functions of the serpinin peptides indicate their physiological importance.

Y. Peng Loh (✉) • N. Cawley • A. Woronowicz
Section on Cellular Neurobiology, Eunice Kennedy Shriver National Institute of Child Health and Human Development, National Institutes of Health, 49, Convent Drive, Bldg. 49, Room 6A-10, Bethesda, MD 20892, USA
e-mail: lohp@mail.nih.gov

J. Troger
Department of Ophthalmology, Medical University of Innsbruck, Anichstraße 35, 6020 Innsbruck, Austria

1 Discovery of Serpinin Peptides: An Historical Prospective

Serpinin is a 26 amino acid peptide derived from proteolytic cleavage of the penultimate and last pair of basic residues at the C-terminus of chromogranin A (CgA). This peptide was discovered through studies on the role of CgA in secretory granule biogenesis using the mouse pituitary endocrine cell line, AtT20 as a model system (see section below). We first found that expression of bovine CgA in 6T3 cells, a variant of the AtT20 cells, which lacks CgA and dense core secretory granules (DCG), induced DCG biogenesis and restored regulated secretion in these cells (Kim et al. 2001). Furthermore, CgA-knockout mice (Mahapatra et al. 2005) and CgA-antisense transgenic mice (Kim et al. 2005) had significantly decreased numbers and size of DCG in chromaffin cells in the adrenal medulla, compared to wild type mice. These studies indicated a major role of CgA in granule biogenesis. To understand the mechanism of action of CgA in granule biogenesis, a DNA microarray analysis was carried out on 6T3 cells with and without CgA transfection. Results indicated a significant increase in mRNA encoding a serine protease inhibitor (in the serpin family), protease Nexin-1 (PN-1) in cells transfected with CgA. PN-1 was localized to the Golgi apparatus and stabilized the degradation of DCG proteins required for DCG biogenesis (Kim and Loh 2006). We hypothesized that the increase in PN-1 expression was mediated by secreted CgA or a smaller processed fragment of it. We then synthesized and tested several peptide fragments of CgA based on predicted paired basic residue cleavage sites. A peptide of <3 kD from the C-terminus of CgA was found to increase PN-1 mRNA expression when added to the medium of 6T3 cells. Concomitantly, a<3 kD fraction from conditioned media of AtT20 cells was found to have the same action. The active fragment from conditioned medium of AtT20 cells was then analyzed by high performance liquid chromatography (HPLC) and enzyme immunoassay (EIA) using an anti-serpinin polyclonal antibody which we developed to detect serpinin (Koshimizu et al. 2011b). Subsequently, the serpinin immunoreactive fractions from the HPLC were subjected to matrix assisted laser desorption/ionization time of flight (MALDI-TOF) analysis to identify the peptides. A peptide identical to the sequence of the 26 amino acid synthetic peptide shown to upregulate PN-1 expression was identified. Hence, this endogenous peptide was named *serpinin*. Subsequent studies showed that serpinin was secreted by AtT20 cells and there are two additional forms, pGlu-serpinin and serpinin-RRG. In this review we describe the tissue and cellular distribution of the serpinin peptides and the distinct biological functions of the different members of this peptide family.

2 Expression and Characterization of Serpinin Peptides in AtT20 Cells

Western blot analysis using an antibody that cross reacts with both CgA and serpinin indicate that AtT20, a mouse cell line express a serpinin size immunoreactive peptide, as well as CgA with an apparent molecular weight of 40 kD (Fig. 1A).

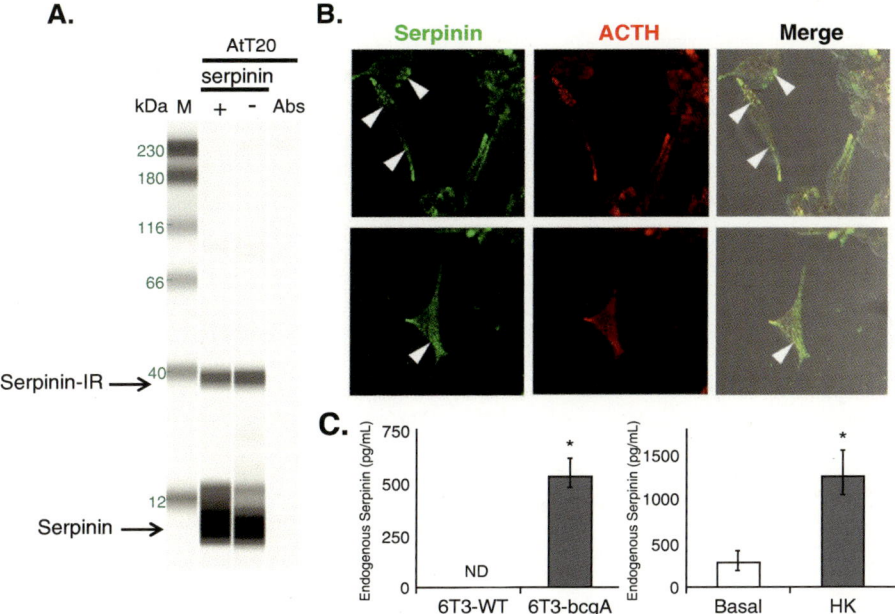

Fig. 1 Analysis of serpinin in AtT20 cells. (**A**) AtT20 cell lysate analysis by Western blot for serpinin-immunoreactivity (*IR*) using the ProteinSimple WES system. Serpinin peptide was added to the lysate as a positive control (+). As a negative control, the immunostaining was eliminated by pre-incubating the serpinin antibody with serpinin peptide as an absorption control (*Abs*). The results show (lane -) the presence of serpinin as well as a serpinin-containing chromogranin peptide at an apparent molecular mass of 40 kDa. M, molecular mass markers. (**B**) Immunocytochemical staining of serpinin-IR in AtT20 cells. Note the punctate staining within the cell body (*arrow heads*) and the co-localization with ACTH accumulated at the tips of the processes (From Koshimizu et al. 2011b). (**C**) Enzyme immunoassay of serpinin-IR in media from 6T3-AtT20 cells devoid of chromogranin A (CgA) and 6T3-AtT20 cells transfected with bovine CgA (*left*). Serpinin-IR was also released from AtT20 cells in a regulated manner by high potassium membrane depolarization (*HK*) (From Koshimizu et al. 2011a)

Immunocytochemistry studies of AtT20 cells showed that serpinin immunoreativity is localized in the cell body, along cell processes and at the tips of the processes where it is co-localized with the hormone, adrenocorticotropin (ACTH), in secretory granules (Fig. 1B; Koshimizu et al. 2011b). Furthermore, serpinin is secreted in a regulated manner with high K^+ stimulation from 6T3 cells transfected with bovine CgA (Fig. 1C; Koshimizu et al. 2011a).

MALDI-TOF analysis of immunoreactive serpinin peptides in secretion medium from AtT20 cells after HPLC purification revealed several peptides with m/z values shown in Fig. 2. Of these, a major peptide with a mass at m/z 2864 was detected and MS/MS fragmentation of this peptide confirmed that the amino acid composition of this peptide matched that of mouse serpinin, and therefore AtT20 cells synthesize and secrete serpinin. Additionally, the analysis revealed a pyro-glutaminated form of serpinin (pGlu-serpinin). Taken together, the identification of the various peptides found in the medium led to a proposed biosynthesis pathway for mouse serpinin

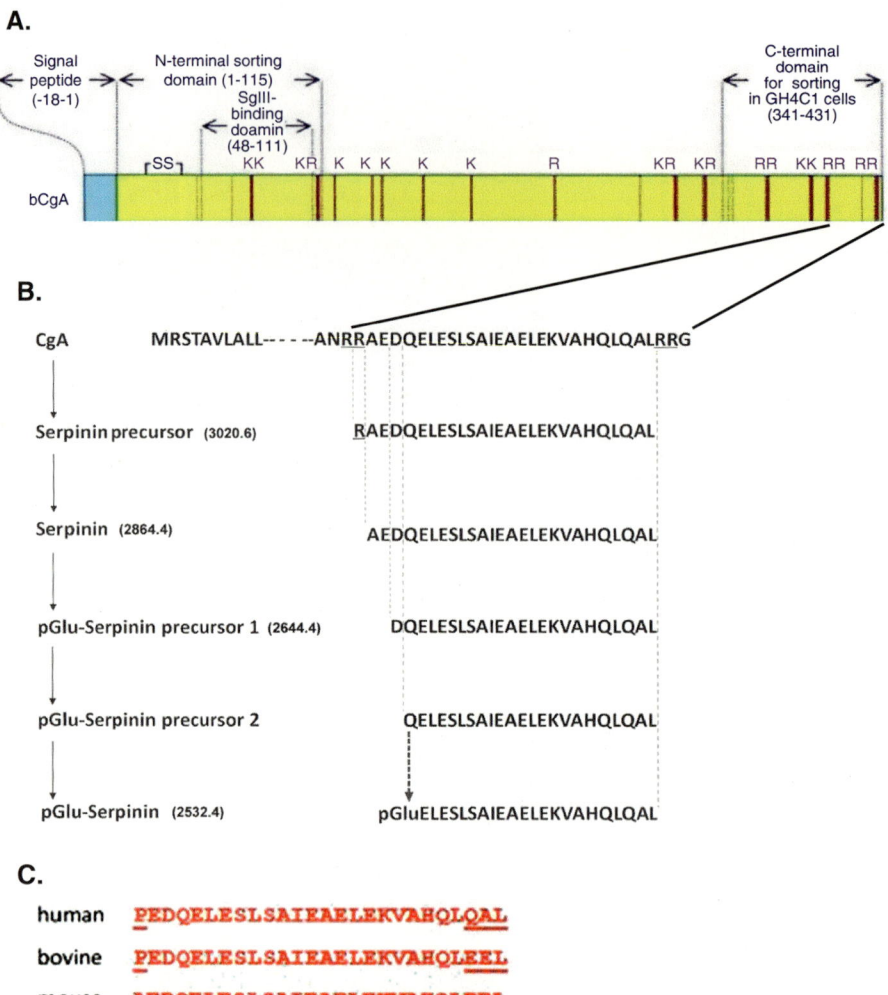

Fig. 2 Schematic of the bovine chromogranin A (bCgA) protein and processing of mCgA at its C-terminus. (**A**) bCgA is a 431 amino acid protein with multiple paired and single basic-residue cleavage sites (K, lysine, R, arginine) that can be used for processing bCgA into smaller peptides by prohormone convertases (From Loh et al. 2012b). (**B**) Schematic representation of the synthetic pathway of mouse pGlu-serpinin (From Koshimizu et al. 2011b). (**C**) Sequence of C-terminus of CgA from human, bovine and mouse. Note that the serpinin sequence (in *red*) differs by 4 amino acids (*underlined*) between mouse, human and bovine

peptides based on potential cleavage sites at paired basic residues in CgA as shown in Fig. 2B (Koshimizu et al. 2011b; Loh et al. 2012b). Based on the HPLC profile of the serpinin peptides in the medium, an earlier peak which eluted in the approximate position of serpinin with the RRG extension in the C-terminus (serpinin-RRG) was also present, suggestive of the presence of this peptide. Studies thus far have only purified and sequenced serpinin from mouse. Of note is the sequence of bovine and human serpinin differs by 4 amino acids from mouse (see Fig. 2C). A bovine CgA fragment consistent with cleavage at the RR↓P site (Fig. 2C) and liberating serpinin has been reported (Wohlfarter et al. 1988). However, cleavage at that site to liberate human serpinin remains to be determined.

3 Expression and Secretion of pGlu-Serpinin in AtT20 Cells

Examination of the cellular localization of pGlu-serpinin using a specific antibody showed that this peptide is expressed in AtT20 cells, but not in 6T3 cells lacking CgA (Fig. 3B; Koshimizu et al. 2011b). pGlu-serpinin was localized primarily in the tips of the processes and co-localized with ACTH in AtT20 cells (Fig. 3A; Koshimizu et al. 2011b). This is not surprising since pyro-glutamination is a late step in the biosynthetic pathway and therefore this peptide resides mainly in

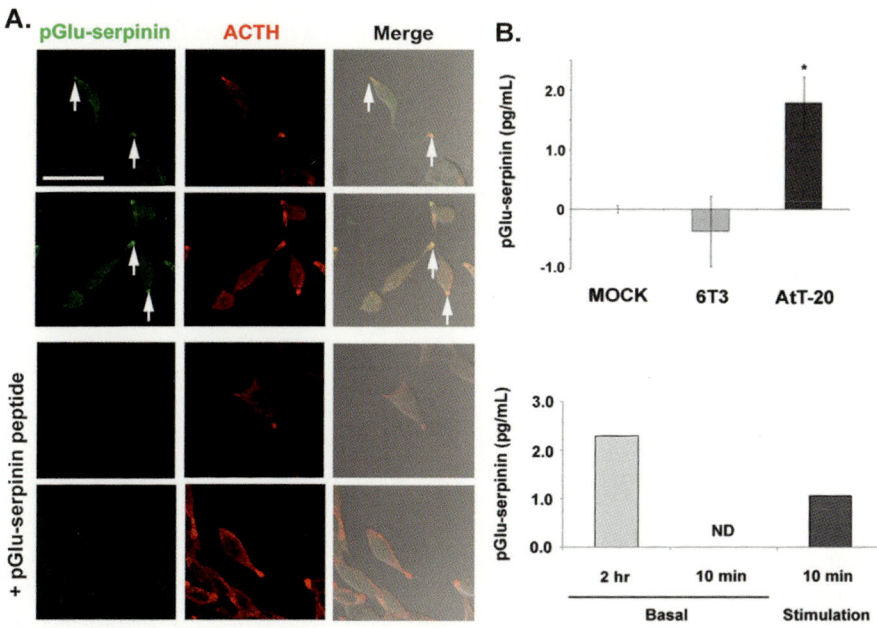

Fig. 3 Analysis of pGlu-serpinin in AtT20 cells. (**A**) Immunocytochemical staining of pGlu-serpinin-immunoreactivity (IR) in AtT20 cells. The punctate staining of pGlu-serpinin-IR co-localized with ACTH accumulated at the tips of the processes. Note the absence of pGlu-serpinin staining when the antibody was pre-absorbed with the pGlu-serpinin peptide. (**B**) Enzyme immunoassay of pGlu-serpinin-IR in 6T3-AtT20 cells devoid of chromogranin A (CgA) and WT AtT20 cells (*top panel*). (**B**, *lower panel*) pGlu-serpinin-IR was also detected in non-stimulated culture media (2 h. basal) of AtT20 cells and was secreted in a stimulated manner by high potassium membrane depolarization (10 min stimulation) (From Koshimizu et al. 2011b)

secretory granules at the tips of processes and poised for secretion. pGlu-serpinin was found to be secreted in a stimulated manner (Fig. 3C; Koshimizu et al. 2011b), suggesting that it may play an extracellular role in signal transduction, as a neurotransmitter or neuromodulator.

4 Distribution of Serpinin Peptides in Organs and the Nervous System

4.1 Serpinin Peptides in the Adrenal Medulla and Heart

Serpinin peptides have been found in the mouse adrenal medulla which synthesizes large amounts of CgA. HPLC analysis revealed the presence of pGlu-serpinin as the major serpinin peptide, with lesser amounts of serpinin in adrenal medulla (Fig. 4A). Serpinin peptides are also expressed in the heart. The major serpinin peptide in rat heart is the C-terminal extended form: serpinin-RRG, with a significant amount of pGlu-serpinin also present (Fig. 4B). When assayed by sandwich ELISA for pGlu-serpinin, the concentration of this peptide reported in rat heart was 103.8 ± 14.7 pg/g rat heart (Tota et al. 2012).

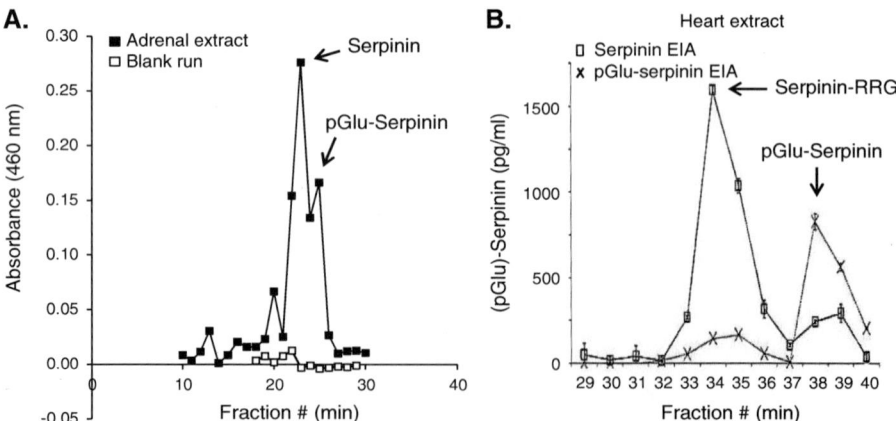

Fig. 4 Analysis of serpinin-immunoreactivity (IR) in tissue. (**A**) High pressure liquid chromatography (HPLC) followed by enzyme-linked immunoassay (*EIA*) analysis of mouse adrenal gland extracts. Note the presence of two peaks consistent with the elution profile for serpinin and pGlu-serpinin standards. (**B**) HPLC-EIA analysis of rat heart extract for serpinin-IR and pGlu-serpinin. Note the presence of two peaks consistent with the elution profile of serpinin-RRG and pGlu-serpinin standards. A shoulder that is evident on the serpinin-RRG peak is indicative of a small amount of serpinin in this tissue. Note that two different HPLC systems were used to analyze these extracts resulting in different elution profiles of the peptides (From Tota et al. 2012)

4.2 Expression and Localization of p-Glu Serpinin in the Central Nervous System

Studies on the expression of serpinin and pGlu-serpinin revealed that there are specific areas in the central nervous system (CNS) where serpinin is present in significant amounts. Using an antibody specific for pGlu-serpinin, immunohistochemistry studies in mouse brain revealed that one of the areas enriched in this peptide is in the pars reticulate of the substantia nigra, (Fig. 5A, *a*). This region is the major source of GABAergic innervation to various brain areas (mainly thalamus) and also the region which receives axons from medium spiny cells from the striatum, as well as GABAergic projections from globus pallidium and glutaminergic projections from subthalamic nucleus (Kanazawa and Toyokura 1975; Tepper and Lee 2007). Indeed, it has been reported that many GABAergic neurons express CgA (Schafer et al. 2010) and pGlu-serpinin are likely in axons of these neurons. Another region of the brain showing expression of pGlu-serpinin in the parafascicular nucleus (Fig. 5A, *b*), which is an intralaminar nucleus of the thalamus, typically considered as part of the ascending activating system and is homologous to the human center median. Parafascicular nucleus is an essential source of thalamostriatal projections (Tsumori et al. 2003; Vercelli et al. 2003). pGlu-serpinin is also present in the nuclei of pontis (Fig. 5B), which are the concentrations of the gray matter of the pons and are involved basically

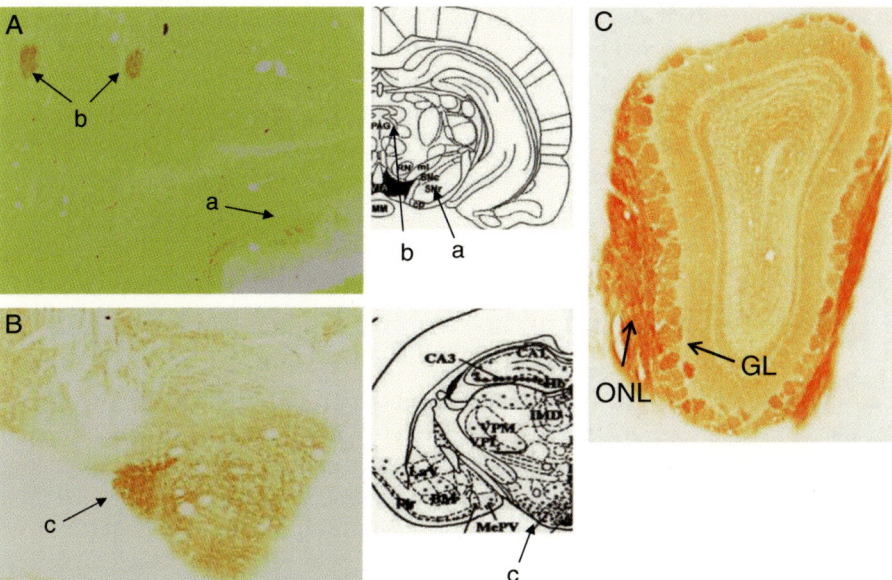

Fig. 5 Immunohistochemical staining for pGlu-serpinin in mouse brain sections. (**A**) Arrows show immuno-staining in the area of the Substantia Nigra (*a*), and the Nucleus Centrum-parafacicularis Thalami (*b*). (**B**) Arrow points to the Nuclei Pontis (*c*). (**C**) The olfactory bulb showing immuno-staining in the Glomerular layer (*GL*) and the olfactory nerve layer (*ONL*)

in the motor activities, by carrying information from the primary motor cortex to the ipsilateral pontine nucleus in the ventral pons via corticopontine fibers and through middle cerebellar peduncule to the contralateral cerebellum (Brodal and Bjaalie 1992; Cicirata et al. 2005). Additionally, pGlu-serpinin has been found in the olfactory bulb (Fig. 5C) primarily in the olfactory nerve layer (ONL) and the glomerular layer (GL) which is enriched in nerve terminals, suggesting that pGlu-serpinin may be released as a neurotransmitter or neuromodulator in the olfactory system.

4.3 Expression and Localization of Serpinin Peptides in the Eye

In addition to the central nervous system, serpinin peptides have been found to be present in the peripheral nervous system. In the rat peripheral nervous system, serpinin immunoreactvity has been detected in abundance in cells in the trigeminal ganglion which is the cranial sensory ganglion (Fig. 6A), indicating that it is a constituent of sensory neurons. Since the peptide is expressed in small to medium-sized cells with a diameter of 20–30 µm in this ganglion, the sensory nerves projecting to target tissues must represent predominantly unmyelinated C-fibers which arise from such cells. In the retina, which is a part of the central nervous system, this peptide is atypically expressed in the innermost part representing glia (Fig. 6B). With respect to the molecular form, only free serpinin has been detected in the trigeminal ganglion (Fig. 6C, *TG*) which indicates that exclusively free serpinin is present in sensory nerves originating from this ganglion and that there exists a pronounced processing of chromogranin A, already at the site of synthesis. Although in lessor amounts, free serpinin also predominates in the retina (Fig. 6C, *RET*). Very small amounts of serpinin-RRG may also be present in the trigeminal ganglion and the retina (Fig. 6C).

5 Role of Serpinin and pGlu-Serpinin in Granule Biogenesis

Early studies showed that AtT20 cells treated with antisense CgA RNA to down-regulate CgA expression (Fig. 7A,B; Kim et al. 2001) resulted in a decrease in dense core secretory granule biogenesis (Fig. 7C; Kim et al. 2001), in a dose dependent manner (Fig. 7D; Kim et al. 2001). To determine the mechanism by which CgA up-regulates secretory granule biogenesis, conditioned medium from AtT20 cells were analyzed. A CgA-derived peptide, serpinin was identified which enhanced secretory granule biogenesis (See Introduction and Expression of serpinin sections). Serpinin and pGlu-serpinin are both found in the secretion medium of AtT20 cells. When these peptides were added to the medium of AtT20 cells, the expression of a protease inhibitor, PN-1 was increased, with pGlu-serpinin being much more potent

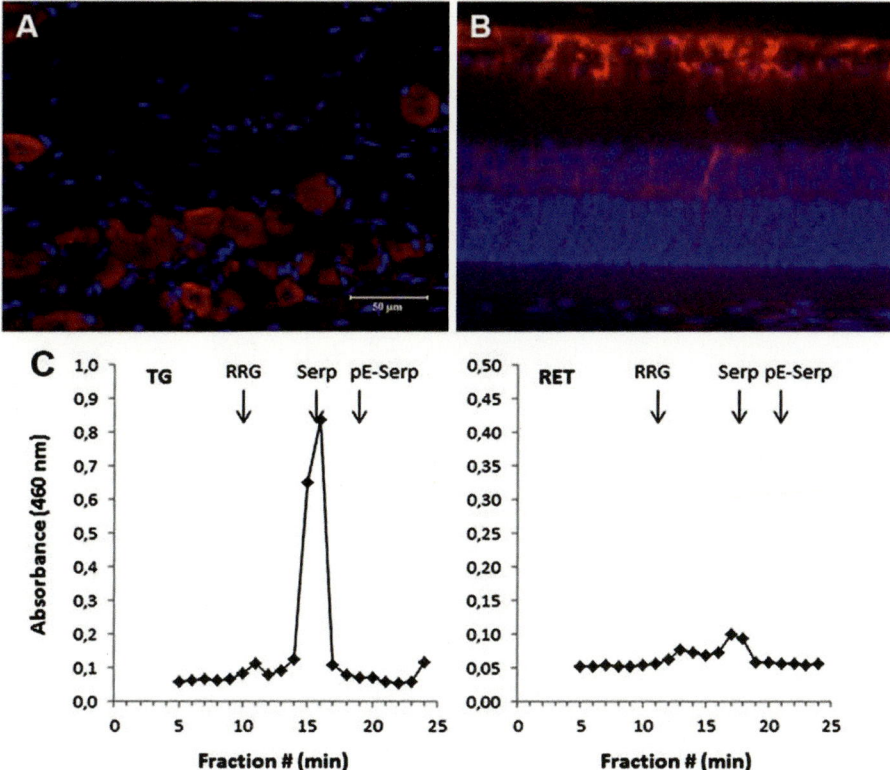

Fig. 6 Analysis of serpinin immunoreactivity (IR) in rat eye. (**A**) Serpinin-IR was found in cells in the trigeminal ganglion, indicative of being a constituent of sensory neurons. (**B**) In the retina, which is a part of the central nervous system, this serpinin-IR is expressed in the innermost part representing glia (red fluorescence). (**C**) Serpinin was detected in the trigeminal ganglion (*TG*) and to a smaller extent in the retina (*RET*) when these tissues we analyzed by HPLC-EIA for serpinin-IR

than serpinin (Koshimizu et al. 2011b). It was demonstrated that the enhanced expression of PN-1 by serpinin was mediated through binding of the peptide to a cognate receptor, which then increased cAMP (Fig. 8B). Addition of 8-bromo-cAMP or forskolin also increased PN-1 expression (Fig. 8A). Enhancement of PN-1 expression by serpinin was inhibited by an inhibitor of protein kinase A (PKA), 622-amide, (Fig. 8C), as well as by mithramycin A, that inhibits the activity of the transcription factor, sp1 (Fig. 8E). Furthermore, serpinin treatment of AtT20 cells resulted in the movement of sp1 into the nucleus of the cells (Fig. 8D). Additionally, luciferase reporter assay indicated that serpinin induced PN-1 expression by binding of sp1 to the promoter (Koshimizu et al. 2011a). Based on these findings, the proposed mechanism of action of serpinin in inducing granule biogenesis in an endocrine cell is summarized in the schematic in Fig. 9 (Kim et al. 2006).

Upon stimulated secretion of an endocrine cell, hormone and CgA-derived serpinin peptides contained in dense-cored secretory granules are released and are

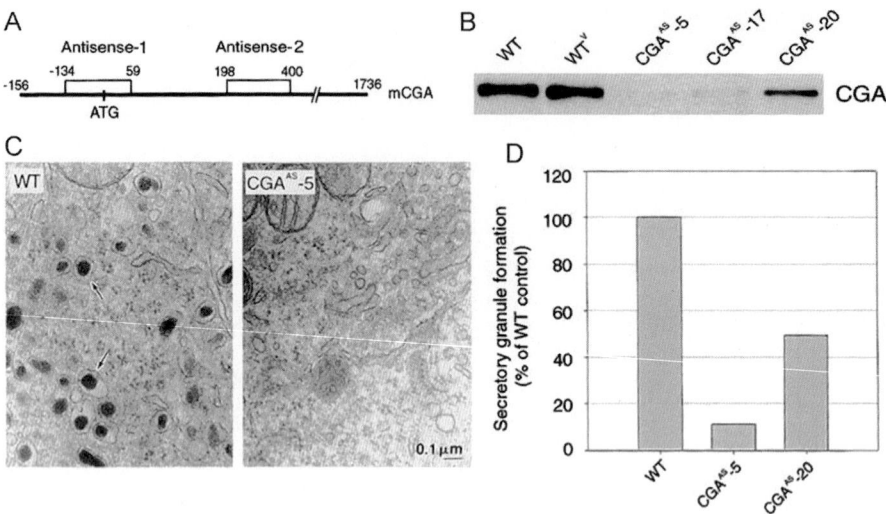

Fig. 7 Chromogranin A (CgA) levels directly affect dense core secretory granule numbers. (**A**) Schematic describing the construction of antisense vectors for silencing CgA expression in AtT20 cells. (**B**) Western blot analysis of CgA in WT AtT20 cells and AtT20 cells treated with antisense vectors. Note the different levels of silencing of CgA expression in the different clones (#5, #17, #20) (**C**) Electron micrograph of WT AtT20 cells and AtT20-CgA antisense clone #5. Note the apparent absence of dense core secretory granules in the antisense cells. (**D**) Quantification of dense core granule formation in WT and CgA antisense AtT20 cells derived from electron micrographs. Note the correlation of CgA levels and dense core granule numbers in the WT and antisense treated clones (From Kim et al. 2001)

subsequently replenished. Our studies found that secreted serpinin and pGlu-serpinin act as a signal to drive granule biogenesis likely through binding to a G-protein coupled receptor (see chapter "Full Length CgA: A Multifaceted Protein in Cardiovascular Health and Disease" by B. Tota) and triggering a cAMP-PKA-sp1 dependent pathway to up-regulate expression of the protease inhibitor, PN-1, at the transcriptional level. PN-1 resides in the Golgi apparatus and stabilizes degradation of secretory granule proteins which are constantly turning over at steady state. As a result, the increased secretory granule protein levels then induce more granule formation.

6 Serpinin and pGlu-Serpinin Peptides Protect against Cell Death

Serpinin and pGlu-serpinin peptides have been found to protect endocrine cells and neurons against oxidative stress and cell death. Both these peptides reduced the induced cytotoxicity in endocrine AtT20 cells, with pGlu-serpinin being 10-fold more potent than serpinin (Fig. 10; Koshimizu et al. 2011b). pGlu-serpinin was also found to be effective in protecting cortical neurons against H_2O_2 induced oxidative

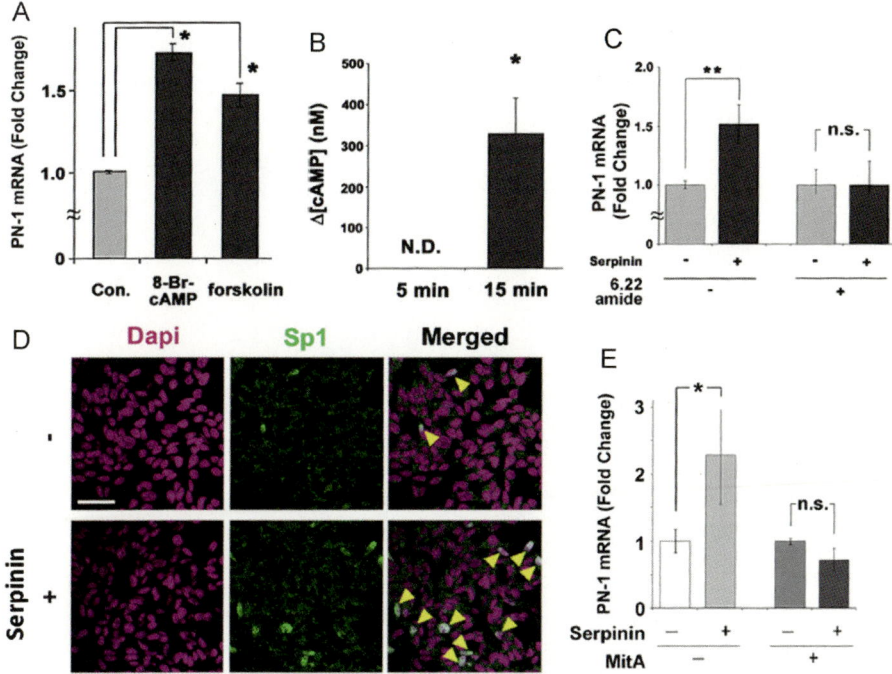

Fig. 8 Serpinin induces PN1 mRNA by a cAMP/PKA/SP1 signaling pathway. (**A**) PN1 mRNA is upregulated in AtT20 cells by 8-Br-cAMP and forskolin. (**B**) Intracellular cAMP increases in AtT20 cells treated with serpinin. (**C**) The PKA inhibitor, 6.22 amide, blocked the serpinin-induced up-regulation of PN1 mRNA. (**D**) Serpinin treatment of AtT20 cells induced the translocation of SP1 into the nucleus. Arrow heads indicate nuclear staining of SP1 immunoreactivity **E**. Mithramycin A (MitA), a specific inhibitor of SP1 binding, inhibits the serpinin-induced expression of PN1 mRNA (From Koshimizu et al. 2011a)

stress. Neurons treated with 10 nM pGlu-serpinin in the presence of H_2O_2 showed no difference in MAP 2 (a neuronal marker) signal compared to non-treated control neurons, but neurons treated with H_2O_2 showed a significant reduction in MAP2 signal (Fig. 11A; Koshimizu et al. 2011b). The neuroprotective effect of pGlu-serpinin is mediated by up-regulating the expression of the pro-survival mitochondrial protein, BCL2, in neurons under H_2O_2 induced oxidative stress (Fig. 11B; Loh et al. 2012a). Additionally, pGlu-serpinin also showed significant cardioprotective effects after ischemic stress of the rat heart ((Pasqua et al. 2015), see also chapter "Full Length CgA: A Multifaceted Protein in Cardiovascular Health and Disease" by B. Tota). Serpinin and pGlu-serpinin peptides have also been found to be positive cardiac β-adrenergic-like inotropes having a powerful effect on enhancing myocardial contractility and relaxation without change in blood pressure (Tota et al. 2012). The cAMP-PKA signaling pathway is also activated by serpinin and pGlu-serpinin to mediate inotropic activity in the heart (Tota et al. 2012, see Tota's chapter "Full LengthCgA: A Multifaceted Protein in Cardiovascular Health and Disease").

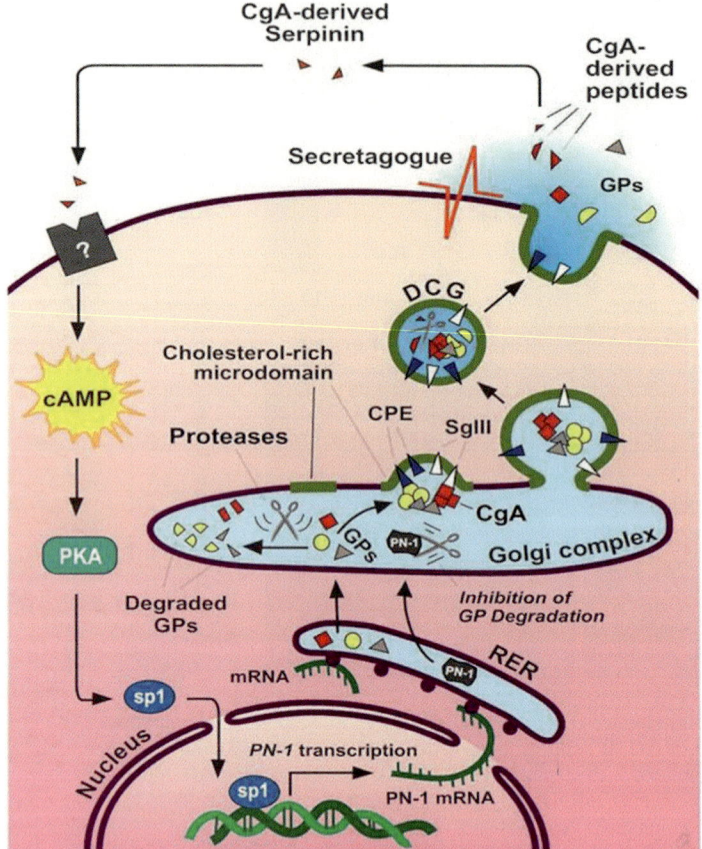

Fig. 9 Schematic of serpinin formation its signaling pathway and PN-1-dependent biogenesis of secretory granules in (neuro)endocrine cells. CgA is proteolytically cleaved to form serpinin which is secreted in an activity-dependent manner. Secreted serpinin binds to a cognate receptor and induces cAMP elevation followed by PKA activation. Then Sp1 a transcriptional factor translocates into the nucleus to upregulate PN-1 transcription. The increase in PN-1 protein stabilizes the secretory granule proteins at the Golgi apparatus to increase their levels which then promotes biogenesis of DCG (From Kim et al. 2006)

7 Anti-Angiogenesis Effect of Serpinin-RRG

The effect of human serpinin peptides on angiogenesis were assayed using the chick chorioallantoic membrane assay (Ponce and Kleinmann 2003). This assay involves the implantation of a filter paper disk embedded with serpinin peptides on the chorioallantoic membrane (CAM) of fertilized chicken eggs. Four days after implantation, the CAMs are fixed in 4% formaldehyde excised and the blood capillaries recruited by the half disk are counted. Fig. 12A shows that the number of new capillaries induced by bovine fibroblast growth factor (bFGF2) in chick embryos were unaffected by C-terminally truncated forms of CgA [CgA(1–373),

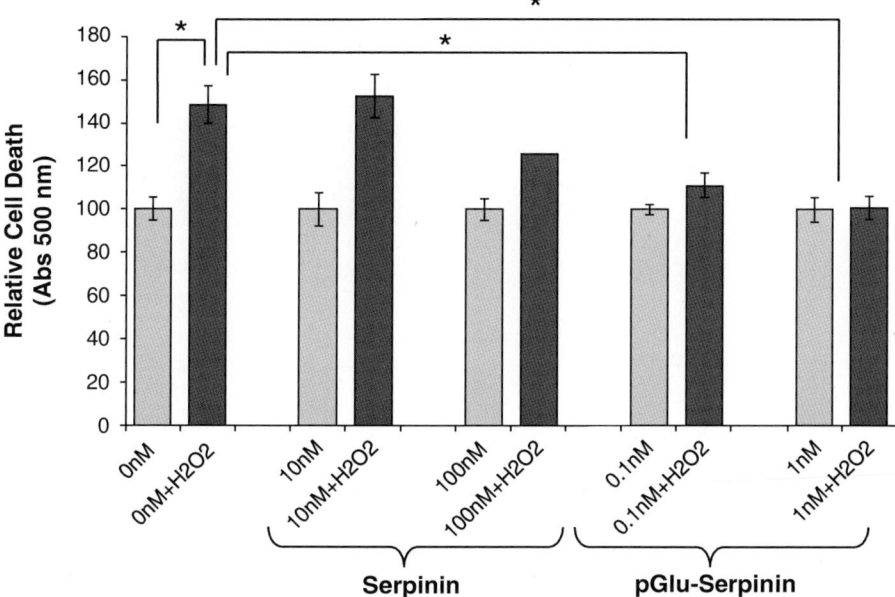

Fig. 10 Anti-apoptotic effects of serpinin and pGlu-serpinin. AtT-20 cells were challenged with 50 μM hydrogen peroxide in the presence or absence of 10 and 100 nM serpinin or 0.1 and 1 nM pGlu-serpinin for 1 day. Cell viability was quantified by the lactate dehydrogenase assay. Serpinin and pGlu-serpinin significantly inhibited hydrogen peroxide-induced cell death (n = 3, *$P < 0.05$) (From Koshimizu et al. 2011b)

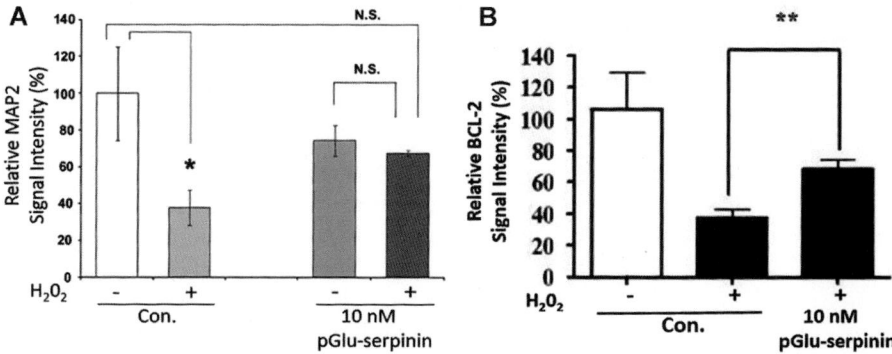

Fig. 11 pGlu-serpinin exhibits anti-apoptotic activity. Cultured rat cortical neurons were challenged with 50 mM hydrogen peroxide (H_2O_2) in the presence or absence of 10 nM pGlu-serpinin for 1 day. (**A**) Note the significant reduction in neuronal cell death, as measured by MAP 2 immunostaining, in the presence of H_2O_2 without pGlu-serpinin treatment (Con.), that was prevented when the neurons were incubated with pGlu-serpinin (From Koshimizu et al. 2011b). (**B**) pGlu-serpinin up-regulated Bcl-2 mRNA expression after H_2O_2-induced oxidative stress. Primary E18 rat cortical neurons were cultured for 5 days and then treated with 10 nM pGlu-serpinin or vehicle for 24 h. Neurons were then treated with 50 μM H_2O_2 to induce cell death. Twenty-four hours later, the neurons were collected for qPCR assay to assess the mRNA level of Bcl-2 in different groups. **$p < 0.01$ (t-test) (From Loh et al. 2012a)

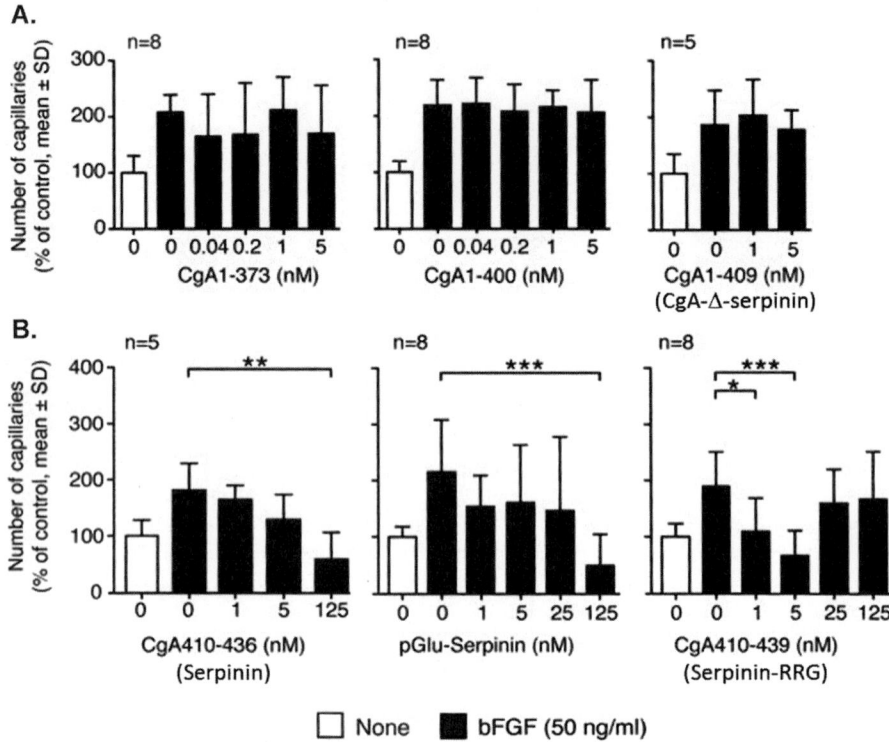

Fig. 12 Serpinin peptides exhibit anti-angiogenic activity. (**A**) The number of new capillaries induced in chick embryos by bFGF were unaffected by C-terminally truncated forms of hCgA (hCgA(1–373), hCgA(1–400) and hCgA(1–409)). (**B**) The hCgA C-terminal peptides, (serpinin, pGlu-serpinin and serpinin-RRG), all reduced the number of new capillaries in the chick embryos induced by bFGF in a dose dependent manner. Note that serpinin-RRG was more effective than serpinin and pGlu-serpinin (From Crippa et al. 2013)

CgA(1–400) and CgA(1–409) which is CgA minus the serpinin sequence]. However, the CgA C-terminal peptides, (serpinin, pGlu-serpinin and serpinin-RRG), all reduced the number of new capillaries in the chick embryos induced by bFGF in a dose dependent manner (Fig. 12B) (Crippa et al. 2013). Interestingly, serpinin-RRG at 5 nM concentration was much more effective than serpinin and pGlu-serpinin, and the dose response was bimodal, which often occurs with peptides (Ojaniemi and Vuori 1997).

8 Conclusions

The serpinin family of peptides is the newest set of biologically active peptides found to be derived from processing of CgA. The cleavage of the penultimate pair of basic residues at the C-terminus to generate this set of peptides appears to occur first in the processing of CgA, since the remaining intact part of the CgA minus the C-terminus was readily isolated in adrenal medulla extract since 1988 (Wohlfarter et al. 1988). Nevertheless, the liberated C-terminal peptide fragment was not isolated till 2011 (Koshimizu et al. 2011a). While serpinin, pGlu-serpinin and serpinin-RRG are the forms that we have found differentially expressed in various tissues, there may be still other members of the serpinin family that have biological activity. These three members of serpinin peptides have diverse functions from regulation of granule biogenesis through up-regulation of a protease inhibitor, PN-1; protection against cell death through increasing BCL2 expression; and anti-angiogenesis activity. pGlu-serpinin appears to be the most potent in these functions except in anti-angiogenesis where serpinin-RRG had the strongest effect. The mechanism of action of serpinin-RRG in promoting anti-angiogensis has yet to be explored. The presence of pGlu-serpinin in areas of the CNS enriched in nerve fibers or terminals and the glomerular layer of the olfactory bulb, and in the retina suggest that this peptide could be an important neurotransmitter or neuromodulator in certain systems. Future work should focus on finding the receptor for the serpinin peptides, investigating a possible neurotransmitter/neuromodulator role and exploring new roles of these peptides in the endocrine and neuronal systems, as well as in metabolism. Finally given the strong activity of pGlu-serpinin in enhancing myocardial contractility and relaxation, this peptide could be an excellent therapeutic agent since its effect does not affect blood pressure (Tota et al. 2012).

Acknowledgements This research was supported by the Intramural Research Program of the *Eunice Kennedy Shriver* National Institute of Child Health and Human Development, National Institutes of Health, USA

References

Brodal P, Bjaalie JG (1992) Organization of the pontine nuclei. Neurosci Res 13:83–118

Cicirata F, Serapide MF, Parenti R, Panto MR, Zappala A, Nicotra A, Cicero D (2005) The basilar pontine nuclei and the nucleus reticularis tegmenti pontis subserve distinct cerebrocerebellar pathways. Prog Brain Res 148:259–282

Crippa L, Bianco M, Colombo B, Gasparri AM, Ferrero E, Loh YP, Curnis F, Corti A (2013) A new chromogranin A-dependent angiogenic switch activated by thrombin. Blood 121:392–402

Kanazawa I, Toyokura Y (1975) Topographical study of the distribution of gamma-aminobutyric acid (GABA) in the human substantia nigra. A case study. Brain Res 100:371–381

Kim T, Gondre-Lewis MC, Arnaoutova I, Loh YP (2006) Dense-core secretory granule biogenesis. Physiology (Bethesda) 21:124–133

Kim T, Loh YP (2006) Protease nexin-1 promotes secretory granule biogenesis by preventing granule protein degradation. Mol Biol Cell 17:789–798

Kim T, Tao-Cheng JH, Eiden LE, Loh YP (2001) Chromogranin A, an "on/off" switch controlling dense-core secretory granule biogenesis. Cell 106:499–509

Kim T, Zhang CF, Sun Z, Wu H, Loh YP (2005) Chromogranin A deficiency in transgenic mice leads to aberrant chromaffin granule biogenesis. J Neurosci 25:6958–6961

Koshimizu H, Cawley NX, Kim T, Yergey AL, Loh YP (2011a) Serpinin: a novel chromogranin A-derived, secreted peptide up-regulates protease nexin-1 expression and granule biogenesis in endocrine cells. Mol Endocrinol 25:732–744

Koshimizu H, Cawley NX, Yergy AL, Loh YP (2011b) Role of pGlu-serpinin, a novel chromogranin A-derived peptide in inhibition of cell death. J Mol Neurosci 45:294–303

Loh YP, Cheng Y, Mahata SK, Corti A, Tota B (2012a) Chromogranin A and derived peptides in health and disease. J Mol Neurosci 48:347–356

Loh YP, Koshimizu H, Cawley NX, Tota B (2012b) Serpinins: role in granule biogenesis, inhibition of cell death and cardiac function. Curr Med Chem 19:4086–4092

Mahapatra NR, O'connor DT, Vaingankar SM, Hikim AP, Mahata M, Ray S, Staite E, Wu H, Gu Y, Dalton N, Kennedy BP, Ziegler MG, Ross J, Mahata SK (2005) Hypertension from targeted ablation of chromogranin A can be rescued by the human ortholog. J Clin Invest 115:1942–1952

Ojaniemi M, Vuori K (1997) Epidermal growth factor modulates tyrosine phosphorylation of p130Cas. Involvement of phosphatidylinositol 3′-kinase and actin cytoskeleton. J Biol Chem 272:25993–25998

Pasqua T, Tota B, Penna C, Corti A, Cerra MC, Loh YP, Angelone T (2015) pGlu-serpinin protects the normotensive and hypertensive heart from ischemic injury. J Endocrinol 227:167–178

Ponce, ML, Kleinmann, HK (2003) The chick chorioallantoic membrane as an in vivo angiogenesis model Curr Protoc Cell Biol Chapter 19: Unit 19.5

Schafer MK, Mahata SK, Stroth N, Eiden LE, Weihe E (2010) Cellular distribution of chromogranin A in excitatory, inhibitory, aminergic and peptidergic neurons of the rodent central nervous system. Regul Pept 165:36–44

Tepper JM, Lee CR (2007) GABAergic control of substantia nigra dopaminergic neurons. Prog Brain Res 160:189–208

Tota B, Gentile S, Pasqua T, Bassino E, Koshimizu H, Cawley NX, Cerra MC, Loh YP, Angelone T (2012) The novel chromogranin A-derived serpinin and pyroglutaminated serpinin peptides are positive cardiac beta-adrenergic-like inotropes. FASEB J 26:2888–2898

Tsumori T, Yokota S, Ono K, Yasui Y (2003) Nigrothalamostriatal and nigrothalamocortical pathways via the ventrolateral parafascicular nucleus. Neuroreport 14:81–86

Vercelli A, Marini G, Tredici G (2003) Anatomical organization of the telencephalic connections of the parafascicular nucleus in adult and developing rats. Eur J Neurosci 18:275–289

Wohlfarter T, Fischer-Colbrie R, Hogue-Angeletti R, Eiden LE, Winkler H (1988) Processing of chromogranin A within chromaffin granules starts at C- and N-terminal cleavage sites. FEBS Lett 231:67–70

Action and Mechanisms of Action of the Chromogranin A Derived Peptide Pancreastatin

N.E. Evtikhova, A. Pérez-Pérez, C. Jiménez-Cortegana,
A. Carmona-Fernández, T. Vilariño-García, and V. Sánchez-Margalet

Abstract Pancreastatin (PST), is a biologically active peptide isolated from porcine pancreas in 1986. Soon after PST was found to be contained within the chromogranin A (CGA) sequence, and therefore distributed throughout the neuroendocrine and gastrointestinal systems. This finding started up the consideration of CGA as a source of different biologically active peptides, which were then identified.

Even though many metabolic effects, such as the modulation of secretion of different glands, as well as the general metabolism regulation, have been described, the definitive picture of the physiological role of PST has not yet been established. Nevertheless, the sum of these metabolic effects, forged the name of PST as a dysglycemic peptide, with conterregulatory effects on insulin action. Thus, elevated circulating levels of PST have been found in Type 2 diabetes, gestational diabetes and essential hypertension, suggesting that PST is a negative regulator of insulin sensitivity and glucose homeostasis. The mechanism of action whereby PST could modulate insulin action has been thoroughly studied in various cellular systems (rat liver cells and adipocytes), and G coupled protein nature of the receptor has been established. But, although the purification process is able to yield some amount of PST binding protein, the final characterization of such PST receptor have been elusive so far. On the other hand, a different kind of receptor for PST has been proposed, and it may be the surface chaperone GRP78. Therefore, PST could modulate the energy metabolism through different mechanisms, such as insulin signaling antagonism, and as a protein folding regulator.

N.E. Evtikhova • A. Pérez-Pérez • C. Jiménez-Cortegana • A. Carmona-Fernández
T. Vilariño-García • V. Sánchez-Margalet (✉)
Department of Clinical Biochemistry, Virgen Macarena University Hospital,
School of Medicine, University of Seville, Seville, Spain
e-mail: margalet@us.es

1 Introduction

Pancreastatin (PST) is a regulatory peptide with multiple effects on the general metabolism. And even though the physiological role for PST has not yet been completely elucidated, there is no doubt about its importance in both endocrine and exocrine secretion, as well as in the regulation of energy metabolism. The counter-regulatory action of PST on insulin signaling has been described many times in different systems, such as in rat hepatoma cells or adipocytes.

The present chapter aims to compile the current knowledge about PST, including the biological origin of the peptide, its synthesis and secretion, as well as its action and the underlying molecular mechanisms in liver and adipose tissues.

1.1 Structure, Processing and Secretion

The biologically active peptide PST was first isolated from porcine pancreas in 1986, as a 49 aa (51 kDa) chain with no homology with any other gastrointestinal hormone (Tatemoto et al. 1986). Almost immediately, PST presence was found in neuroendocrine and gastrointestinal tissues, and in 1987 the origin of the peptide was already determined. PST was found to be the proteolytic product of chromogranin A (CGA), an acidic glycoprotein ubiquitously present in the secretory granules of the sympathetic adrenal system (Eiden 1987; Konecki et al. 1987; Schmidt et al. 1988; Tatemoto and Mutt 1978; Huttner and Benedum 1987; Winkler and Fischer-Colbrie 1992). Today it is well-known that CGA processing produces a variety of regulatory peptides, but PST was the first one to be described (Taupenot et al. 2003).

1.1.1 Pancreastatin Structure

Primary Structure

Despite of the fact that PST is a unique peptide with no evident homology to other peptide family, it shares the poly-glutamate sequence with gastrin and the COOH-terminal Arg-Gly-NH2 with vasopressin, being the amidated C-terminus a common feature among many regulatory peptides from the neuroendocrine system (Tatemoto et al. 1986). In fact, this part of the molecule is the responsible of its biological activity (Tatemoto et al. 1986; Zhang et al. 1990).

Another interesting point is that among all the CGA derived peptides, PST is the only one conserved exclusively in mammals, with no homology with any other vertebrates. In addition, a 41.5% homology exists between the human and the Tasmanian devil, and being these species so highly separated in the phylogenetic tree, it may indicate its importance in early mammalian evolution. This homology is even higher when the C-terminal fragment is compared. This is not surprising considering that it is the biological active part of the peptide. Indeed, this fragment was shown to inhibit the first and second phases of glucose-stimulated insulin release in a dose dependent way, being residues 35–36 (Glu-Glu) essential for this activity (Zhang et al. 1990).

Molecular Forms

Regarding the existing PST isoforms, a variety of them has been described in human tumors and blood (Tamamura et al. 1990, Kitayama et al. 1994, Funakoshi et al. 1989a) as well as several rat tissues (Hakanson et al. 1995; Curry et al. 1990). These forms clearly differ in the length of the peptide chain, but they still share the biologically active C terminus. Thus, we can find fragments between 29 and 186 aa, being the PST-52 the most abundant in the human plasma (Kitayama et al. 1994).

Finally, diverse kinds of phosphorylated PST forms have been described. This differences in phosphate residues are correlated with the physical location. Moreover, there is a correlation between the degree of phosphorylation of the CGA and it's processing in different tissues. Thus, CGA and PST are highly phosphorylated at pancreas, where the most mature form of PST is secreted, while, in the ilium, phosphorylation is poor and there is no processing of PST (Watkinson et al. 1993).

Pancreastatin Synthesis and Secretion

The cleavage of PST from CGA is specific for each specie and tissue, and occurs both inside and outside the cell, mainly in the pituitary gland, the endocrine pancreas and stomach (Watkinson et al. 1991; Simon and Aunis 1989; Metz-Boutigue et al. 1993; Leduc et al. 1990; Curry et al. 1991). It is in the stomach and endocrine pancreas, were the more complete processing of CGA is carried out, being the PST peptide its major product. Although many different molecular forms of PST can arise in the process, all of them share the C-terminus, the biologically active sequence (Tamamura et al. 1990; Schmidt et al. 1988; Sekiya et al. 1994; Hakanson et al. 1995; Funakoshi et al. 1989b; Curry et al. 1990). The best known processing and secretion system of PST is, certainly, that studied in entherocromaffin cells (ECL) of the gastric antrum. These cells respond to the gastrin stimulation increasing CGA expression, and consequently, the plasma PST levels. This upward effect has been achieved in vivo (rodent model), either directly by gastrin infusion or by suppressing acid secretion, whereas the drop of gastrin levels was due to fasting or antrectomy, resulting in a decreased plasmatic PST levels. According to this results, gastrinoma patients present increased plasma PST, which proceed from both, tumor secretions as well as normal gastrin stimulation (Hakanson et al. 1995). It is therefore not surprising that higher PST secretion takes place in human insulinoma cells, compared with primary islets (Hakanson et al. 1992).

2 Biological Effects

As explained above, pancreastatin is an important mediator in many biological functions such as autocrine, paracrine or endocrine secretion (see the Fig. 1). Thus, in pancreas and parathyroid the hormonal release is modulated by PST in an autocrine manner, whereas in the gastric mucosa occurs a paracrine effect, when ECL cells release PST modulating gastric secretion (Sanchez-Margalet et al. 1996).

Probably, the best described PST effect is the endocrine modulation of insulin action in the liver, and paracrine/endocrine in the adipocytes (Sanchez-Margalet et al. 1996). In this way, PST results particularly important in physiological studies of homeostasis of blood glucose and insulin, as well as pathological conditions such as diabetes mellitus (Sanchez-Margalet et al. 1996). For instance, plasma pancreastatin concentration was elevated in type 2 diabetes about 3.7 times, being unchanged in insulin resistance related obesity. Furthermore, this elevation was resistant to weight loss by diet (O'Connor et al. 2005). In the essential hypertension, a symptom related to metabolic syndrome, plasma levels also appears elevated, and thus PST could contribute to the insulin resistance that often accompanies this condition (Sanchez-Margalet et al. 1995).

In the next sections we will explore the physiological role of PST as a global regulator of exocrine and endocrine secretion, as well as energy metabolism or antagonism of insulin action.

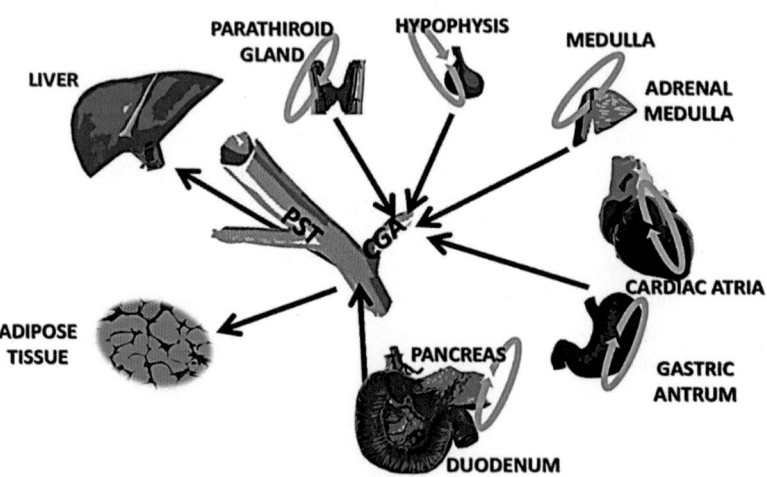

Fig. 1 Target tissues of pancreastatin action

2.1 Effects on Glandular Secretions

2.1.1 Endocrine Secretion

PST Inhibits Glucose Stimulated Insulin Secretion in Endocrine Pancreas

Since its isolation (Tatemoto et al. 1986) PST has been designated as an inhibitor of glucose stimulated insulin secretion, especially the first phase (Tatemoto et al. 1986). This effect was achieved both in vivo, in porcine pancreas, as well as in vitro, in beta cell lines. Thus, the cell line RIN m5F, a model of Langerhans islet beta cells, under PST treatment, showed significantly inhibition of insulin secretion stimulated by glyceraldehyde, carbachol, or ionophore A23187 (Hertelendy et al. 1996). Moreover, the inhibition of insulin secretion by PST seems to keep close relationship with elevation of cytosolic Ca^{2+} levels (Sanchez-Margalet et al. 1992). Even more, PST didn't inhibit insulin secretion stimulated by mastoparan and phorbol myristate acetate ester (PMA), being those mechanisms not dependent on Ca^{2+}. In general, all inhibitory effects of pancreastatin have been shown to be pertussis toxin sensitive, suggesting that PST could perform its action through a G-protein regulated Ca^{2+}-dependent mechanism, a common feature of other physiological inhibitors of insulin secretion (Hertelendy et al. 1996). Insulin release seems to be affected by PST in the presence of multiple physiological stimuli, such as glucose, arginine (Efendic et al. 1987), hormones GIP, VIP, CCK-8 (Peiro et al. 1989) or glucagon (Efendic et al. 1987), and some drugs like 3-isobutyl-1-methylxanthine or sulphonylurea (Schmidt and Creutzfeldt 1991). However in rat perfused pancreas, PST did not affect the somatostatin and glucagon release (both insulin mediated processes), while in canine pancreas and pig (Ohneda et al. 1989) rat PST had no effect on glucose-induced insulin release, suggesting that PST mediated effects could be species dependent.

PST Restricts Parathyroid Hormone Secretion

PST also exerts inhibitory effects on thyroid-parathyroid axis, being the calcitonin-producing C cells which take care of the active PST production, since the parathyroid gland, despite of being an important producer of CGA, doesn't yield PST (Cohn et al. 1982). So, PST produced by C-cells inhibits secretion of parathyroid hormone (PTH) in porcine, bovine and cultured parathyroid cells under low calcium condition or phorbol ester stimulation (Fasciotto et al. 1989; Drees and Hamilton 1992), while, enhanced PTH secretion in parathyroid cells has been observed when PST peptide was blocked with specific antibodies (Fasciotto et al. 1990). Moreover, PST also inhibited the transcription of the PTH and CGA genes and decreased the stability of the respective mRNAs (Zhang et al. 1994). Since the parathyroid cells don't produce PST, it seems to be unlikely any autocrine regulation of PTH, while paracrine/endocrine regulation via C cells should not be excluded.

2.1.2 Exocrine Secretion

PST Stimulates Gastric Secretion

Despite of some in vitro studies in isolated parietal rabbit cells, where PST has been shown to inhibit gastrin secretion (Lewis et al. 1988), in vivo it clearly enhances gastric activity. This is the case of conscious dogs stimulated with peptone food, phenylalanine or glucose (Hashimoto et al. 1990). Moreover, ECL cells of the gastric antrum have been shown to display high anti PST immunoreactivity, suggesting its important role in the paracrine regulation of gastric acid secretion through PST (Hakanson et al. 1995).

The mechanism behind gastric stimulation has a clear relationship with the inhibition of parietal cells stimulation by histamine. Such inhibition is interrupted in the presence of the pertussis toxin, suggesting the existence of a cAMP-dependent mechanism (Lindstrom et al. 1997; Lewis et al. 1988).

PST Inhibits the Exocrine Pancreas Secretion

Most studies affirms that PST has a negative effect on exocrine pancreatic secretion, which has been seen in physiological assays with food stimulation, vagal nerve arousal, and with CCK-8 treatment (Miyasaka et al. 1989). Only very few studies have been able to show an increased amylase secretion under PST treatment (Arden et al. 1994).

Inhibition of pancreatic secretion was reached by human synthetic PST and by bovine, porcine, and rat PST, or using only the carboxy-terminal fragment. These effects appear to be mediated by the presynaptic modulation of acetylcholine release from the vagal system. Therefore, PST could be a new islet-acinar axis mediator (Herzig et al. 1992).

2.2 PST Exhibits Negative Effects on Cell Proliferation

In HTC rat hepatoma cells, PST presents inhibitory effects on both cell growth and proliferation, decreasing protein and DNA synthesis. In this case, the underlying mechanism is the PST mediated nitric oxide (NO) and cyclic GMP (cGMP) production. On the other hand, PST can also activate mitogen activated protein kinase (MAPK), which is a growth signal. In this way, PST is able to increase NO levels in a dose depend manner, which overtake the MAPK pathway effect. However, if NO production is blocked by using pharmacological NOS inhibitors, PST action changes into a growth promoting effect. Therefore, PST effect on cell growth is depending on the NO availability (Sanchez-Margalet et al. 2001).

2.3 PST Increase Hepatic Glycogen Catabolism and Inhibits Insulin-Stimulated Glucose Uptake

In vivo, PST seems to exert direct effects on the liver metabolism stimulating glycogenolysis, while the basal levels of insulin and glucagon remained unchanged. The mechanism mediating the glycogenolytic effect of PST was studied in rat isolated hepatocytes. The glycogenolytic effect of PST turned out to be mediated by a cAMP-independent and Ca^{2+-}dependent mechanism. Moreover, the glycogenolytic effect of PST is related to the increase of cytosolic calcium concentration in a dose-dependent manner (Sanchez et al. 1992).

On the other hand, insulin stimulated glycogen synthesis was also inhibited by PST, in a similar way to the effect of glucagon, increasing the ratio of glycolysis on insulin-stimulated hepatocytes to 25%. But unlike glucagon, PST did not affect the basal rate of glycolysis. In addition, PST inhibits glucagon-stimulated insulin release enhancing the net hyperglycemia. However, only high concentrations of PST and insulin were able to induce insulin-stimulated glycolysis, which proposes that PST could play some role in insulin resistance (Sanchez-Margalet and Goberna 1994a, b). As a result, PST causes hyperglycemia by increasing glucose release from the liver (Sanchez et al. 1990).

Using CGA knockout mice, more recent animal studies have found that the lack of CGA expression reduces hepatic gluconeogenesis, improves insulin sensitivity, and results in low glycemia, despite of elevated plasma catecholamines and corticosterone levels. The mice has also hypertension, which could result from the loss of the catestatin fragment of CGA. These results may indicate that in CGA KO mice, the lack of PST could increase insulin sensitivity, contributing to maintain euglycemia by relieving inhibition from IRS1/2-PI 3-kinase-Akt signaling (achieved through suppression of cPKC and NOS activity) leading to increased suppression of hepatic gluconeogenesis. Thus, in normal condition, PST boosts the gluconeogenesis, through the inhibition of insulin signaling (Gayen et al. 2009; Valicherla et al. 2013).

Taken together, these findings indicate that PST could be an important modulator of hepatic glucose metabolism, an effect that may be mediated by an inhibitory cross-talk with insulin signaling pathway.

2.4 In Adipocyte, PST Enhances Energy Metabolism and Stimulates Protein Synthesis

In adipocytes PST has a similar role than in hepatocytes, exhibiting an increment of the energy expenditure and anti-insulin properties. In isolated white adipocytes we found a dose-dependent inhibition of both basal and insulin-stimulated glycogen synthesis, lactate production and glucose transport (Sanchez-Margalet and Gonzalez-Yanes 1998).

Regarding the lipid metabolism, PST causes a threefold rise of free fatty acids and glycerol release, although this effect is reversed by insulin co-stimulation (Sanchez-Margalet and Gonzalez-Yanes 1998). These results are consistent with the Chromogranin A knock-out mice model where it has been found a diminished expression of PPAR-γ and Srebp 1c, two of the most important lipogenic genes. However no changes were seen in hepatic lipids levels, probably due to an enhancement of fatty acid oxidation, since there is an increase in ketone bodies levels, UCP2 expression and enhanced acetil CoA carboxylase phosphorylation (Gayen et al. 2009; Valicherla et al. 2013). In conclusion, PST seems to stimulate lipolysis rather than lipogenesis (Carmen and Victor 2006).

PST has been found to act as a stimulator of protein synthesis, as well as does insulin. Thus, insulin can induce an increment of 40% in basal protein synthesis, and PST stimulation is able to produce a 30% increase. The mechanism used by PST to regulate these effects seems to be the activation of ERK1/2 signaling pathway (Sanchez-Margalet and Gonzalez-Yanes 1998).

In summary, PST has similar effects on glycogen, lipid metabolism and protein synthesis stimulation than other insulin contrarregulatory hormones, like catecholamines, calcium dependent hormones or growth hormone. But, in addition, PST has also been found to affect the expression of key genes in adipose tissue regulation. In this line, PST is able to inhibit leptin expression up to 60%, an effect that is completely reversed if protein synthesis inhibitors are employed. In the same way PST seems to enhance UCP2 expression, although the expression of other important adipocyte genes, such PPAR gamma, or UCP-1 were not affected (Gonzalez-Yanes and Sanchez-Margalet 2003).

3 Mechanisms of Action

3.1 *In Hepatocytes*

Molecular mechanisms of PST action in hepatocytes have been extensively studied by our group. Thus, a cAMP-independent and a Ca^{2+}-dose-dependent mechanism has been observed in rat hepatocytes (Sanchez et al. 1992). The generation of inositol 1,4,5-triphosphate (IP3) is responsible for the mobilization of intracellular calcium through a pertussis toxin-insensitive mechanism, while the activation of calcium influx implicates a pertussis toxin-sensitive mechaism. In this line, phospholipase C (PLC) has been found to be stimulated by PST in a dose-dependent manner, producing an increase in the levels of IP3 and diacylglycerol (DAG) in rat hepatocytes, leading to the activation of protein kinase C (PKC) (Sanchez-Margalet and Goberna 1994a).

PST action is also mediated by cGMP production through a pertussis toxin-sensitive G protein. Although the physiological role of cGMP in liver metabolism is not yet fully understood, there are some studies confirming cGMP as a mediator of

PLC inhibition, stimulated by PST, probably mediating a negative feed-back of PST signaling (Sanchez-Margalet and Goberna 1994b).

Using different approaches (GTP-γ-S binding, GTP-azido photolabeling, and GTPase activity), in rat liver membranes, our group has demonstrated that a pertussis toxin-insensitive stimulation of PLC by PST, is mediated by the activation of a G protein of the $G_{\alpha q/11}$ family, while the PST activated pertussis toxin-insensitive G protein is a part of the $G_{\alpha 1,2}$ family (Sanchez-Margalet 1999; Santos-Alvarez and Sanchez-Margalet 1998). Therefore, the activation of PLC by PST is mediated by $G_{\alpha 11}$ rather than $G_{\alpha q}$, and more precisely, PLC-β3 is the isoform activated in rat liver membranes by PST (see the Fig. 2) (Santos-Alvarez and Sanchez-Margalet 1998).

In recent CGA knock-out studies, it has been found that the lack of CGA expression leads to a decreased hepatic gluconeogenesis, a diminished glycaemia and an enhanced insulin sensitivity, despite of a high levels of corticosterone, and catecholamines, which may cause the observed hypertension. The increase in catecholamines could be a consequence of the loss of the CGA-derived peptide catestatin. The interpretation of these data is that PST may decrease insulin sensitivity, and the lack of PST may help to maintain euglycemia, since the inhibitory effect of PST on IRS1/2-PI3K-AKT signaling (reached through NOS and cPKC inactivation) would be not present, and therefore, insulin would generate an increased inhibition of hepatic gluconeogenesis. Thus, in normal conditions PST should enhance the

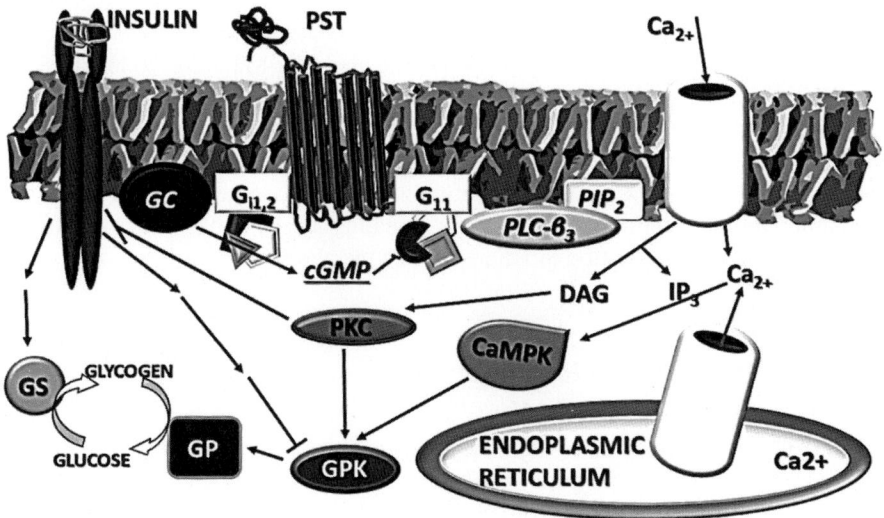

Fig. 2 Model of PST mechanism of action in hepatocyte. PST binds to its putative receptor at the plasma membrane, activating signaling through a $G_{\alpha 11}$ protein. The activation of a $G_{\alpha 11}$ stimulates PLC-β3 increasing the production of IP3 and DAG, and finally, the activated PKC and calcium dependent kinases mediate the metabolic effects. PKC phosphorylates the insulin receptor inhibiting its activity and counterregulating insulin action. $G_{\alpha i1,2}$ appears to mediate the production of cGMP inhibiting PLC activity to downregulate PST signaling

gluconeogenesis, by the inhibition of the insulin signaling (Gayen et al. 2009; Valicherla et al. 2013).

All these data suggest that PST could be an important regulator of hepatic glucose metabolism, by cross-talk with the insulin signaling pathway.

3.2 In Adipocytes

In a similar way to the mechanism of PST action described in hepatocytes, PST action in rat adipocytes is also mediated by the activation of PLC-β3, mainly through the $G_{\alpha q/11}$ family of G proteins (Gonzalez-Yanes et al. 1999), and to a lesser extent via the $G_{\alpha i1,2}$ protein (Sanchez-Margalet et al. 2010). Therefore, trough PLC-β3 pathway signaling, PST stimulates the translocation of classical PKC isoform to the plasma membrane. Then, PKC may mediate MAPK activation (Gonzalez-Yanes and Sanchez-Margalet 2000). In this way, the calcium-PKC signaling is responsible for the PST inhibition of glucose transport, glycogen synthesis, leptin expression, and the activation of lipolysis (Gonzalez-Yanes and Sanchez-Margalet, 2000, 2001, 2003), whereas the PST stimulation of protein synthesis in the adipocyte is mediated by the MAPK pathway, and the activation the translation initiation machinery (Gonzalez-Yanes and Sanchez-Margalet 2002). In this context, PKC activated through $G_{q/11}$ has been shown to stimulate MAPK by a Ras-independent, Raf-1 dependent, Ser-phosphorylation (Kolch et al. 1993). Consequently, we determined that PST stimulates both basal and insulin stimulated p42/44-MAPK activity trough Tyr/Thr phosphorylation (Gonzalez-Yanes and Sanchez-Margalet 2000). On the other hand, the initiation factor eIF-4E and 4E–BP1, which helps to liberate eIF-4E, are phosphorylated and activated in presence of PST (Gonzalez-Yanes and Sanchez-Margalet 2002). In conclusion, PST signaling in adipocytes, has two important pathways: the PLC- β3 signaling regulating the energy storage and MAPK/eIF-4E regulating protein synthesis. Thus, PLC signaling is the molecular mechanism underlying the PST regulation of adipose tissue metabolism.

4 The Pancreastatin Receptor

There is no final conclusion to identify the specific receptor through which PST is working. We have try out to purify this receptor from liver cells with no success. A possible explanation could be that the interaction between ligand and receptor is too labile to be maintained during the current elution process. But, in rat liver plasma membranes, we have carry out the purification of the active PST receptor by covalent cross linking and gel filtration study, identifying a ~85 kDa glycoprotein that specifically bound with the wheat-germ agglutinin (WGA) lectin. In addition, the fact that the eluted protein has sensitivity to Pertussis Toxin (PT) suggest that the

PST receptor could belong to the G protein coupled receptors family (Sanchez-Margalet and Santos-Alvarez 1997).

Therefore, our group performed a two-step purification method to obtain an 80 KDa monomeric glycoprotein, partially sensitive to PT and associated with $G_{\alpha q/11}$, to demonstrate the G coupled protein and glycoprotein nature of the receptor; although a final identification and complete sequencing have to be carry out, possibly solved by a microsequencing technique (Sanchez-Margalet et al. 2003, Santos-Alvarez and Sanchez-Margalet 2000).

4.1 G-Proteins Coupled to Pancreastatin Receptors

G protein-coupled receptors play a crucial role in signal transduction of vertebrate organisms. They belong to the most extensive family of receptors with more than 800 different GPCRs known so far. The structure of all of them has a common seven membrane-spanning domains. Their mechanism of action is also common. Signal starts with the ligand binding to the extracellular region that causes a conformational change in the seven-transmembrane domains, which is transmitted to the third loop and the C-terminal domain, activating the associated G-protein. This heterotrimeric G-protein is composed of three different subunits: α, β, and γ, which can offer a large number of permutations between them. In addition, many different ligands are associated with diverse classes of α subunits to mediate the activation of different signaling pathways. For example, $G_{\alpha q/11}$ can activate IP3 signaling pathway through phospholipase C beta, while $\alpha i1, 2$ causes the inhibition of cAMP levels (Culhane et al. 2015; Tuteja 2009).

As explained above, PST seems to bind a G-protein coupled receptor, but there is not a direct evidence for this yet, only some indirect binding and photolabelling studies combining with pertussis toxin and some with antibodies which block different alpha subunits. We suppose that PST is mainly coupled to a G-protein of the $G_{\alpha q/11}$ family, and secondly with the $G_{\alpha i1}, 2$ family. The specific G-protein coupled to putative PST receptor has been assessed in rat liver membranes, having the $G_{\alpha q}$ protein a better functional coupling than $G_{\alpha 11}$ protein (Santos-Alvarez and Sanchez-Margalet 1999, 2000; Santos-Alvarez et al. 1998). Furthermore, it has been found, only in heart cells, a different member of the alpha q/11 family ($G_{\alpha 16}$) that is coupled to the PST receptor.

4.2 Crosstalk of Pancreastatin Receptor with Insulin Receptor Signaling

The cross-talk between PST and insulin signaling has been largely studied in adipocyte and hepatoma cells. So, it was shown that PST can impair the insulin-stimulated auto-phosphorylation of the beta subunit of insulin receptor, in a dose-dependent manner, (Sun et al. 1991). Moreover, PST was also found to interfere other steps in insulin tyrosine kinase signaling: such as the tyrosine phosphorylation of insulin receptor substrate-1/2 (IRS-1/2) or the blocking of the association between p60–70 and the regulatory subunit of phosphatidyl inositol 3-kinase (PI3K): p85, causing an inhibition of PI3K activity (Sanchez-Margalet 1999; Gonzalez-Yanes and Sanchez-Margalet 2000) and consequently all the downstream events, such as the prevention of the activation of protein kinase B (PKB) and S6. Thus, PST addition in insulin stimulated adipocytes decreases PKB activity by 15% and 20% compared with basal activity (Gonzalez-Yanes and Sanchez-Margalet 2001).

In this context, PST has been found to block insulin pathway through Ser-phosphorylation of IR beta subunit and IRS1, generating a separation of the IR-IRS1-PI3K complex and a downregulation of PKB activity. Thus, a 3–4 fold raise of Ser phosphorylated IR/IRS1 has been found when PST was used (Gonzalez-Yanes and Sanchez-Margalet 2000).

Furthermore, PKC is the key factor to mediate all PST effects on insulin signaling. Thus, blocking PKC activity has been found to prevent IR/IRS1 Ser phosphorylation, and PI3K inhibition in hepatocytes and adipocytes. Finally, It has also been found that this PKC blocking can reduce other physiological actions of PST, such as insulin stimulated glucose transport or glycogen synthesis (Gonzalez-Yanes and Sanchez-Margalet 2000).

All these findings confirm the anti-insulin effects of PST, and points to the underlying mechanism, which seems to be the PKC mediated Ser-phosphorylation, impairing the Tyr-phosphorylation in insulin receptor signaling.

4.3   More than One Pancreastatin Receptor?

Various studies showed that PST displays its physiological effects in nM range of concentration (Gayen et al. 2009; O'Connor et al. 2005). These findings suggest, that PST receptor should be a typical one for biologically active peptides. As mentioned in previous points, this receptor seems to be a glycoprotein which has Pertussis Toxin sensitive nature, associated with $G_{\alpha q/11}$ proteins (Santos-Alvarez and Sanchez-Margalet 2000; Sanchez-Margalet et al. 2000; Gonzalez-Yanes et al. 1999, 2001; Sanchez-Margalet et al. 1994a, b). These facts strongly suggest that the PST receptor should be a GPCR. Nevertheless, due to the insufficient amount of protein purified, it was not possible to finally demonstrate this hypothesis. More recent studies discovered the interaction between PST and GRP78, a stress response chaperone, which regulates the protein half-life in stressful situations. Not only physical coupling exists between both molecules, moreover, PST has been shown to inhibit the ATPase activity of GRP78 (Biswas et al. 2014). But being GRP78 an intracellular protein, the question would be how is it possible its interaction with PST on the cell membrane surface. The answer is that a cell surface form of GRP78 exists (Arap et al. 2004; Delpino and Castelli 2002). In this line, it may also act as a receptor for several tumor proliferation-related peptides. Moreover, evidence exist about the GRP78 mediated AKT activation in cell surface, which is downstream pathway of insulin signaling (Arap et al. 2004; Delpino and Castelli 2002; Misra et al. 2006; Misra and Pizzo 2010; Shani et al. 2008; Wang et al. 2009; Zhang et al. 2010). Although, ligand-affinity experiments show some evidence for a GPCR-PST binding, it is conceivable that an additional high-affinity receptor for PST should exist. In conclusion, GRP78, could be a probable additional mechanism of PST cross-talk with insulin receptor whereby PST may interfere with insulin action.

5 PST and Inflammation Signaling

Studies in chromogranin A knockout-mice have brought to light the development of an anti-inflammatory environment in adipose tissue, which involves the downregulation of pro-inflammatory genes and cytokines and an upregulation of inflammatory genes in WAT and peritoneal macrophages. In parallel to anti-inflammatory events, the insulin resistance produced by high fat diet was prevented by CGA gene ablation. Thus, anti-inflammatory environment and enhanced insulin sensitivity, have been reversed by PST administration. Furthermore, it has been demonstrated that the insulin resistance in obese individuals is manifested only if PST is present, while in its absence obesity dissociates from it. The anti-inflammatory environment keeps a close relationship with altered PI3-K/Akt /Foxo1 signaling in CGA-KO individuals, suggesting that this may be the underlying mechanism in the pro-inflammatory activity of PST. By contrast, application of the truncated, variant of PST peptide (PST v1) simulates KO anti-inflammatory phenotype in WT obese individuals and simultaneously improving insulin sensitivity. Thus it seems that PSTv1 could operate as a PST competitive inhibitor, but further studies will be necessary to prove whether its administration to insulin resistant subjects could improve insulin sensitivity by reducing PST induced inflammation (Bandyopadhyay et al. 2015).

6 PST as a Stress Peptide

Considering the variety of biological actions ascribed to PST, the physiological role of this peptide could be a local modulator of secretion in glands where this peptide is actually processed and secreted. Hence, we can speculate that PST operates as an autocrine and paracrine regulatory peptide of endocrine and exocrine secretion attributing a significant physiological role for PST

The fact is that, PST levels are increased under the stressful stimuli when CGA is co-secreted with catecholamines from the sympathetic system. Higher PST levels leads to reduced insulin sensitivity in the liver and adipose tissues, which results in increased glycogenolysis in the liver and lipolysis in the adipose tissue, providing an extra energy to the whole organism in stressful conditions (Sanchez-Margalet et al. 2010). High affinity PST receptors have been characterized by our group and their signal transduction in liver and adipose tissue, so, the basis for the molecular mechanisms of the PST effects observed in glucose and lipid metabolism has been elucidated. Therefore it's foreseeable that PST displays its endocrine actions in stressful conditions when circulating PST levels are high enough to interact with specific receptors in target cells.

The mechanisms by which PST controls stress metabolism seems to be the inhibitory crosstalk with insulin receptor pathway. In this way, in CGA knockout mouse, the inhibition of IRS-1/2-PI3K-Akt signaling pathway by PST has been confirmed, which is consistent with previous in vitro studies with hepatocytes and adipocytes (Sanchez-Margalet et al. 2000; Gonzalez-Yanes et al. 1999). Nevertheless, despite these findings, some questions about PST signaling are still awaiting for further elucidation. Although active PST receptors were purified from rat liver membranes and shown association with a $G_{\alpha q/11}$ protein (Santos-Alvarez and Sanchez-Margalet 2000) the complete characterization and sequencing of such receptor was not possible. Even so, PST physiological actions could have behand them the G_q-PLC-calcium-PKC and protein kinase- $Ca^{2+/}$ NO mediated signaling pathway activation (Gayen et al. 2009; Santos-Alvarez and Sanchez-Margalet 1999; Sanchez-Margalet and Goberna 1994a, b; Sanchez-Margalet et al. 1994a, b).

In this physiological stress context, a novel PST target has been discovered recently in liver tissue, the adaptive UPR chaperone GRP78 also known as "Glucose-Regulated Protein". In summary, GRP78 may function in stress conditions as binding and degrading undesirable proteins. PST was found not only to bind GRP78 (in pH-dependent manner), but also to inhibit its ATPase enzymatic activity, resulting in enhanced G6P-ase expression and reduced glucose uptake (Biswas et al. 2014).

In this way it seems possible that PST could regulate the energy availability, in a global manner, through very different mechanisms, such as insulin signaling antagonism, as well as regulation of protein folding, and other remaining unknown possible ways.

7 PST Natural Occurring Variants and Its Implications

Finally, recent studies of the Indian population revealed different biological potency among natural PST peptide variants. Thus, two potent PST naturally occurring variants have been discovered: a more frequent (G297S) and an unusual one (PST287K), with more potent activity than WT-PST. This is not surprising, since both variants differ in the amino acid content of the C-terminus fragment, the responsible of PST biological activity. Thus, both variants have been shown to have higher activity than WT-PST in a variety of biological effects: decreased insulin stimulated glucose uptake, increased intracellular NO and Ca^{2+} levels, and enhanced gluconeogenesis enzymes expression. This enlarged activity keeps close relationship with a more ordered secondary structure of the resultant peptide, which results in a higher α-helicoidal content. In this way, the quantity of alpha-helices structures is higher in

PST-297S than PST 287 K, and this one, in turn, higher than WT-PST. These differences in secondary structure, have a direct relation with the plasma glucose and gluconeogenic genes expression, being highest in case of PST-297S. These findings could shed light on the inter-individual variations in glucose homeostasis, which opens new doors to a better diagnostics and personalized therapies of glucose/insulin disorders (Allu et al. 2014).

8 Final Remarks

The isolation of the PST peptide from porcine pancreas 30 years ago, reporting a biological effect, started up the era of the extracellular function of CGA as a precursor of peptides with regulatory functions. Different autocrine, paracrine and endocrine effects of PST have been described, and we have proposed that metabolic effects of PST could be useful in the physiology of stress, and could also participate in pathophysiological conditions such as insulin resistance. Different mechanisms of PST action have been proposed so far, and new mechanisms may be revealed in the future. Nevertheless, the precise role of PST in the context of CGA function in health and disease deserves further investigation, and we are positive that it will help to find the role of CGA as a precursor of biologically active peptides.

Acknowledgement We acknowledge the financial support of the Consejería de Innovacion, Ciencia y Empresa, Junta de Andalucía, Spain (Proyecto de Excelencia 08-CTS-4329), funded in part by FEDER.

References

Allu PK, Chirasani VR, Ghosh D, Mani A, Bera AK, Maji SK, Senapati S, Mullasari AS, Mahapatra NR (2014) Naturally occurring variants of the dysglycemic peptide pancreastatin: differential potencies for multiple cellular functions and structure-function correlation. J Biol Chem 289:4455–4469

Arap MA, Lahdenranta J, Mintz PJ, Hajitou A, Sarkis AS, Arap W, Pasqualini R (2004) Cell surface expression of the stress response chaperone GRP78 enables tumor targeting by circulating ligands. Cancer Cell 6:275–284

Arden SD, Rutherford NG, Guest PC, Curry WJ, Bailyes EM, Johnston CF, Hutton JC (1994) The post-translational processing of chromogranin A in the pancreatic islet: involvement of the eukaryote subtilisin PC2. Biochem J 298(Pt 3):521–528

Bandyopadhyay GK, Lu M, Avolio E, Siddiqui JA, Gayen JR, Wollam J, Vu CU, Chi NW, O'Connor DT, Mahata SK (2015) Pancreastatin-dependent inflammatory signaling mediates obesity-induced insulin resistance. Diabetes 64:104–116

Biswas N, Friese RS, Gayen JR, Bandyopadhyay G, Mahata SK, O'Connor DT (2014) Discovery of a novel target for the dysglycemic chromogranin A fragment pancreastatin: interaction with the chaperone GRP78 to influence metabolism. PLoS One 9:e84132

Carmen GY, Victor SM (2006) Signalling mechanisms regulating lipolysis. Cell Signal 18:401–408

Cohn DV, Zangerle R, Fischer-Colbrie R, Chu LL, Elting JJ, Hamilton JW, Winkler H (1982) Similarity of secretory protein I from parathyroid gland to chromogranin A from adrenal medulla. Proc Natl Acad Sci U S A 79:6056–6059

Culhane KJ, Liu Y, Cai Y, Yan EC (2015) Transmembrane signal transduction by peptide hormones via family B G protein-coupled receptors. Front Pharmacol 6:264

Curry WJ, Johnston CF, Shaw C, Buchanan KD (1990) Distribution and partial characterisation of immunoreactivity to the putative C-terminus of rat pancreastatin. Regul Pept 30:207–219

Curry WJ, Johnston CF, Hutton JC, Arden SD, Rutherford NG, Shaw C, Buchanan KD (1991) The tissue distribution of rat chromogranin A-derived peptides: evidence for differential tissue processing from sequence specific antisera. Histochemistry 96:531–538

Delpino A, Castelli M (2002) The 78 kDa glucose-regulated protein (GRP78/BIP) is expressed on the cell membrane, is released into cell culture medium and is also present in human peripheral circulation. Biosci Rep 22:407–420

Drees BM, Hamilton JW (1992) Pancreastatin and bovine parathyroid cell secretion. Bone Miner 17:335–346

Efendic S, Tatemoto K, Mutt V, Quan C, Chang D, Ostenson CG (1987) Pancreastatin and islet hormone release. Proc Natl Acad Sci U S A 84:7257–7260

Eiden LE (1987) Is chromogranin a prohormone? Nature 325:301

Fasciotto BH, Gorr SU, DeFranco DJ, Levine MA, Cohn DV (1989) Pancreastatin, a presumed product of chromogranin-A (secretory protein-I) processing, inhibits secretion from porcine parathyroid cells in culture. Endocrinology 125:1617–1622

Fasciotto BH, Gorr SU, Bourdeau AM, Cohn DV (1990) Autocrine regulation of parathyroid secretion: inhibition of secretion by chromogranin-A (secretory protein-I) and potentiation of secretion by chromogranin-A and pancreastatin antibodies. Endocrinology 127:1329–1335

Funakoshi S, Tamamura H, Ohta M, Yoshizawa K, Funakoshi A, Miyasaka K, Tateishi K, Tatemoto K, Nakano I, Yajima H (1989a) Isolation and characterization of a tumor-derived human pancreastatin-related protein. Biochem Biophys Res Commun 164:141–148

Funakoshi A, Miyasaka K, Kitani K, Tamamura H, Funakoshi S, Yajima H (1989b) Bioactivity of synthetic C-terminal fragment of rat pancreastatin on endocrine pancreas. Biochem Biophys Res Commun 158:844–849

Gayen JR, Saberi M, Schenk S, Biswas N, Vaingankar SM, Cheung WW, Najjar SM, O'Connor DT, Bandyopadhyay G, Mahata SK (2009) A novel pathway of insulin sensitivity in chromogranin A null mice: a crucial role for pancreastatin in glucose homeostasis. J Biol Chem 284:28498–28509

Gonzalez-Yanes C, Sanchez-Margalet V (2000) Pancreastatin modulates insulin signaling in rat adipocytes: mechanisms of cross-talk. Diabetes 49:1288–1294

Gonzalez-Yanes C, Sanchez-Margalet V (2001) Pancreastatin, a chromogranin-A-derived peptide, inhibits insulin-stimulated glycogen synthesis by activating GSK-3 in rat adipocytes. Biochem Biophys Res Commun 289:282–287

Gonzalez-Yanes C, Sanchez-Margalet V (2002) Pancreastatin, a chromogranin A-derived peptide, activates protein synthesis signaling cascade in rat adipocytes. Biochem Biophys Res Commun 299:525–531

Gonzalez-Yanes C, Sanchez-Margalet V (2003) Pancreastatin, a chromogranin A-derived peptide, inhibits leptin and enhances UCP-2 expression in isolated rat adipocytes. Cell Mol Life Sci 60:2749–2756

Gonzalez-Yanes C, Santos-Alvarez J, Sanchez-Margalet V (1999) Characterization of pancreastatin receptors and signaling in adipocyte membranes. Biochim Biophys Acta 1451:153–162

Gonzalez-Yanes C, Santos-Alvarez J, Sanchez-Margalet V (2001) Pancreastatin, a chromogranin A-derived peptide, activates Galpha(16) and phospholipase C-beta(2) by interacting with specific receptors in rat heart membranes. Cell Signal 13:43–49

Hakanson R, Tielemans Y, Chen D, Andersson K, Ryberg B, Mattsson H, Sundler F (1992) The biology and pathobiology of the ECL cells. Yale J Biol Med 65:761–774

Hakanson R, Ding XQ, Norlen P, Chen D (1995) Circulating pancreastatin is a marker for the enterochromaffin-like cells of the rat stomach. Gastroenterology 108:1445–1452

Hashimoto T, Kogire M, Lluis F, Gomez G, Tatemoto K, Greeley GH Jr, Thompson JC (1990) Stimulatory effect of pancreastatin on gastric acid secretion in conscious dogs. Gastroenterology 99:61–65

Hertelendy ZI, Patel DG, Knittel JJ (1996) Pancreastatin inhibits insulin secretion in RINm5F cells through obstruction of G-protein mediated, calcium-directed exocytosis. Cell Calcium 19:125–132

Herzig KH, Louie DS, Tatemoto K, Chung OY (1992) Pancreastatin inhibits pancreatic enzyme secretion by presynaptic modulation of acetylcholine release. Am J Phys 262:G113–G117

Huttner WB, Benedum UM (1987) Chromogranin A and pancreastatin. Nature 325:305

Kitayama N, Tateishi K, Funakoshi A, Miyasaka K, Shimazoe T, Kono A, Iwamoto N, Matsuoka Y (1994) Pancreastatin molecular forms in normal human plasma. Life Sci 54:1571–1578

Kolch W, Heidecker G, Kochs G, Hummel R, Vahidi H, Mischak H, Finkenzeller G, Marme D, Rapp UR (1993) Protein kinase C alpha activates RAF-1 by direct phosphorylation. Nature 364:249–252

Konecki DS, Benedum UM, Gerdes HH, Huttner WB (1987) The primary structure of human chromogranin A and pancreastatin. J Biol Chem 262:17026–17030

Leduc R, Hendy GN, Seidah NG, Chretien M, Lazure C (1990) Fragmentation of bovine chromogranin A by plasma kallikrein. Life Sci 46:1427–1433

Lewis JJ, Zdon MJ, Adrian TE, Modlin IM (1988) Pancreastatin: a novel peptide inhibitor of parietal cell secretion. Surgery 104:1031–1036

Lindstrom E, Bjorkquist M, Boketoft A, Chen D, Zhao CM, Kimura K, Hakanson R (1997) Release of histamine and pancreastatin from isolated rat stomach ECL cells. Inflamm Res 46(Suppl 1):S109–S110

Metz-Boutigue MH, Garcia-Sablone P, Hogue-Angeletti R, Aunis D (1993) Intracellular and extracellular processing of chromogranin A. Determination of cleavage sites. Eur J Biochem 217:247–257

Misra UK, Pizzo SV (2010) Modulation of the unfolded protein response in prostate cancer cells by antibody-directed against the carboxyl-terminal domain of GRP78. Apoptosis 15:173–182

Misra UK, Deedwania R, Pizzo SV (2006) Activation and cross-talk between Akt, NF-kappaB, and unfolded protein response signaling in 1-LN prostate cancer cells consequent to ligation of cell surface-associated GRP78. J Biol Chem 281:13694–13707

Miyasaka K, Funakoshi A, Nakamura R, Kitani K, Shimizu F, Tatemoto K (1989) Effects of porcine pancreastatin on postprandial pancreatic exocrine secretion and endocrine functions in the conscious rat. Digestion 43:204–211

O'Connor DT, Cadman PE, Smiley C, Salem RM, Rao F, Smith J, Funk SD, Mahata SK, Mahata M, Wen G, Taupenot L, Gonzalez-Yanes C, Harper KL, Henry RR, Sanchez-Margalet V (2005) Pancreastatin: multiple actions on human intermediary metabolism in vivo, variation in disease, and naturally occurring functional genetic polymorphism. J Clin Endocrinol Metab 90:5414–5425

Ohneda A, Koizumi F, Ohneda M (1989) Effect of porcine pancreastatin on endocrine function of canine pancreas. Tohoku J Exp Med 159:291–298

Peiro E, Miralles P, Silvestre RA, Villanueva ML, Marco J (1989) Pancreastatin inhibits insulin secretion as induced by glucagon, vasoactive intestinal peptide, gastric inhibitory peptide, and 8-cholecystokinin in the perfused rat pancreas. Metabolism 38:679–682

Sanchez V, Calvo JR, Goberna R (1990) Glycogenolytic effect of pancreastatin in the rat. Biosci Rep 10:87–91

Sanchez V, Lucas M, Calvo JR, Goberna R (1992) Glycogenolytic effect of pancreastatin in isolated rat hepatocytes is mediated by a cyclic-AMP-independent Ca(2+)-dependent mechanism. Biochem J 284(Pt 3):659–662

Sanchez-Margalet V (1999) Modulation of insulin receptor signalling by pancreastatin in HTC hepatoma cells. Diabetologia 42:317–325

Sanchez-Margalet V, Goberna R (1994a) Pancreastatin activates pertussis toxin-sensitive guanylate cyclase and pertussis toxin-insensitive phospholipase C in rat liver membranes. J Cell Biochem 55:173–181

Sanchez-Margalet V, Goberna R (1994b) Pancreastatin inhibits insulin-stimulated glycogen synthesis but not glycolysis in rat hepatocytes. Regul Pept 51:215–220

Sanchez-Margalet V, Gonzalez-Yanes C (1998) Pancreastatin inhibits insulin action in rat adipocytes. Am J Phys 275:E1055–E1060

Sanchez-Margalet V, Santos-Alvarez J (1997) Solubilization and molecular characterization of active pancreastatin receptors from rat liver membranes. Endocrinology 138:1712–1718

Sanchez-Margalet V, Lucas M, Goberna R (1992) Pancreastatin increases cytosolic Ca2+ in insulin secreting RINm5F cells. Mol Cell Endocrinol 88:129–133

Sanchez-Margalet V, Lucas M, Goberna R (1994a) Pancreastatin activates protein kinase C by stimulating the formation of 1,2-diacylglycerol in rat hepatocytes. Biochem J 303(Pt 1):51–54

Sanchez-Margalet V, Valle M, Goberna R (1994b) Receptors for pancreastatin in rat liver membranes: molecular identification and characterization by covalent cross-linking. Mol Pharmacol 46:24–29

Sanchez-Margalet V, Valle M, Lobon JA, Maldonado A, Escobar-Jimenez F, Olivan J, Perez-Cano R, Goberna R (1995) Increased plasma pancreastatin-like immunoreactivity levels in non-obese patients with essential hypertension. J Hypertens 13:251–258

Sanchez-Margalet V, Lucas M, Goberna R (1996) Pancreastatin: further evidence for its consideration as a regulatory peptide. J Mol Endocrinol 16:1–8

Sanchez-Margalet V, Gonzalez-Yanes C, Santos-Alvarez J, Najib S (2000) Characterization of pancreastatin receptor and signaling in rat HTC hepatoma cells. Eur J Pharmacol 397:229–235

Sanchez-Margalet V, Gonzalez-Yanes C, Najib S (2001) Pancreastatin, a chromogranin A-derived peptide, inhibits DNA and protein synthesis by producing nitric oxide in HTC rat hepatoma cells. J Hepatol 35:80–85

Sanchez-Margalet V, Santos-Alvarez J, Diaz-Troya S (2003) Purification of pancreastatin receptor from rat liver membranes. Methods Mol Biol 228:187–194

Sanchez-Margalet V, Gonzalez-Yanes C, Najib S, Santos-Alvarez J (2010) Metabolic effects and mechanism of action of the chromogranin A-derived peptide pancreastatin. Regul Pept 161:8–14

Santos-Alvarez J, Sanchez-Margalet V (1998) Pancreastatin activates beta3 isoform of phospholipase C via G(alpha)11 protein stimulation in rat liver membranes. Mol Cell Endocrinol 143:101–106

Santos-Alvarez J, Sanchez-Margalet V (1999) G protein G alpha q/11 and G alpha i1,2 are activated by pancreastatin receptors in rat liver: studies with GTP-gamma 35S and azido-GTP-alpha-32P. J Cell Biochem 73:469–477

Santos-Alvarez J, Sanchez-Margalet V (2000) Affinity purification of pancreastatin receptor-Gq/11 protein complex from rat liver membranes. Arch Biochem Biophys 378:151–156

Santos-Alvarez J, Gonzalez-Yanes C, Sanchez-Margalet V (1998) Pancreastatin receptor is coupled to a guanosine triphosphate-binding protein of the G(q/11)alpha family in rat liver membranes. Hepatology 27:608–614

Schmidt WE, Creutzfeldt W (1991) Pancreastatin–a novel regulatory peptide? Acta Oncol 30:441–449

Schmidt WE, Siegel EG, Kratzin H, Creutzfeldt W (1988) Isolation and primary structure of tumor-derived peptides related to human pancreastatin and chromogranin A. Proc Natl Acad Sci U S A 85:8231–8235

Sekiya K, Haji M, Fukahori M, Takayanagi R, Ohashi M, Kurose S-n, Oyama M, Tateishi K, Funakosh A, Nawata H (1994) Pancreastatin-like immunoreactivity of cerebrospinal fluid in patients with Alzheimer type dementia: evidence of aberrant processing of pancreastatin in Alzheimer type dementia. Neurosci Lett 177:123–126

Shani G, Fischer WH, Justice NJ, Kelber JA, Vale W, Gray PC (2008) GRP78 and Cripto form a complex at the cell surface and collaborate to inhibit transforming growth factor beta signaling and enhance cell growth. Mol Cell Biol 28:666–677

Simon JP, Aunis D (1989) Biochemistry of the chromogranin A protein family. Biochem J 262:1–13

Sun XJ, Rothenberg P, Kahn CR, Backer JM, Araki E, Wilden PA, Cahill DA, Goldstein BJ, White MF (1991) Structure of the insulin receptor substrate IRS-1 defines a unique signal transduction protein. Nature 352:73–77

Tamamura H, Ohta M, Yoshizawa K, Ono Y, Funakoshi A, Miyasaka K, Tateishi K, Jimi A, Yajima H, Fujii N (1990) Isolation and characterization of a tumor-derived human protein related to chromogranin A and its in vitro conversion to human pancreastatin-48. Eur J Biochem 191:33–39

Tatemoto K, Mutt V (1978) Chemical determination of polypeptide hormones. Proc Natl Acad Sci U S A 75:4115–4119

Tatemoto K, Efendic S, Mutt V, Makk G, Feistner GJ, Barchas JD (1986) Pancreastatin, a novel pancreatic peptide that inhibits insulin secretion. Nature 324:476–478

Taupenot L, Harper KL, O'Connor DT (2003) The chromogranin-secretogranin family. N Engl J Med 348:1134–1149

Tuteja N (2009) Signaling through G protein coupled receptors. Plant Signal Behav 4:942–947

Valicherla GR, Hossain Z, Mahata SK, Gayen JR (2013) Pancreastatin is an endogenous peptide that regulates glucose homeostasis. Physiol Genomics 45:1060–1071

Wang M, Wey S, Zhang Y, Ye R, Lee AS (2009) Role of the unfolded protein response regulator GRP78/BiP in development, cancer, and neurological disorders. Antioxid Redox Signal 11:2307–2316

Watkinson A, Jonsson AC, Davison M, Young J, Lee CM, Moore S, Dockray GJ (1991) Heterogeneity of chromogranin A-derived peptides in bovine gut, pancreas and adrenal medulla. Biochem J 276(Pt 2):471–479

Watkinson A, Rogers M, Dockray GJ (1993) Post-translational processing of chromogranin A: differential distribution of phosphorylated variants of pancreastatin and fragments 248-313 and 297-313 in bovine pancreas and ileum. Biochem J 295(Pt 3):649–654

Winkler H, Fischer-Colbrie R (1992) The chromogranins A and B: the first 25 years and future perspectives. Neuroscience 49:497–528

Zhang T, Mochizuki T, Kogire M, Ishizuka J, Yanaihara N, Thompson JC, Greeley GH Jr (1990) Pancreastatin: characterization of biological activity. Biochem Biophys Res Commun 173:1157–1160

Zhang JX, Fasciotto BH, Darling DS, Cohn DV (1994) Pancreastatin, a chromogranin A-derived peptide, inhibits transcription of the parathyroid hormone and chromogranin A genes and decreases the stability of the respective messenger ribonucleic acids in parathyroid cells in culture. Endocrinology 134:1310–1316

Zhang Y, Liu R, Ni M, Gill P, Lee AS (2010) Cell surface relocalization of the endoplasmic reticulum chaperone and unfolded protein response regulator GRP78/BiP. J Biol Chem 285:15065–15075

Chromogranins and the Quantum Release of Catecholamines

Leandro Castañeyra, Michelle Juan-Bandini, Natalia Domínguez, José David Machado, and Ricardo Borges

Abstract Chromogranins (Cgs) are the most abundant intravesicular proteins of chromaffin granules. Using Cgs knockout mice, we found that the lack of chromogranin A (CgA), chromogranin B (CgB) or both drastically reduce the vesicular content of catecholamines (CA), impair its accumulation in granules and largely affect the kinetics of exocytosis. Conversely, the overexpression of CgA induces the genesis of vesicles, increases their quantal content and even transforms non-secretory in cells capable to secrete substances. We conclude that Cgs contribute to a highly efficient system that directly mediates monoamine accumulation and regulates the exocytotic process.

Abbreviations

CA	Catecholamines
CgA	Chromogranin A
CgA&B	Chromogranins A and B
CgB	Chromogranin B
Cgs	Chromogranins
KO	Knockout
LDCV	Large dense core vesicle
SgII	Secretogranin II

We dedicate this review to Prof. Hans-Hermann Gerdes who passed away in August 18, 2013. He largely contributed to enhance our knowledge of the functional roles of chromogranins.

L. Castañeyra • M. Juan-Bandini • J.D. Machado • R. Borges (✉)
Unidad de Farmacología, Facultad de Medicina, Universidad de La Laguna,
E-38320-La Laguna, Tenerife, Spain
e-mail: rborges@ull.es

N. Domínguez
Unidad de Farmacología, Facultad de Medicina, Universidad de La Laguna,
E-38320-La Laguna, Tenerife, Spain

VU Medical Center, Clinical Genetics,
De Boelelaan 1085, 1081 HV Amsterdam, Netherlands

VMAT   Vesicular mono-amine transporter
VNUT   Vesicular nucleotide transporter
WT   Wild type animals

Chromogranins are the main protein component of chromaffin granules, a similar organelle to the large dense core vesicles (LDCVs) found in many neuroendocrine cells and in some neurons. Chromogranin A (CgA) was described in the mid 60s (Blaschko et al. 1967) being the first of a series of acidic proteins known as granins, of which currently has 9 members (Huttner et al. 1991; Winkler and Fischer-Colbrie 1992; Taupenot et al. 2003). Chromogranins are characterized by highly hydrophilic and acidic primary amino acid sequences (Huttner et al. 1991), as well as the presence of multiple paired basic residues that form cleavage sites in pro-hormones to generate bioactive peptides (Helle et al. 2007; Lee and Hook 2009). They also undergo a multitude of post-translational modifications. CgA and CgB share a tendency to self-aggregate at acidic pH values and high Ca^{2+} concentrations, conditions typical of the lumen of the trans-Golgi network and of secretory granules (Huttner et al. 1991; Rosa and Gerdes 1994). The most important chromogranins in chromaffin granules are CgA and CgB, and to a lesser extent secretogranin II (SgII). Aggregated granins provide the physical driving force to induce budding of trans-Golgi network membranes, resulting in the formation of dense core granules (Koshimizu et al. 2010; Tooze and Huttner 1990).

The physiological roles attributed to Cgs are:

(i) **To promote granulogenesis.** Down-regulation of CgA (Kim et al. 2001) and CgB (Huh et al. 2003) provokes a loss of secretory granules in PC12 cells, while overexpression induces the biogenesis of structures resembling secretory granules in non-endocrine cells, including CV-1, NIH3T3, COS-7 or HEK293 cells. Indeed, these granule-like structures are able to release/secrete their contents (Kim et al. 2001; Huh et al. 2003; Beuret et al. 2004; Stettler et al. 2009; Dominguez et al. 2014). However, secretory granules can be formed independently of CgA expression in PC12 (Day and Gorr 2003) and mouse chromaffin cells (Mahapatra et al. 2005; Hendy et al. 2006; Montesinos et al. 2008). The same assumption can be applied to CgB as granulogenesis is still observed in CgB knockout mice (Obermuller et al. 2010; Diaz-Vera et al. 2010). Indeed, granule biogenesis and calcium-evoked secretory responses are present in chromaffin cells when both chromogranins A and B are absent (Diaz-Vera et al. 2010, 2012).

(ii) **To sort vesicles towards secretory pathway.** Cgs can act as chaperones for prohormone-mediated sorting and packaging of neuropeptides in granules within the trans-Golgi network (Iacangelo and Eiden 1995; Rosa et al. 1985; Natori and Huttner 1996; Courel et al. 2006; Montero-Hadjadje et al. 2009).

(iii) **To serve as prohormones.** Cgs constitute a source of biologically active peptides, These granins are secreted during regulated exocytosis and they may fulfil hormonal, autocrine and paracrine activities through their peptide derivatives (Taupenot et al. 2003; Montero-Hadjadje et al. 2008; Zhao et al. 2009; Helle 2004).

(iv) **To facilitate the storage of amines.** Cgs are the main component of the dense core of granules, facilitating the storage of catecholamines (Dominguez et al. 2014; Nanavati and Fernandez 1993; Helle et al. 1985). Granins exhibit pH-buffering capacities and thus, they help concentrate soluble products for secretion. This was the first function attributed to Cgs and is one of main interests of our research group.

The ability of secretory vesicles to actively accumulate enormous concentrations of solutes has intrigued scientists for decades. This process is crucial in cells whose primary function is to efficiently secrete substances such as neurotransmitters and hormones, as few exocytotic events can provoke sufficiently large secretory responses. The strong accumulation of solutes into the vesicles requires the interaction of intravesicular species to reduce the osmotic pressure. In the limited space of secretory vesicles: amines, nucleotides and Ca^{2+} are the mobile components whose concentration gradients relative to the cytosol are maintained by transporters (VMAT, VNUT). By contrast, Cgs constitute the immobile components that form the dense core of LDCV that aggregates the majority of solutes (Fig. 1).

These granins are currently considered to be high capacity and low affinity buffers. For example, CgA binds 32 mol adrenaline per mol with a Kd of 2.1 mM

Fig. 1 Estimated composition of the vesicular cocktail of chromaffin granules. In the vesicular components, mobile solutes (catecholamines, calcium, ATP, ascorbate) can be distinguished from immobile species such as Cgs and enzymes. Calcium free is about 40 µM whereas the estimated bounded fraction is very high. While catecholamines are efficiently packaged in normal vesicles, some room for the uptake of newly synthesized catecholamines remains

(Videen et al. 1992), and depending on the granin type, chromogranins can bind ≈50 mol Ca^{2+} per mol with a Kd of 1.5–4 mM (Yoo 2010). The ability of CgA and CgB to form dimers or hetero-tetramers with one another has been studied to further elucidate the interactions of Cgs with Ca^{2+} (Yoo 1996; Yoo and Albanesi 1991). Similar interactions with soluble species such as catecholamines and ATP are also likely to occur, as the presence of multiple dibasic groups in the chromogranin structure increases their ability to concentrate solutes (Yoo 1996; Yoo and Albanesi 1990; Park et al. 2002). CgA and CgB are the most abundant soluble proteins in LDCVs and thus, they are the main candidates for facilitating the condensation of soluble species to generate the functional matrix (Helle et al. 1985). This matrix probably corresponds to the electron-dense core observed in electron microscopy images (Ehrhart et al. 1986; Crivellato et al. 2008).

Protons are a crucial component of vesicles and they are concentrated by a specific V-ATPase to maintain an inner pH of 5.5, approximately coinciding with the isoelectric point of Cgs. As the association of Cgs with other solutes is pH-dependent (Helle et al. 1985), vesicular pH may also regulate the ability of CgA to form aggregates (Taupenot et al. 2005), thereby playing a functional role in the dynamics of vesicular Ca^{2+}, ATP and catecholamines.

Two CgA-KO mice have been developed using distinct strategies (Mahapatra et al. 2005), and a CgB-KO mouse was developed later (Obermuller et al. 2010). By crossbreeding these two strains, we recently developed the first double CgA/B-KO mouse, which was viable and fertile in homozygosis (Diaz-Vera et al. 2012). These three strains constitute valuable tools to analyse the role of Cgs in cargo concentration and exocytosis in chromaffin vesicles.

Consequences of the Lack of CgA on the Exocytosis of Catecholamine The absence of CgA appears to trigger compensatory mechanisms that include the overexpression of CgB (Mahapatra et al. 2005; Montesinos et al. 2008). However, the number of LDCVs seems to be decreased either in norepinephrine- and epinephrine-containing cells (Pasqua et al. 2016) the redistribution of Cgs has drastic effects on the storage and release of catecholamines from the LDCVs of adrenal chromaffin cells. Using amperometry, we showed CgA-KO cells released ≈30% less catecholamines than wild-type cells upon stimulation (Fig. 2a), which is due to a reduction in the net catecholamine quantum content (Fig. 4a). These kinetic changes mainly affected the later (descending) portion of the spikes. Taken together, it appears that in the absence of CgA, the LDCV matrix is less capable of concentrating and retaining catecholamines, resulting in more rapid exocytosis (Montesinos et al. 2008).

The capacity of LDCVs to concentrate their cargo can be explored using the catecholamine precursor L-DOPA. L-DOPA penetrates the chromaffin cell membrane and it is rapidly converted to dopamine, which is then taken up by LDCVs and converted to noradrenaline by dopamine-β-hydroxylase. Thus, the usual effect of L-DOPA incubation is a notable increase in vesicular catecholamine content (Colliver et al. 2000; Sombers et al. 2007; Gong et al. 2003), as we observed in WT cells. By contrast, no increase in amine uptake was detected in the LDCVs of

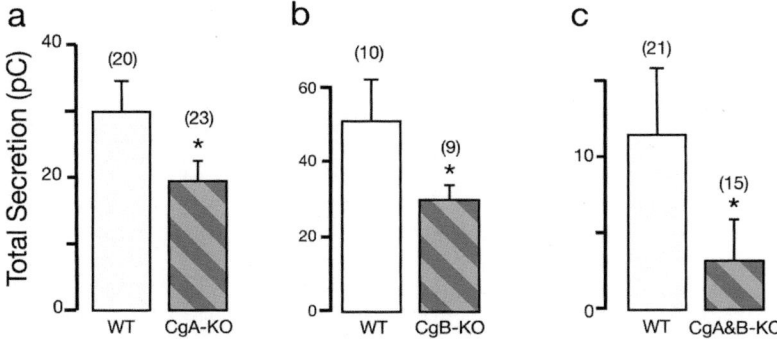

Fig. 2 The secretory response in chromaffin cells from Cgs-KO mice. Data were obtained from mouse chromaffin cells by carbon fibre amperometry. Average number of secretory spikes counted over 2 min following a 5 s pulse of 5 mM $BaCl_2$ from CgA-KO (**a**), CgB-KO (**b**) and the double KO CgA&B (**c**). Data are compared with their isogenic controls (WT) by alternating WT and KO cells from the same culture day and using the same calibrated electrode. Note the quantitative differences obtained within the experiments. $*p < 0.05$, $**p < 0.01$, Mann-Whitney test. Cell number is indicated in brackets (Modified from (Montesinos et al. 2008; Diaz-Vera et al. 2010) and from (Diaz-Vera et al. 2012))

CgA-KO chromaffin cells. Conversely, the overexpression of CgA causes, both in PC12- and in HEK293-cells, a drastic increase in secretion (Fig. 5).

To determine whether the deficit in catecholamine uptake was due to a reduction in the availability of cytosolic catecholamines, we analysed the intracellular electrochemistry in the presence of the monoamine oxidase inhibitor, pargyline. The technique used was a modified version of patch-amperometry using the whole-cell configuration, thereby allowing a carbon fibre electrode to contact the cytosol (Mosharov et al. 2003). Chromaffin cells from KO animals contained less free cytosolic catecholamines than their WT counterparts. However, a dramatic increase in free cytosolic amines was observed in CgA-KO mice after incubation with L-DOPA (100 µM for 90 min) when compared with the WT controls. This finding suggests that saturation of the LDCV matrix prevents the uptake of newly synthesized catecholamines (Montesinos et al. 2008).

The storage and release properties of LDCVs lacking CgA were studied in more detail using patch-amperometry in the cell-attached configuration, simultaneously monitoring vesicle size (capacitance) and catecholamine release (amperometry) in the same vesicle (Montesinos et al. 2008; Albillos et al. 1997). The results revealed a decrease in vesicular catecholamine concentration from 870 mM in WT to 530 mM in CgA-KO mice. Taken together, these findings indicate a dramatic reduction in the capacity of chromaffin cell LDCVs to concentrate catecholamines in the absence of CgA, despite the apparent compensatory overexpression of CgB.

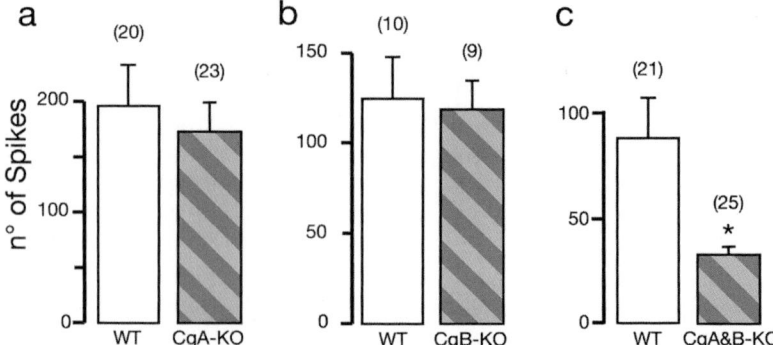

Fig. 3 The frequency of exocytotic events in chromaffin cells from Cgs-KO mice. Data were obtained from mouse chromaffin cells by carbon fibre amperometry. Average number of secretory spikes counted over 2 min following a 5 s pulse of 5 mM $BaCl_2$ from CgA-KO (**a**), CgB-KO (**b**) and the double KO CgA&B (**c**). Data are compared with their isogenic controls (WT) by alternating WT and KO cells from the same culture day and using the same calibrated electrode. Note the quantitative differences obtained within the experiments. *$p < 0.05$, **$p < 0.01$, Mann-Whitney test. Cell number is indicated in brackets (Modified from (Montesinos et al. 2008; Diaz-Vera et al. 2010) and from (Diaz-Vera et al. 2012))

1 Catecholamine Exocytosis in the Absence of Chromogranin B

The absence of CgB in CgB-KO mice was confirmed by immunohistochemistry and western blotting, also revealing the overexpression of CgA (Diaz-Vera et al. 2010). Hence, the secretory characteristics of chromaffin cells in these mice were then analyzed as described for the CgA-KO strain. Despite of critical role of CgB in the genesis and sorting of LDCVs, sustained exocytotic catecholamine release was described in chromaffin cells from CgB-KO mice (Natori and Huttner 1996; Glombik et al. 1999; Kromer et al. 1998). We observed no differences in the frequency of secretory events in chromaffin cells from CgA-KO or CgB-KO mice comparing with their controls (Fig. 3a, b). However, chromaffin cells from CgA&B-KO exhibited a lower spike firing (Fig. 3c). The total catecholamine release from CgB-KO cells was 33% lower than from control cells (Fig. 2b), roughly coinciding with the reduction observed in the amount released per quanta (Fig. 4b). Careful analysis of the kinetic properties of secretory spikes showed that the reduction in exocytosis primarily affected the initial (ascending) portion of the spikes, in contrast to the pattern observed in CgA-KO mice (Diaz-Vera et al. 2010). Moreover, L-DOPA overloading revealed that LDCVs in CgB-KO cells were unable to take up more catecholamines, with excess amines remaining in the cytosol.

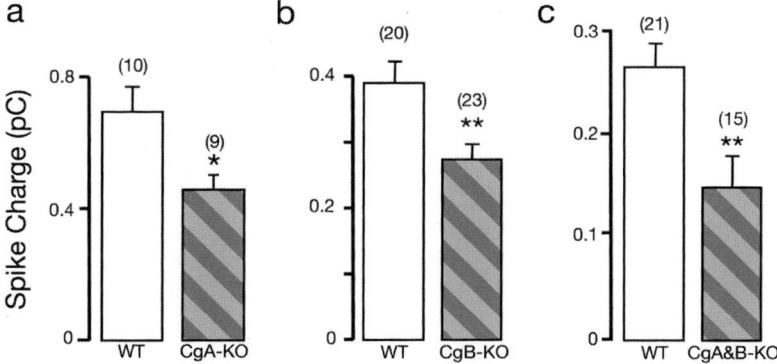

Fig. 4 The quantum size of secretory spikes in chromaffin cells from Cgs-KO mice. Data were obtained from mouse chromaffin cells by carbon fibre amperometry. Average number of secretory spikes counted over 2 min following a 5 s pulse of 5 mM $BaCl_2$ from CgA-KO (**a**), CgB-KO (**b**) and the double KO CgA&B (**c**). Data are compared with their isogenic controls (WT) by alternating WT and KO cells from the same culture day and using the same calibrated electrode. Note the quantitative differences obtained within the experiments. *$p < 0.05$, **$p < 0.01$, Mann-Whitney test. Cell number is indicated in *brackets* (Modified from (Montesinos et al. 2008; Diaz-Vera et al. 2010) and from (Diaz-Vera et al. 2012))

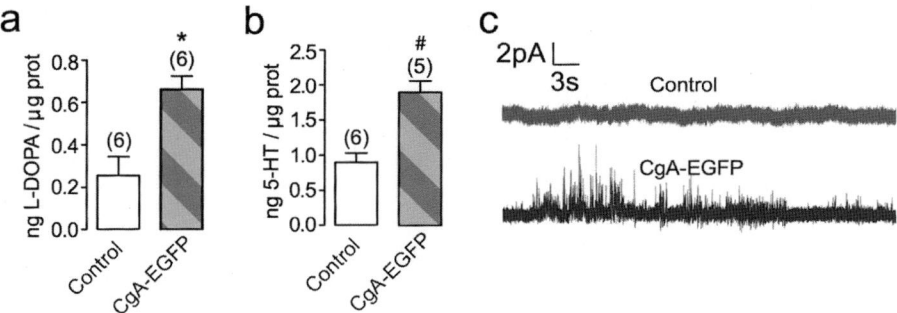

Fig. 5 Chromogranins as promoter of the neoformation of functional LDCV. (**a**) HEK293 cells were incubated for 60 min with L-DOPA (100 µM, 60 min) and the lysates were prepared for HPLC analysis. Control: non-transfected cells; CgA-EGFP cells expressing CgA-EGFP. Bar graphs normalized the L-DOPA accumulation (measured as ng L-DOPA/µg protein): *$p = 0.0022$; (Mann–Whitney U test). (**b**) As in **a** but incubating the cells with 5-HT (100 µM for 60 min) #$p = 0.0043$; (Mann–Whitney U test). (**c**) Amperometric recording showing exocytotic events from HEK293 cells incubated for 90 min with 1 mM of L-DOPA. This secretory activity is only detected in cells that express CgA-EGFP (Modified from (Dominguez et al. 2014))

2 Catecholamine Exocytosis in the Absence of Chromogranins A and B

The generation of double CgA&B-KO mice allowed us to analyse catecholamine secretion in the absence of both chromogranins. Electron microscopy images of the adrenal medulla revealed the presence of giant granules with little or no vesicular matrix (Diaz-Vera et al. 2012). The large vesicular size is likely the result of osmotic decompensation and it may explain the dramatic reduction in the frequency of exocytotic firing in CgA&B KO mice, which was not observed in the absence of CgA or CgB alone. Granule membranes were usually broken, indicating a high susceptibility to the osmotic changes associated with the fixation procedure.

Total amine secretion was strongly reduced (Fig. 2c) in CgA&B-KO mice due to a combination of low spike firing (Fig. 3c) and the small quantum size (Fig. 4c). When determined by amperometric spikes, the kinetics of exocytosis differed clearly from those of control mice and they bore a greater resemblance to CgB-KO rather than CgA-KO cells. Indeed, the Imax value was halved and the slope of the ascending region of the spikes was not as steep as in the wild-type controls. This apparent general slowdown of exocytosis may have been influenced by the very low catecholamine concentration. However, the kinetic changes observed appear to have been produced more by a combination of the limited amounts of amines and the very large size of the secretory vesicles (Montesinos et al. 2008; Diaz-Vera et al. 2010, 2012).

Incubation of cells with L-DOPA showed that the uptake of newly synthesized catecholamines granules was impaired in CgA/B-KO cells. No increase in the net charge of granules was detected after incubation with L-DOPA, although the free cytosolic catechols increased as granules cannot easily remove the catecholamines from cytosol (Diaz-Vera et al. 2012).

While the concentration of catecholamines accumulated in CgA&B-KO chromaffin vesicles was significantly reduced, it remained above that required to reach isotonicity with the cytosol. As such, we cannot rule out the possibility that other components of the vesicular cocktail, such as ATP (Kopell and Westhead 1982) and/or H^+ (Camacho et al. 2006, 2008), contribute to the maintenance of amine accumulation.

To determine whether other granins could fulfil the role of Cgs in forming the dense matrix, we performed a proteomic analysis of the enriched LDCV fraction from the adrenal medulla of the CgA&B-KO mouse (Diaz-Vera et al. 2012). While no significant changes in the amount of SgII or other granins were observed, and no other protein appears to be capable of fulfilling the role of Cgs as a matrix-condenser for soluble intravesicular components (Diaz-Vera et al. 2010, 2012).

We have also studied the role of CgA in the accumulation and exocytosis of catecholamines in cells when the levels of CgA were increased (Dominguez et al. 2014). We overexpressed CgA in non-secretory HEK293 and in secretory PC12 cells in order to study the genesis, movement and exocytosis of newly formed granules by evanescent wave microscopy. We also analysed the association of Cgs with catecholamines by HPLC and amperometry, and their role in the accumulation and exocytosis of amines, both under resting conditions and after L-DOPA overloading. The CgA overexpression in PC12 cells doubles the accumulation of amines in secretory vesicles and a significant increase in the quantal size. Moreover, the overexpression of CgA converts a non-secretory-, like HEK293, in a secretory-cell capable to make granules that accumulate L-DOPA or serotonin and release it by exocytosis (Fig. 5).

Large secretory vesicles are still accumulating CA even in the absence of Cgs. Although amine concentration resulted halved (400–500 mM), it is still hypertonic to cytosol. As the only intravesicular species present inside LDCV is ATP and considering that ATP is present almost in all secretory vesicles from all animal cells (Borges 2013), we decided to test the role of ATP as a colligative agent. This ability of ATP to bind CA to reduce the osmotic pressure was demonstrated in vitro (CITA). We were successful showing that reduction of vesicular ATP, assessed by siVNUT treatment, was accompanied with a reduction in vesicular ATP. In addition, the overexpression of VNUT resulted in an increase of vesicular CA (Estevez-Herrera et al. 2016a). This later results offers a new view of the physic-chemical interaction between the solutes present in the vesicular cocktail (Estevez-Herrera et al. 2016b).

3 Concluding Remarks

Since their discovery, Cgs have captivated the attention of scientists and they have been implicated in several processes, including granule biogenesis and sorting, catecholamine storage and release, as prohormones for the production of bioactive peptides, tumour marking, and the pathophysiology of neurodegenerative diseases. Data from Cg-KO mice are providing direct evidence implicating Cgs in vesicular storage and in the exocytotic release of catecholamines. While the exocytosis process is maintained, even in the complete absence of Cgs, the absence of Cgs impairs vesicular catecholamine accumulation. Hence, although LDCV biogenesis does not appear to be affected, the saturation of vesicular storage capacity might well be. Protein analysis of the secretory vesicle fraction revealed the compensatory overexpression of CgA in the absence of CgB, and *vice versa*. Unexpectedly, other proteins that are apparently unrelated with secretion were only present in the adrenomedullary tissue of CgB-KO animals. In conclusion, in accordance with earlier hypotheses and findings, Cgs are highly efficient direct mediators of monoamine accumulation, influencing the kinetics of exocytosis in LDCVs.

Acknowledgments LC is the recipient of a fellowship from the Fundación CajaCanarias. This work is partially funded by the grant BFU2013-45253-P from the MINECO (Spain) to RB and JDM.

References

Albillos A, Dernick G, Horstmann H, Almers W, Alvarez de Toledo G, Lindau M (1997) The exocytotic event in chromaffin cells revealed by patch amperometry. Nature 389:509–512

Beuret N, Stettler H, Renold A, Rutishauser J, Spiess M (2004) Expression of regulated secretory proteins is sufficient to generate granule-like structures in constitutively secreting cells. J Biol Chem 279:20242–20249

Blaschko H, Comline RS, Schneider FH, Silver M, Smith AD (1967) Secretion of a chromaffin granule protein, chromogranin, from the adrenal gland after splanchnic stimulation. Nature 215:58–59

Borges R (2013) The ATP or the natural history of neurotransmission. Purinergic Signal 9:5–6

Camacho M, Machado JD, Montesinos MS, Criado M, Borges R (2006) Intragranular pH rapidly modulates exocytosis in adrenal chromaffin cells. J Neurochem 96:324–334

Camacho M, Machado JD, Alvarez J, Borges R (2008) Intravesicular calcium release mediates the motion and exocytosis of secretory organelles: a study with adrenal chromaffin cells. J Biol Chem 283:22383–22389

Colliver TL, Pyott SJ, Achalabun M, Ewing AG (2000) VMAT-mediated changes in quantal size and vesicular volume. J Neurosci 20:5276–5282

Courel M, Rodemer C, Nguyen ST, Pance A, Jackson AP, O'Connor DT et al (2006) Secretory granule biogenesis in sympathoadrenal cells: identification of a granulogenic determinant in the secretory prohormone chromogranin A. J Biol Chem 281:38038–38051

Crivellato E, Nico B, Ribatti D (2008) The chromaffin vesicle: advances in understanding the composition of a versatile, multifunctional secretory organelle. Anat Rec 291:1587–1602

Day R, Gorr SU (2003) Secretory granule biogenesis and chromogranin A: master gene, on/off switch or assembly factor? Trends Endocrinol Metab 14:10–13

Diaz-Vera J, Morales YG, Hernandez-Fernaud JR, Camacho M, Montesinos MS, Calegari F et al (2010) Chromogranin B gene ablation reduces the catecholamine cargo and decelerates exocytosis in chromaffin secretory vesicles. J Neurosci 30:950–957

Diaz-Vera J, Camacho M, Machado JD, Dominguez N, Montesinos MS, Hernandez-Fernaud JR et al (2012) Chromogranins A and B are key proteins in amine accumulation, but the catecholamine secretory pathway is conserved without them. FASEB J 26:430–438

Dominguez N, Estevez-Herrera J, Borges R, Machado JD (2014) The interaction between chromogranin A and catecholamines governs exocytosis. FASEB J 28:4657–4667

Ehrhart M, Grube D, Bader MF, Aunis D, Gratzl M (1986) Chromogranin A in the pancreatic islet: cellular and subcellular distribution. J Histochem Cytochem 34:1673–1682

Estevez-Herrera J, Dominguez N, Pardo MR, Gonzalez-Santana A, Westhead EW, Borges R et al (2016a) ATP: the crucial component of secretory vesicles. Proc Natl Acad Sci U S A 113:E4098–E4106

Estevez-Herrera J, Gonzalez-Santana A, Baz-Davila R, Machado JD, Borges R (2016b) The intravesicular cocktail and its role in the regulation of exocytosis. J Neurochem 137:897–903

Glombik MM, Kromer A, Salm T, Huttner WB, Gerdes HH (1999) The disulfide-bonded loop of chromogranin B mediates membrane binding and directs sorting from the trans-Golgi network to secretory granules. EMBO J 18:1059–1070

Gong LW, Hafez I, Alvarez de Toledo G, Lindau M (2003) Secretory vesicles membrane area is regulated in tandem with quantal size in chromaffin cells. J Neurosci 23:7917–7921

Helle KB (2004) The granin family of uniquely acidic proteins of the diffuse neuroendocrine system: comparative and functional aspects. Biol Rev Camb Philos Soc 79:769–794

Helle KB, Reed RK, Pihl KE, Serck-Hanssen G (1985) Osmotic properties of the chromogranins and relation to osmotic pressure in catecholamine storage granules. Acta Physiol Scand 123:21–33

Helle KB, Corti A, Metz-Boutigue MH, Tota B (2007) The endocrine role for chromogranin A: a prohormone for peptides with regulatory properties. Cell Mol Life Sci 64:2863–2886

Hendy GN, Li T, Girard M, Feldstein RC, Mulay S, Desjardins R et al (2006) Targeted ablation of the chromogranin a (Chga) gene: normal neuroendocrine dense-core secretory granules and increased expression of other granins. Mol Endocrinol 20:1935–1947

Huh YH, Jeon SH, Yoo SH (2003) Chromogranin B-induced secretory granule biogenesis: comparison with the similar role of chromogranin A. J Biol Chem 278:40581–40589

Huttner WB, Gerdes HH, Rosa P (1991) The granin (chromogranin/secretogranin) family. Trends Biochem Sci 16:27–30

Iacangelo AL, Eiden LE (1995) Chromogranin A: current status as a precursor for bioactive peptides and a granulogenic/sorting factor in the regulated secretory pathway. Regul Pept 58:65–88

Kim T, Tao-Cheng JH, Eiden LE, Loh YP (2001) Chromogranin A, an "on/off" switch controlling dense-core secretory granule biogenesis. Cell 106:499–509

Kopell WN, Westhead EW (1982) Osmotic pressures of solutions of ATP and catecholamines relating to storage in chromaffin granules. J Biol Chem 257:5707–5710

Koshimizu H, Kim T, Cawley NX, Loh YP (2010) Chromogranin A: a new proposal for trafficking, processing and induction of granule biogenesis. Regul Pept 160:153–159

Kromer A, Glombik MM, Huttner WB, Gerdes HH (1998) Essential role of the disulfide-bonded loop of chromogranin B for sorting to secretory granules is revealed by expression of a deletion mutant in the absence of endogenous granin synthesis. J Cell Biol 140:1331–1346

Lee JC, Hook V (2009) Proteolytic fragments of chromogranins A and B represent major soluble components of chromaffin granules, illustrated by two-dimensional proteomics with NH2-terminal edman peptide sequencing and MALDI-TOF MS. Biochemistry 48:5254–5262

Mahapatra NR, O'Connor DT, Vaingankar SM, Hikim AP, Mahata M, Ray S et al (2005) Hypertension from targeted ablation of chromogranin A can be rescued by the human ortholog. J Clin Invest 115:1942–1952

Montero-Hadjadje M, Vaingankar S, Elias S, Tostivint H, Mahata SK, Anouar Y (2008) Chromogranins A and B and secretogranin II: evolutionary and functional aspects. Acta Physiol (Oxf) 192:309–324

Montero-Hadjadje M, Elias S, Chevalier L, Benard M, Tanguy Y, Turquier V et al (2009) Chromogranin A promotes peptide hormone sorting to mobile granules in constitutively and regulated secreting cells: role of conserved N- and C-terminal peptides. J Biol Chem 284:12420–12431

Montesinos MS, Machado JD, Camacho M, Diaz J, Morales YG, Alvarez de la Rosa D et al (2008) The crucial role of chromogranins in storage and exocytosis revealed using chromaffin cells from chromogranin A null mouse. J Neurosci 28:3350–3358

Mosharov EV, Gong LW, Khanna B, Sulzer D, Lindau M (2003) Intracellular patch electrochemistry: regulation of cytosolic catecholamines in chromaffin cells. J Neurosci 23:5835–5845

Nanavati C, Fernandez JM (1993) The secretory granule matrix: a fast-acting smart polymer. Science 259:963–965

Natori S, Huttner WB (1996) Chromogranin B (secretogranin I) promotes sorting to the regulated secretory pathway of processing intermediates derived from a peptide hormone precursor. Proc Natl Acad Sci U S A 93:4431–4436

Obermuller S, Calegari F, King A, Lindqvist A, Lundquist I, Salehi A et al (2010) Defective secretion of islet hormones in chromogranin-B deficient mice. PLoS One 5:e8936

Park HY, So SH, Lee WB, You SH, Yoo SH (2002) Purification, pH-dependent conformational change, aggregation, and secretory granule membrane binding property of secretogranin II (chromogranin C). Biochemistry 41:1259–1266

Pasqua T, Mahata S, Bandyopadhyay GK, Biswas A, Perkins GA, Sinha-Hikim AP et al (2016) Impact of Chromogranin A deficiency on catecholamine storage, catecholamine granule morphology and chromaffin cell energy metabolism in vivo. Cell Tissue Res 363:693–712

Rosa P, Gerdes HH (1994) The granin protein family: markers for neuroendocrine cells and tools for the diagnosis of neuroendocrine tumors. J Endocrinol Investig 17:207–225

Rosa P, Hille A, Lee RW, Zanini A, De Camilli P, Huttner WB (1985) Secretogranins I and II: two tyrosine-sulfated secretory proteins common to a variety of cells secreting peptides by the regulated pathway. J Cell Biol 101:1999–2011

Sombers LA, Maxson MM, Ewing AG (2007) Multicore vesicles: hyperosmolarity and L-DOPA induce homotypic fusion of dense core vesicles. Cell Mol Neurobiol 27:681–685

Stettler H, Beuret N, Prescianotto-Baschong C, Fayard B, Taupenot L, Spiess M (2009) Determinants for chromogranin A sorting into the regulated secretory pathway are also sufficient to generate granule-like structures in non-endocrine cells. Biochem J 418:81–91

Taupenot L, Harper KL, O'Connor DT (2003) The chromogranin-secretogranin family. N Engl J Med 348:1134–1149

Taupenot L, Harper KL, O'Connor DT (2005) Role of H+−ATPase-mediated acidification in sorting and release of the regulated secretory protein chromogranin A: evidence for a vesiculogenic function. J Biol Chem 280:3885–3897

Tooze SA, Huttner WB (1990) Cell-free protein sorting to the regulated and constitutive secretory pathways. Cell 60:837–847

Videen JS, Mezger MS, Chang YM, O'Connor DT (1992) Calcium and catecholamine interactions with adrenal chromogranins. Comparison of driving forces in binding and aggregation. J Biol Chem 267:3066–3073

Winkler H, Fischer-Colbrie R (1992) The chromogranins A and B: the first 25 years and future perspectives. Neuroscience 49:497–528

Yoo SH (1996) pH- and Ca(2+)-dependent aggregation property of secretory vesicle matrix proteins and the potential role of chromogranins A and B in secretory vesicle biogenesis. J Biol Chem 271:1558–1565

Yoo SH (2010) Secretory granules in inositol 1,4,5-trisphosphate-dependent Ca2+ signaling in the cytoplasm of neuroendocrine cells. FASEB J 24:653–664

Yoo SH, Albanesi JP (1990) Ca2(+)-induced conformational change and aggregation of chromogranin A. J Biol Chem 265:14414–14421

Yoo SH, Albanesi JP (1991) High capacity, low affinity Ca2+ binding of chromogranin A. Relationship between the pH-induced conformational change and Ca2+ binding property. J Biol Chem 266:7740–7745

Zhao E, Zhang D, Basak A, Trudeau VL (2009) New insights into granin-derived peptides: evolution and endocrine roles. Gen Comp Endocr 164:161–174

Index

A
Acute myocardial infarction (AMI), 172, 181
Adrenergic drugs, 102
Adrenocorticotropic hormone (ACTH), 41, 215, 217
African American Study of Kidney Disease and Hypertension (AASK) trial cohort, 203
Angiogenesis
    chromogranin A
        proteases, 88
        tumors, 9, 90–92
        VEGF-and FGF-2, RAR assay, 87–88
    secretoneurin, 30–31
Anti-adrenergic drug therapy, 102
Antibiotics, 50, 57–58
Antimicrobial peptides (AMPs)
    CGA-derived peptides
        antibiotics, 57–58
        bacterial proteases, 56–57
        biomaterials, 62–63
        catestatin, 53, 55, 56
        chromofungin, 53, 55–56
        location, sequence and net charge, 53, 54
        SIRS and acute circulatory failure criteria, 59–61
        VS-I, 53–56
    innate and adaptive immunity, 5, 50, 51
Anti-Thy1.1 antibody, 92
Arfaptin 1, 42, 43
Arlequin, 197
Atrial natriuretic peptide (ANP), 104, 161
Autophagy, 172

B
Bacterial proteases, 56–57
Basic fibroblast growth factor (bFGF), 162
β-adrenergic receptors (β-ARs), 116
Bin/Amphiphysin/Rvs (BAR) domains, 42, 43
Bovine chromogranin A (bCgA) protein, 216, 217
Bovine fibroblast growth factor (bFGF2), 224
Brain natriuretic peptide (BNP), 104, 105, 161, 171

C
Carboxypeptidase B, 88
Carboxypeptidase E (CPE), 42
Carcinoids, 89
Cardioprotection, CgA derived peptides
    catestatin, 181–184
    chromofungin, 184
    ischemia/reperfusion injury
        apoptosis, 172, 175
        autophagy, 172
        Ca^{2+} overload phenomenon, 174
        complex biochemical and mechanical mechanisms, 172
        inflammatory process, 175
        mPTP, 174
        myocardial ischemia, 173
        necrosis, 172, 175
        NFκB, 175
        nitric oxide, 173–175
        no-reflow phenomenon, 175
        oxidative stress, 173, 174
        peroxynitrite, 174
        ROS/RNS, 173, 175

Cardioprotection, CgA derived peptides (*cont.*)
  ischemic preconditioning, 175–178
  natriuretic peptide, 171
  neuroendocrine regulation, 171
  postconditioning, 179–180
  serpinin, 184–185
  vasostatin 1, 180–181
Catecholamines (CA) exocytosis
  chromogranin A, absence of, 252–253, 255
  chromogranin B, absence of, 253–255
  chromogranins A and B, absence of, 253–257
Cateslytin, 100
Catestatin (CST)
  antimicrobial activity, 53, 55, 56
  basic fibroblast growth factor, 162
  biologic variants, 140
  cardiac effects of, 143–145, 151, 156, 164–165
    in aged hypertrophic heart, 154
    basal and ISO-stimulated conditions, 140–141
    $\beta 2$-AR/$G_{i/o}$-dependent inhibitory pathway, 122
    CgA352-372, 121
    ET-1, 123, 124
    Frank-Starling response, 154
    frog and eel heart, 154
    G364S-CST, and P370L-CST, 121–122
    histamine-dependent positive inotropism, 122
    isolated cardiomyocytes, cardioprotective effect, 162–163
    isolated rat heart, basal conditions, 153
    loading stimulated conditions, 142–143
    mechanisms of action, 141–142
    myocardial stretch, 124–125
    NO production and $P^{Ser1179}$eNOS, 123
  cardioprotection, 181–184
  definition, 140
  endothelial-mediated effects, 157–160
  functional genetic polymorphisms
    cardiovascular function, 206–208
    catecholamine secretion, 206
    discovery of, 206
  histamine, 161
  physiological functions, 85
Caveolin 1 (Cav1), 158–160
Cell adhesion, 85
Cell penetrating peptides (CPPs), 159
Central nervous system (CNS), 219–220
Cerebral ischemia, 32
Chorioallantoic membrane (CAM), 224
Chromaffin cell, 2

Chromaffin granules, 2–3
Chromofungin (Chr)
  antimicrobial activity, 53, 55–56
  cardiac effects of, 151, 156
    AKT/eNOS/cGMP/PKG pathway, 155
    dose-dependent negative inotropic effects, 154–155
    VS-1 domain, 120–121
  cardioprotection, 184
Chromogranin A (CgA), 4–5, 12–13
  angiogenesis
    proteases, 88
    tumors, 9, 90–92
    VEGF-and FGF-2, RAR assay, 87–88
  ANP, BNP and TNFα, 161–162
  antimicrobial peptides (*see* (Antimicrobial peptides (AMPs)))
  biologically active peptides, 99–100
  calcium homeostasis and N-terminus of, 4
  cardioprotection (*see* (Cardioprotection, CgA derived peptides))
  catecholamines exocytosis, 252–253, 255
  cell adhesion, 85
  circulating CgA
    carcinoids, 89
    diabetic retinopathy, 89–90
    ELISA, 89
    full-length CGA, cardiovascular system (*see* (Full length chromogranin A, cardio-circulatory system))
    heart failure and rheumatoid arthritis, 89
    immunoassays, 89–90
    as marker for inflammatory disease, 10
    multiple myeloma, 90
    neuroendocrine tumors, 89
    physiological role for, 10
  definition, 134
  endothelial barrier integrity, 85–87
  essential hypertension, 44
  fibroblast adhesion, 85
  glucose homeostasis, pancreastatin and prohormone, 3
  hormone aggregation, 41
  inflammatory diseases, 44
  IP_3R/Ca^{2+} channel interaction
    (*see* (Inositol 1,4,5-trisphosphate receptor (IP_3R)/Ca^{2+} channel, chromogranins A and B))
  metabolic disorders, 44
  phylogenetic conservation, 134
  physiological functions, 84–85
  plasma CGA, 50–51
  post-translational modifications, 100

processing of, 115
  bovine CGs, 51
  DV vs. NDV, 52–53
  human CGA, endogenous cleavage sites of, 53
  PMNs, 51–52
receptors, 10–11
secretory granule biogenesis
  hormone sorting, TGN membrane, 42–43
  regulators of, 40–41
structure and expression, 84
Chromogranin B (CgB)
catecholamines exocytosis, 253–255
IP$_3$R/Ca^{2+} channel interaction
  (see (Inositol 1,4,5-trisphosphate receptor (IP$_3$R)/Ca^{2+} channel, chromogranins A and B))
type-2 diabetes, 44
Chromogranin C. See Secretogranin II (SgII)
Chromogranins (Cgs), 45
hormone aggregation, 41
hypersecretory endocrine pathologies, 43–44
immobile components, 251
physiological roles, 250–251
secretory granule biogenesis
  hormone sorting, TGN membrane, 42–43
  regulators of, 40–41
structural properties, 40, 41
vesicular components, 251–252
Circular dichroism (CD), 159
Coronary pressure (CP), 116
CST. See Catestatin (CST)
Cyclic AMP response element (CRE), 27
Cytosolic proteins, 43

D
Dense core secretory granules (DCG), 214
Diabetic retinopathy, 89
Diazoxide, 176, 180
Doxorubicin, 91

E
Electropherogram, 205
EM66, 43–44
Endocardial endothelium (EE), 139, 141–143, 152
Endothelial nitric oxide synthase (eNOS), 86–87, 108, 117–118
Endothelin-1 (ET-1), 116, 117

Endothelin-1 A subtype receptor (ETAR), 141–142
Endothelin-1 B subtype receptor (ETBR), 141–142
End-stage renal disease (ESRD), 203
Entherocromaffin cells (ECL), 231
Enzyme immunoassay (EIA), 214
Enzyme-linked immunosorbent assays (ELISA), 89, 105
Essential hypertension, 43, 44, 90
Ezrin-radixin-moesin binding phosphoprotein 50 (EBP50), 85

F
Fibroblast adhesion, 85
Fibroblast growth factor (FGF)-2, 87, 90
Frank–Starling response, 138, 142–143, 154
Full length chromogranin A, cardio-circulatory system
  intracardiac localization and myocardial processing, 104–105
  normotensive and SHR heart
    intracardiac CGA processing, 108–109
    myocardial and coronary actions, 106–107
    obligatory endothelium-NO involvement, 107–108
    plasma CGA levels, 103–104
    region-specific processing-dependent analysis, 103
    SAN overactivation, 101–102
    serological determinations, 103
    sites of interventions, 110

G
General adaptation syndrome, 101
Glomerular filtration rate (GFR), 203
Glomerular layer (GL), 220
Golgi apparatus, 214, 222
G-protein coupled receptor (GPCR), 31
G-proteins, 11–12
Guanylyl cyclase (GC), 176

H
Half time relaxation (HTR), 116
Heart failure (HF), 28
  chronic SAN activation, 102
  circulating CgA and fragments, 89
Heparan sulfate proteoglycans (HSPGs), 157, 159
High-mobility group box (HMGB)-1, 86

High performance liquid chromatography (HPLC), 214
Histamine, 161
Host defense peptides (HDPs), 50, 51
Human chromogranin A (CHGA) gene, SNPs
  autonomic activity and blood pressure
    CHGA promoter SNPs, 201–202
    CHGA 3'-UTR, 202–203
  catecholamines, 196
  functional genetic polymorphisms
    CST domain, 206–208
    PST domain, 204–205
  functional genetic variations, 197, 201
  gene regulatory regions, discovery of, 197–199
  haplotype distribution, 200
  hypertensive nephrosclerosis, 203–204
  resequencing strategy and identified variants in, 198
  systematic polymorphism, 197
Human umbilical vein endothelial cells (HUVECs), 86, 108
Hypertension
  catestatin, 151
  human *CHGA* gene, SNPs in (*see* (Single nucleotide polymorphisms (SNPs), human *CHGA* gene))
Hypertensive nephrosclerosis, 203–204
Hypothalamus-pituitary-adrenal (HPA) axis, 101

I
Implantable medical devices, 62
Innate immune system, 5, 50, 51
Inositol 1,4,5-trisphosphate receptor (IP$_3$R)/ Ca^{2+} channel, chromogranins A and B
  endoplasmic reticulum, 70, 76, 77
  intraluminal L3-2 loops
    conserved amino acid sequences, 70–73
    near C-terminal end, 70–72
    pH-dependent interaction, 70, 74–76
  IP$_3$-induced Ca^{2+} release in
    cytoplasm, 78
    nonsecretory NIH3T3 cells, nucleus of, 78–79
  nucleoplasmic Ca^{2+} store vesicles, 70, 77, 79
  organelles, 77
  secretory granules, 70, 76, 77
Irritable bowel syndrome (IBS), 44

Ischemia/reperfusion injury (IRI), CgA derived peptides
  apoptosis, 172, 175
  autophagy, 172
  Ca^{2+} overload phenomenon, 174
  complex biochemical and mechanical mechanisms, 172
  inflammatory process, 175
  mPTP, 174
  myocardial ischemia, 173
  necrosis, 172, 175
  NFκB, 175
  nitric oxide, 173–175
  no-reflow phenomenon, 175
  oxidative stress, 173, 174
  peroxynitrite, 174
  ROS/RNS, 173, 175
Ischemic preconditioning (IP), 175–178, 183
Isoproterenol (ISO), 116–118, 136–137

J
Janus Kinase (JAK), 178

L
Langerhans islet beta cells, 233
Large dense core vesicles (LDCVs), 252–257
Left ventricular pressure (LVP), 116
Lethal myocardial reperfusion injury, 172
Linkage disequilibrium (LD), 199
Luciferase reporter assay, 221
Lymphoma, 91

M
Mastoparan, 233
Matrix assisted laser desorption/ionization time of flight (MALDI-TOF) analysis, 214, 215
Mediation phase, 177
Memory phase, 177
Mercaptopropionyl glycine (MPG), 176
Metabolic disorders, 44
Microsequencing technique, 239
Mitochondrial membrane potential (MMP), 163
Mitochondrial permeability transition pore (mPTP), 173
Mitogen activated protein kinase (MAPK), 234, 238
Molecular heterosis, 202
Monocyte chemoactractant protein (MCP)-1, 86

Multidrug antibiotic resistance, 50
Multiple myeloma, 90
Myocardial ischemia/reperfusion injury, 173

N
National Institute of Diabetes and Digestive
 and Kidney Diseases (NIDDK), 203
Natriuretic peptides (NPs), 105
Neuroendocrine tumors, 43–44, 89, 92
Neurohormones, 39
Non-diabetic vitreous (NDV), 52–53
Non small cell lung cancer, 92
Nuclear factor kappa B (NFκB), 173
Nuclear magnetic resonance (NMR)
 spectroscopy, 159

O
Olfactory nerve layer (ONL), 220

P
Pancreastatin (PST), 3, 4
 action, target tissues of, 232
 in adipocyte, energy metabolism and
  protein synthesis, 235–236
 biological functions, 232
 cell proliferation, negative effects on, 234
 functional genetic polymorphisms
  discovery of, 204
  glucose homeostasis, 204–205
  Gly297Ser variant, biochemical
   parameters, 205
 glandular secretions
  exocrine pancreatic secretion, 234
  gastric secretion, 234
  glucose stimulated insulin secretion, 233
  PTH, 233
 glucose and lipid metabolism, regulation
  of, 85
 G protein-coupled receptors, 239
 GRP78 mediated AKT activation, 240
 hepatic glycogen catabolism and insulin-
  stimulated glucose uptake, 235
 and inflammation signaling, 241
 insulin receptor signaling, 239–240
 insulin resistance, 232
 ligand-affinity experiment, 241
 mechanisms of action
  in adipocytes, 238
  in hepatocytes, 236–238
 natural occurring variants and implications,
  242–243

 neuroendocrine and gastrointestinal
  tissues, 230
 as stress peptide, 241–242
 structure
  molecular forms, 231
  primary structure, 230–231
  synthesis and secretion, 231
Parathyroid hormone (PTH) secretion, 4
Pertussis toxin (PTX), 7, 11
Pheochromocytomas, 43–44
Phorbol myristate acetate ester (PMA), 233
Phosphatidylinositol triphosphate (PIP3)
 levels, 178
2 Phosphodiesterases (PDE2), 157
Phospholipase C (PLC), 177, 236
PICK1, 42, 43
Plasmin, 88
Polyelectrolyte multilayer (PEM) films, 62–63
Prohormone convertases (PCs), 23
Prohormones, 3
Proopiomelanocortin (POMC), 29, 41, 42
Prostate cancer, 44
Prostate tumors, 92
Proteases, 88
Proton pump inhibitors, 89
Pyroglutaminated serpinin (pGlu-serpinin)
 anti-apoptotic activity, 225
 AtT20 cells, 217–218
 cell death, 222–223
 CNS, 219–220
 granule biogenesis, 220–222
 HPLC analysis, 218

R
Rat aortic rings (RAR) angiogenesis assay,
 87–88
Rate pressure product (RPP), 116
Reactive nitrogen species (RNS), 173
Reactive oxygen species (ROS), 173
Receptor tyrosine kinase (RTK), 31
Renin-angiotensin system (RAS), 101
Reperfusion Injury Salvage Kinase (RISK)
 pathway, 177, 179, 180, 184
Reperfusion therapy, 172
RE-1 silencing transcription factor (REST), 27
Rheumatoid arthritis, 89

S
Sarcoplasmic reticulum Ca^{2+}-ATPase
 (SERCA2a), 123
SCG10, 43
SCLIP, 43

Secretogranin II (SgII)
    biomarker and disease
        CNS diseases, 28
        SN serum levels, heart failure, 28
        tumors, 27–28
    gene organisation and regulation, 27
    hormone secretion and tumorigenesis, 44
    prostate cancer, 44
    proteolytic processing of, 23
    secretoneurin
        angiogenesis and vasculogenesis, 30–31
        cardiomyocytes, 32
        cerebral ischemia, 32
        high vascular risk, animal models of, 32–33
        immune system, 30
        nervous system, 29–30
        phylogenetic conservation of, 25–26
        wound healing, 31
    secretory granule biogenesis, 29, 40
    secretory proteins, sorting and release of, 29
    structure and posttranslational modifications, 22–23
    subcellular localisation, LDVs, 24
    tissue expression and secretion from cell, 24–25
Secretoneurin (SN) peptide
    phylogenetic conservation of, 25–26
    physiological functions of
        angiogenesis and vasculogenesis, 30–31
        cardiomyocytes, 32
        cerebral ischemia, 32
        high vascular risk, animal models of, 32–33
        immune system, 30
        nervous system, 29–30
        wound healing, 31
Serine protease inhibitor, 214
Serpinin (Serp), 9, 224
    AtT20 cells
        expression and characterization, 214–217
        pGlu-serpinin, expression and secretion of, 217–218
    cardiac effects of, 156
        AKT/eNOS/cGMP/PKG pathway, 155
        β-adrenergic-like cardiac modulators, 155
        β1-AR/AC/cAMP/PKA pathway, 126, 155
        ERK1/2 and GSK3β phosphorylation, 127

Serp-Ala29Gly and pGlu-Serp, 125–127
cardioprotection, 184–185
cell death, 222–223
discovery of, 214
in granule biogenesis, 220–222
organs and nervous system
    adrenal medulla and heart, 218
    CNS, pGlu-serpinin, 219–220
    eye, expression and localization of, 220
physiological functions, 85
serpinin-RRG, anti-angiogenesis effect of, 224, 226
Serum-response element (SRE), 27
SgII. See Secretogranin II (SgII)
Signal transducer and activator of transcription 3 (STAT3), 178, 180
Single nucleotide polymorphisms (SNPs), human *CHGA* gene
    autonomic activity and blood pressure
        CHGA promoter SNPs, 201–202
        CHGA 3'-UTR, 202–203
    catecholamines, 196
    functional genetic polymorphisms
        CST domain, 206–208
        PST domain, 204–205
    functional genetic variations, 197, 201
    gene regulatory regions, discovery of, 197–199
    haplotype distribution, 200
    hypertensive nephrosclerosis, 203–204
    resequencing strategy and identified variants in, 198
    systematic polymorphism, 197
Small bowel Crohn's disease, 44
Soluble tumor necrosis factor receptors (sTNF-Rs), 89
Spontaneously hypertensive rats (SHR) heart, 106–109
Stathmins, 42, 43
Survivor Activating Factor Enhancement (SAFE) pathway, 177, 179, 180
Sympatho-adrenal neuroendocrine (SAN) activity, 99, 101–102
Systemic inflammatory response syndrome (SIRS), 59–61

T
Thrombin, 88
Transcriptome/proteome-wide approach, 43
Trans-Golgi network (TGN) membrane, 39–40, 42–43, 45
Tumor necrosis factor α (TNFα), 86, 91, 161, 180

Tumors, 28
    chromogranin A, 84, 90–92
    secretogranin II, 27–28
Type-2 diabetes, 44

V

Vascular endothelial growth factor (VEGF), 30, 31, 86, 87, 90, 104
Vasculogenesis, 30–31
Vasoconstriction-inhibiting factor (VIF), 7
Vasostatin-1 (VS-1)
    antimicrobial activity, 53–56
    as cardiac stabilizer
        chromofungin, cardiac properties of, 119–121
        isolated and Langendorff perfused rat heart, 116
        ISO-mediated stimulation, 117–118
        negative inotropic and lusitropic effects, 116–117
        physiological pathways, 117–119
    cardioprotection, 180–181
    physiological functions, 84–85
Vasostatins (VSs), 135–136, 150–151
    cardiac effects, 143–145, 156, 164–165
        basal and ISO-stimulated conditions, 136–137
        frog and eel heart, 151–152
        isolated rat heart, negative inotropic and lusitropic effects, 152
        mechanisms of action, 137–139
        PI3K-Akt-NO pathway, 153
        rCgA1–64SH and rCgA1–64OX, 152–153
    catestatin and serpinin, 7–9
    endothelial-mediated effects, 157–158
    and vascular endothelium, 6–7
Vitreous of diabetic patients (DV), 52–53

W

Wheat-germ agglutinin (WGA) lectin, 238
Wound healing, SN gene therapy, 31

Printed by Printforce, the Netherlands